MANAGING THE HUMAN IMPACT ON THE NATURAL ENVIRONMENT

MANAGING THE HUMAN IMPACT ON THE NATURAL ENVIRONMENT: PATTERNS AND PROCESSES

Edited by Malcolm Newson
with
Jeremy Barnes
Alan Davison
Terry Douglas
Ian Fells
David Harte
Mark Macklin
Bill Myers
Stan Openshaw
Paul Read
Anthony Stevenson

JOHN WILEY & SONS
Chichester • New York • Brisbane • Toronto • Singapore

© Editors and contributors, 1992

First published in Great Britain in 1992 by
Bellhaven Press (a division of Pinter Publishers Limited)
25 Floral Street, London WC2E 9DS

Published in 1995 by John Wiley & Sons Ltd.,
  Baffins Lane, Chichester,
  West Sussex PO19 1UD, England

  Telephone National   (01243) 779777
     International   (+44) 1243 779777

All rights reserved.

No part of this book may be reproduced by any means,
or transmitted, or translated into a machine language
without the written permission of the publisher.

Other Wiley Editorial Offices

John Wiley & Sons, Inc., 605 Third Avenue,
New York, NY 10158-0012, USA

Jacaranda Wiley Ltd, 33 Park Road, Milton,
Queensland 4064, Australia

John Wiley & Sons (Canada) Ltd, 22 Worcester Road,
Rexdale, Ontario M9W 1LI, Canada

John Wiley & Sons (SEA) Pte Ltd, 37 Jalan Pemimpin #05-04.
Block B, Union Industrial Building, Singapore 2057

ISBN 0 471 94733 4 (paper)

Filmset by Mayhew Typesetting, Rhayader, Powys
Printed and bound in Great Britain by SRP Ltd, Exeter

Contents

List of plates viii

List of figures ix

List of tables xi

List of contributors xiv

Preface xvii

Editor's acknowledgements xxi

Part 1 Principles

1 Introduction: managing the natural environment — why and how 3
  Malcolm Newson
  1.1 Background to environmental concern 3
  1.2 People and environment: scales of interaction 8
  1.3 Knowledge and institutions: defining the breadth of policy 10

2 The geography of pollution 14
  Malcolm Newson
  2.1 Global patterns framing environmental management 14
  2.2 Contamination and pollution 20
  2.3 Definitions of pollution in terms of toxicity 22
  2.4 Sources, pathways, targets and control 28
  2.5 Pollution monitoring: a measure of concern 30

3 The geography of conservation 37
  Anthony Stevenson
  3.1 Biodiversity and environmental change 37
  3.2 Species conservation and pollution: relationships 38
  3.3 Biogeography and habitat conservation 41
  3.4 Biogeography and nature reserve design 49
  3.5 Historical perspective 50
  3.6 Monitoring success/failure 52

4 Environmental law 56
  David Harte
  4.1 The relationship of law to other disciplines and its role as a tool in managing the natural environment 56
  4.2 Legal systems and legal sources 59

4.3	From the protection of private property to an integrated public environmental law	63
4.4	Law as an instrument for limiting environmental harm	69
4.5	Practical aspects of enforcement	72
4.6	Law as an instrument for promoting environmental improvement	75
4.7	The future of environmental law: public rights and public involvement	77

5 Environmental economics, resources and commerce — 80
Malcolm Newson

5.1	Economics: principles	80
5.2	Attempts to value the environment	87
5.3	National and international environmental accounts	92
5.4	Commerce and the environment	98
5.5	Principles of sustainability	101

Part 2 Practices

6 Patterns of air pollution: critical loads and abatement strategies — 109
Alan Davison and Jeremy Barnes

6.1	Traditional approaches to air pollution control	109
6.2	Point sources: control of fluoride emissions — a success	111
6.3	Diffuse sources of air pollution	112
6.4	Abatement strategies on an international scale	120
6.5	Conclusions	127

7 Patterns of freshwater pollution — 130
Malcolm Newson

7.1	Sources, pathways and targets in the freshwater environment	130
7.2	Chemical and biological monitoring and control of water pollution	139
7.3	Towards geographical patterns of management: river basin scale	145

8 Patterns of land, water and air pollution by waste — 150
Terry Douglas

8.1	The waste 'stream': sources, types, volumes	150
8.2	Disposal technology: the alternatives — landfill, incineration, treatment and recycling	154
8.3	Emissions from waste disposal: leachate from landfill, air pollution from landfill and incineration	157
8.4	Case studies in a developing arena	163
8.5	Towards environmental protection in the waste industry	169

9 Metal pollution of soils and sediment: a geographical perspective — 172
Mark Macklin

9.1	Patterns of pollution in time and space	172
9.2	Nature and history of metal pollution	173
9.3	Sources, pathways and targets	174
9.4	The toxicology of heavy metals	186
9.5	Geographical audits of metal-contaminated soil and sediment systems	187
9.6	Conclusions	192

10	Emergency planning and pollution *Bill Myers and Paul Read*	196
	10.1 Chemical emergencies and their environmental effect	196
	10.2 Major incidents: chemical emergencies and the planning response	198
	10.3 Major incidents: some case-studies	198
	10.4 Emergency planning legislation in the United Kingdom	200
	10.5 Real-time environmental information for chemical emergencies: the CHEMET scheme	204
	10.6 Moving hazards and other considerations	209

Part 3 Futures

11	Radiation and the environment: types, sources, impacts and management *Stan Openshaw*	213
	11.1 A geographical view of radiation and the environment	216
	11.2 Non-ionizing radiation	216
	11.3 Ionizing radiation	218
	11.4 Nuclear power	221
	11.5 Nuclear waste	225
	11.6 Nuclear disease?	227
	11.7 Ultrasound	228
	11.8 Conclusions	229
12	Global environmental implications for future energy supply and use *Ian Fells*	232
	12.1 History and geography of energy development and energy-related pollution	232
	12.2 Taking stock: energy reserves and use	233
	12.3 Energy demands and development	235
	12.4 Energy supply	236
	12.5 World trade in energy	236
	12.6 Fuel technology	238
	12.7 Energy generation: pollution and environment	240
	12.8 The environment and future energy supply patterns	240
13	Natural environments of the future: adapting, conserving, restoring *Anthony Stevenson and Malcolm Newson*	242
	13.1 How we got to here: humans and their environment	242
	13.2 The living planet: species diversity and future environments	246
	13.3 Global capacities: wilderness, commons and indigenous peoples	251
	13.4 The end of nature or recovery, restoration, enhancement?	253
14	Planning, control or management? *Malcolm Newson*	258
	14.1 Knowledge bases for future environmental interventions	258
	14.2 Environmental planning	261
	14.3 Environmental control	270
	14.4 Managing the environment	274

Index   280

List of plates

4.1	Development in Exeter: luxury housing, the industrial surroundings	64
4.2	Mallam Water, Gloucestershire	73
7.1(a)	A point of source pollution, Causey Barn, County Durham	132
7.1(b)	Downstream from the outfall of Stanley works	133
7.2	Sources of agricultural pollution if spillage occurs	135
8.1	Domestic waste tip for an island community	153
8.2	Municipal solid waste being tipped at Keeble Valley Landfill	154
8.3	The Byker Reclamation Plant, Newcastle	158
8.4	Leachate collection lagoon, North Carolina, USA	160
8.5	Newly established golf course on an old landfill, Charlotte, North Carolina	162
8.6	Abandoned house in the Love Canal Emergency Declaration Area, Niagara Falls, New York State	167

List of figures

i	An interdisciplinary approach to environmental management	xvi
2.1(a)	World methane emissions	16
2.1(b)	Patterns of rainfall acidification	17
2.2	Resource location and world trade	19
2.3	Pollution as a system diagram	21
2.4(a)	World industrial chemical production, 1930–90	23
2.4(b)	The dose–response curve	24
2.4(c)	Segregation of the toxicological curve	25
2.5(a)	Car exhaust emission: the synergistic formation of pollutant gas	26
2.5(b)	The food chain and pesticide pollution	27
2.6(a)	UK water quality monitoring stations	33
2.6(b)	Problems of sampling river water quality	34
3.1	Species extinctions	37
3.2	Species rarity	43
3.3	The waterfowl site at Garaet el Ichkeul	46
3.4	Wetlands: their interrelationships	48
3.5	Peat erosion in the Round Loch, south-west Scotland	51
4.1(a)	Archer's application to abstract water	57
4.1(b)	Judgemental process of Archer's application	58
4.2	The centralization of pollution control	68
4.3	The definition of sites of special scientific interest	77
5.1	Market forces and scarcity	82
5.2	Defining resources	83
5.3	Elemental resources and the economy	86
5.4	The Human Development Index	95
5.5	Political economy of soil erosion	96
5.6	Environmental assessment and analysis by agencies	97
5.7	Environmental awareness and GDP	98
5.8	The consumer and 'green' products	99
6.1	Effects of the London smogs, early 1950s	110
6.2	The spread of pollution in the USA	112
6.3(a)	SO_2 and NO_x concentrations in the UK, 1978–88	113
6.3(b)	Ammonia concentration in the UK, 1950–80	113
6.4	Transport of pollutants across international borders	115
6.5(a)	Acid deposition in lakes and rivers	116
6.5(b)	Acidity in lakes and rivers	116
6.6	Ozone concentration over the UK	118
6.7(a)	The critical load factor	120
6.7(b)	Application of the critical load	121
6.8(a)	1985 sulphur deposition in Europe	124

List of figures

6.8(b)	Relative sensitivity to acid deposition	125
7.1	Rivers and the natural process of waste disposal	131
7.2	The sewerage system	134
7.3(a)	Agricultural pollution ratios	136
7.3(b)	Rural point-source pollution	136
7.4	Cultural eutrophication	137
7.5	Effluents from sewage treatment works and total flow	140
7.6	River water quality and polluting outfall	142
7.7	The effects of point-source pollution on a river basin	146
7.8	Pollution control successes	148
8.1	MSW volumes, 1960–2010	155
8.2	Groundwater protection: the 'dilute and attenuate' philosophy	159
8.3	The effects of methane migration from a landfill, 1986	161
8.4	The Tyne and Wear incinerator system, 1971–88	163
8.5	Leakage from a closed toxic waste site, Love Canal, USA	165
9.1	Anthropogenic perturbation of geochemical cycles	175
9.2	Ground pattern of metal contamination at a metal refinery, South Wales	180
9.3	Lead concentrations in roadside soils	180
9.4	Contaminated overbank deposits of the Geul River, the Netherlands	182
9.5	Pb levels in amenity soils, Tyneside	188
9.6	Pb levels in river sediments, Tyne Basin	191
10.1	Radioactivity release from Chernobyl, 26 April–8 May 1986	201
10.2	Reducing the effects of major industrial disasters	202
10.3	The Zone of Consequence	204
10.4	A chemical site layout	205
10.5	Involvement during the course of a chemical emergency	207
11.1	Infeasible sites for nuclear power stations	223
11.2	Geographical Information Systems (GIS)	227
11.3	Statistically significant clusters	229
12.1	World distribution of oil reserves	237
12.2	Carbon dioxide emissions	239
12.3	Cutting the UK's carbon dioxide emissions: contribution towards target	239
13.1	A hypothesis for 'Pleistocene overkill' sweeping through the North American continent, with the initial colonization occurring from the north	243
13.2	Schematic representation of the displacement of species range limits beyond the boundaries of a biological reserve under changing climatic conditions	249
14.1	Environmental assessment	264
14.2	EA: procedures and interactions	266
14.3	The course of EA simplified	267
14.4	The process of environmental decision-making	274
14.5	Operating environmental management policies: two models	275

List of tables

1.1	Robert Arvill's candidate environmental ethics	5
1.2	Significant individuals and events in modern environmental management	7
1.3	Niches in the pattern of policy opportunities	8
2.1	Attitudes in environmental management	18
2.2	Possible calculations of a 'suitable' dose of LSD to give to an elephant	24
2.3	Some critical pathways and critical groups identified in the UK	30
2.4	UK Environmental monitoring register	32
3.1	Current rates of loss of wildlife habitat	38
3.2	Heavy metal inputs, excluding atmospheric inputs, by the UK to the North Sea, 1985–86	40
3.3	A classification of critiera used to evaluate and justify the conservation of a site	42
3.4	Heilinghenhafen criteria used to select wetlands of international importance as used by the Ramsar convention, 1971–87	44
3.5	Revised criteria used to identify wetlands of international importance as adopted by the Ramsar conference in Regina, Canada, 1987	45
3.6	Comparison of economic benefits of water releases into the Ichkeul National Park compared to the use of water on the agricultural scheme	47
5.1	Neo-classical assumptions	81
5.2	How the automobile has altered our lives (1895 to present)	85
5.3	Schools of thought within green economies	87
5.4(a)	The benefits of pollution control in the USA, 1978	88
5.4(b)	Summary of employment figures created by nature conservation, by category	89
5.5	Pollution damage costs	91
5.6	Types of pollution charge systems	92
5.7	National environmental accounting	94
5.8	Costs of watershed degradation, Java	95
5.9	1987 estimates of national environmental protection markets	101
5.10	Ten principles of sustainable development	103
5.11	Relative costs of command and control and efficient policy instruments	104
6.1	Comparison of critical loads for deposited nitrogen	123
6.2	Critical levels for ozone when present as a single pollutant, estimated from dose–response of sensitive species	
7.1	The aims of sewage treatment in chemical terms	134
7.2	Trophic status of river systems	138
7.3	Chemical and biological monitoring of river pollution: comparative advantages	144
7.4	River and canal water quality classification applied to quinquennial surveys in England and Wales	147
8.1	Waste arisings in England and Wales	150

List of tables

8.2	Waste per employee for selected industries in Tyne and Wear	151
8.3	Composition of household waste in the UK	152
8.4	Selected MSW statistics for various countries	152
8.5	MSW disposal by incineration for European countries	156
8.6	Siting criteria for Alberta hazardous waste management facility	168
8.7	Average disposal costs for MSW in England and Wales, 1988–89	169
9.1	The more important heavy metals with their commonly accepted densities and crustal abundances	174
9.2	Global annual contributions of trace elements from natural and anthropogenic high-temperature processes, 1983	176
9.3	Anthropogenic inputs of trace metals into the aquatic ecosystems	177
9.4	Worldwide emissions of trace metals into soils	177
9.5	Emissions of selected metals into the Hudson–Raritan river basin	179
9.6	Percentage particulate-bound metal of total discharge	181
9.7	Land-based and atmospheric contributions to heavy metals in the waters of the North Sea	182
9.8	Background concentrations of heavy metals in sediments and soils	183
9.9	Statistically based threshold values for lead in soils	184
9.10	Tentative trigger concentrations for selected metal contaminants	185
9.11	Guidelines for toxic element trigger concentrations in minespoil-contaminated 'soils'	185
9.12	Signal values of soil pollution according to Dutch guidelines	185
9.13	Signal values for metal concentrations in the soil for various types of agricultural land-use/products	186
9.14	Inventory of estimated average daily amounts of lead absorbed into blood of adult humans in prehistoric times	186
9.15	Percentage of Tyneside population living in wards where measured values exceed Pb thresholds, and percentages of soils sampled exceeding thresholds	188
10.1	The long-term heath effect of certain categories of chemical incidents	197
10.2	Some major chemical emergencies since 1917	199
10.3	CIMAH plan outline	206
11.1	Principal types of electromagnetic radiation	216
11.2	Half-lives for selected radionuclides	221
11.3	Risks of death in Britain	222
11.4	Populations near to UK reactor sites, 1981	224
11.5	Feasible nuclear power sites in Britain in terms of demographic criteria	225
11.6	Sources of total projected nuclear wastes	226
11.7	Feasible deep repository radwaste sites in Britain	228
12.1	World energy reserves and consumption, 1987	234
12.2	Reserves/current consumption	235
12.3	Fuel scenarios for 2020	235
13.1	Holocene environmental changes in north-west Europe	245
13.2	Recorded extinctions, 1600 to present	247
13.3	Wildlife habitats on the Afrotropical and Indomalayan Realms, 1986	247
13.4	Predicted percent loss of tropical forest species due to extinction	247
13.5	O'Riordan's environmental ideologies	248
14.1	Types of uncertainty and styles of problem-solving	260
14.2	Approximate consultation distances for major hazard sites as specified by the Health and Safety Executive	263
14.3	European Community guidance on the selection of developments for EA	265

14.4	A classification of assessment methods by task	268
14.5	Major activities and types of personnel involved in the EA process	268
14.6	Patterns in accuracy of impact predictions	269
14.7	Comparison of contamination potentials for three types of disposal	271
14.8	Number of works and processes under IPC inspection	272
14.9	Main UK pollution control arrangements	273
14.10	A decision protocol to promote learning from experience	276
14.11	Land-use planning: policy processes in contemporary British public administration	277

List of contributors

Malcolm Newson
Professor of Physical Geography, University of Newcastle

David Harte
Lecturer in Law, University of Newcastle

Alan Davison
Senior Lecturer in Biology, University of Newcastle

Jeremy Barnes
Royal Society Fellow, Department of Agriculture and Environmental Sciences, University of Newcastle

Terry Douglas
Senior Lecturer in Geography, Newcastle Polytechnic

Anthony Stevenson
Lecturer in Geography, University of Newcastle

Mark Macklin
Lecturer in Geography, University of Newcastle

Stan Openshaw
Professor of Quantitative Geography, Centre for Urban and Regional Development Studies, University of Newcastle (recently appointed Professor of Quantitative Geography, University of Leeds)

Paul Read
Chief Emergency Planning Officer, Tyne and Wear

Bill Myers
Deputy Chief Emergency Planning Officer, Tyne and Wear

Ian Fells
Professor of Energy Conversion, University of Newcastle

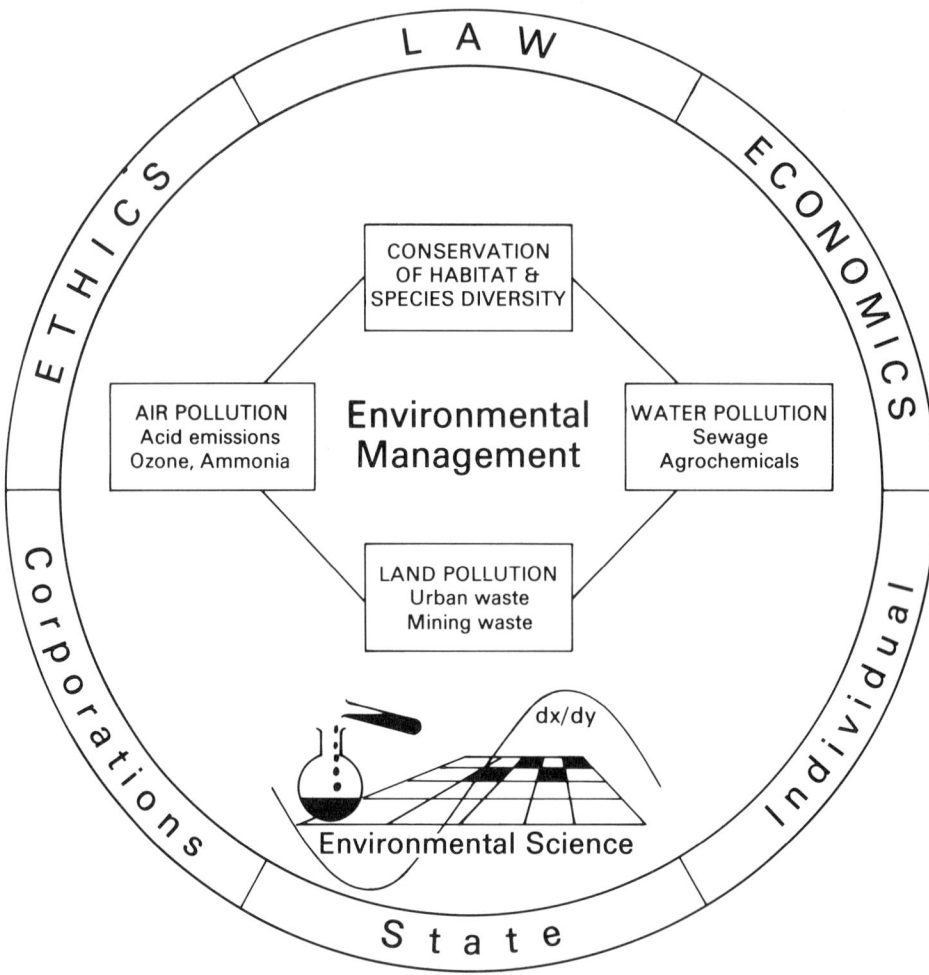

Figure i The context of environmental management, the issues treated by this book, science, institutions and scales

Preface

An essential feature of environmental problems is their geographical pattern; most geographers use a combination of spatial and temporal scales as a methodological thread for their investigations. Environmental problems are not new and the addition of time (or, stratigraphically, of depth) is critical to understanding 'where we have been' in the environment: how the problem in question has evolved. Time is of particular relevance in the consideration of our response to environmental problems. Space is an essential element of our understanding and control of pollution; spatial patterns are widespread in both popular and professional depictions of environmental problems. Another feature of geographical enquiries is that they operate around the physical science/social science boundary; for this reason we are able here to consider both environmental measurement (in space and through time) and the social institutions and systems involved in management of environmental problems. Figure i illustrates the relationships we incorporate in this interdisciplinary view, with environmental management surrounded by problems, supported by science but circumscribed by principles and the institutions operating those principles at various scales.

Long before human numbers and needs produced a common perception of widespread environmental problems many philosophers pondered and pronounced on mankind's relationship with nature; biblical notions of stewardship of the earth abound. However, in pre-biblical times civilizations had already perished as the result of environmental changes, some of them artificially exacerbated. This part of human ethical debate has resurfaced strongly during the last century or so, beginning (in the geographical field) with George Perkins Marsh's book *Man and Nature* (1864). Marsh subtitled his book *Physical Geography as modified by Human Action*. In Marsh's world human action was essentially pioneer exploitation of resources; he could not reconcile deforestation and erosion with good stewardship.

In this century and in the developed world, the ethical and cultural arguments have tended to subside beneath scientific claims and counterclaims, not so much about resource exploitation but about pollution. Environmental pollution has inevitably taken over from exploitation as the most commonly appreciated sign of 'poor stewardship'. The take-over is inevitable because pollution of some types proves an immediate threat to humans and, as a species, they have become threatened by the waste products of their own success.

This book, as a product of its time, is therefore dominated by pollution: of air, land and water. We deal with the geographical triple — 'source/pathway/target' — and with the spatial and temporal processes which help nature deal with pollution: dilution, dispersion, decay.

Nevertheless, the ethical stewardship argument lives on in newsworthy form: violent protests over animal rights have been commonplace in recent years. The much wider modern concern with conservation of species and their critical habitats has a distinct geographical dimension with 'island biogeography' at the heart of theoretical approaches to nature conservation. Conservation is therefore an aspect of biogeography; since it was Charles Darwin's use of biogeography which first led us to appreciate the significance of species diversity it is appropriate to include measurement and management in the field of conservation here too.

Finally we make for ourselves a strict rule: we do not deal with conservation and pollution phenomena in a restricted, reductionist, scientific fashion — each aspect of measurement is dealt with in the context of management. This is critical for a geographical audience and has been a strong reason for the success of the undergraduate course which first brought the authors together. Neither do we fall for the trap of attempting to shock and horrify merely because we are interfacing scientifically derived, often startling results, with a policy system which we hope will be impressed!

The management sections of the book are never far from the measurement sections but two major blocks of the management system — law (both planning and judicial) and economics — deserve separate attention. Readers are referred to Ron Johnston's book in this series (Johnston, 1989) for further work on the structures through which environmental management policies may be routed, i.e. those of the nation state.

At the core of this book is the question of the sustainability of our management of the environment; mankind cannot now resign from stewardship, nor fail critically to assess the ethics of our relationship with the natural environment. Since the spatial and temporal aspects of this relationship are critical to rational yet humane intervention we trust our readers will become convinced of the relevance of the holistic geographical perspective.

A guide to the organization of the book

The history of our concern for the natural environment and of the perceived need for management is an essential prerequisite for the assessment of modern problems and of our cultural response. Chapter 1 therefore sets up a series of 'environmental epochs' during which the prevailing attitude was relatively stable and for which there often turn out to be key events and key actors. The progress of scientific concepts, particularly of evolution (and latterly the Gaia hypothesis) provides an illuminating backcloth.

Just as patterns of attitudes to the management of the natural environment emerge in time, so are they also well developed in space. Part of the reason is the strung-out developmental sequence: East Europe, for example, still has aspects of industrial production (and therefore pollution) redolent of Victorian Britain. Chapter 2 addresses the spatial patterns which are produced by the process of pollution and attempts to summarize the toxicological and monitoring data upon which we base our environmental concern. We clearly need a formal appreciation of these patterns before setting up improved monitoring programmes and, thereafter, a policy and management response.

In Chapter 3, Tony Stevenson sets out the spatial and temporal aspects of the conservation of resources, principally the resource which is genetic diversity. This is, therefore, nature conservation but its fundamental ecological principles apply equally to the balance needed to exploit other resources. It is often assumed that the conservation voice is one of the many protest voices in environmental management process: Chapter 3 establishes the scientific credentials of — and priorities for — conservation of all biological resources.

The final two chapters of the first part of our volume establish the major social science principles of our relationship with the natural environment. Laws and the actions of law rose to prominence in our field towards the end of the Industrial Revolution largely because of the links made between industrial processes and a range of 'nuisances'; David Harte provides the crucial links between the progress made by case law in this era and the more recent evolution of statutory approaches to environmental management. Although pollution control might be considered a relatively 'simple case' for legislators, it is not; David reveals that the operation of legal processes through property rights makes countryside issues of the environment more straightforward (if less democratic).

At the time of writing this book there is a prevailing political view that environmental economics may prove more effective than laws in bringing our whole production system into a more stable relationship with the environment. It is, however, essential to know the rather limited armoury of economic techniques available. This Chapter 5 attempts to do, concluding that 'long-termism' is the most difficult of our requirements of environmental economics to codify.

In Part 2 our authorship is chosen to provide the typical researcher's concentration on one or two key processes in the environmental pollution field; this concentration is crucial to the effective design of both management and monitoring systems. The three elements, air, water and land, are treated separately (but see Chapter 14 for integrated control) and with an emphasis on their spatial dynamics. Alan Davison and Jeremy Barnes write with the benefit of both active field research programmes and participation in the formulation of European policy. The concept of 'critical loads' which they describe are a perfect translation into policy of geographical principles of pollution control. Critical load concepts will eventually be applied to freshwater pollution; in the interim, Chapter 7 describes a complex contemporary mixture of emission controls and water quality 'objectives'. Terry Douglas, an earth scientist with other interests in 'holes in the ground', presents a clear case that, even without toxic effects, solid waste is a limiting factor in the development process.

To conclude the section we contrast the very long-term pollution 'incident' — the accumulation of heavy metals in the environment — with the 'desperately' short-term — emergency spillages. Mark Macklin's research on metals in river systems has allowed him the ability to draw the essential spatial and temporal contrasts between what constitutes contamination and pollution. In a field where 'standards' are notoriously variable he is able to provide a geographical justification for stricter controls; these were not applied to the Victorian lead-mining which gives him so much of his field data in Northern England.

Paul Read and Bill Myers also have pollutant transport problems — on a scale of minutes! With chemical emergencies they need to know how many minutes it will take to set up emergency controls and how many more minutes to move people from their homes. Once again, the basic concepts of control such as the zone of consultation are geographical. In this chapter we collapse the space and time scales of much of the rest of the book whilst reminding ourselves of the potential death toll and damage from 'accidents'.

Stan Openshaw's work on aspects of the nuclear industry is well known for taking an explicit geographical/statistical approach to the location of nuclear power stations and of nuclear waste repositories. He has also published influential statements on the statistical distribution of childhood leukemia in relation to nuclear sites. In this chapter, Professor Openshaw broadens his outlook to cover the whole field of radiation, which as a pollutant has many distinguishing features in human perception and therefore management approaches.

Part 3 of the book is an attempt to assess the present priorities and future trends in environmental management. Energy-use patterns by human systems represent the largest point of departure from natural systems. Ian Fells places the energy field centrally and pivotally to all arguments over the natural environment, not the least the geographical variation of energy availability and use. The tension between the developed and developing worlds is one aspect of environmental management futures; nuclear power options may be the critical debating topic in both 'worlds'.

Rural activities are no longer 'in tune with

nature'. In terms of both pollution threats off-site and reduction of biological diversity on-site, our other fundamental global need — for food — forces us into the same fundamental viewpoint as does energy use. Chapter 13 examines the influences of the 'Gaia hypothesis' on our concepts of global management. An optimistic scenario of recovery and restoration can be developed for *some* natural systems.

Finally the future options for controlling pollution, carrying out environmental planning and marshalling the essential information and knowledge are examined in relation to the range of management problems which now confronts us. The main aspects of an holistic approach are drawn out.

References

Johnson, R.J. (1989), *Environmental problems: nature, economy and state*, Belhaven, London, 211 pp.

Editor's acknowledgements

This volume is a typical Newcastle effort, coming together in a 'dirty old town', striving to be an 'environment city', set in wonderful countryside; the authorship comes together to teach in the Geography Department at Newcastle University on a course started by Mark Macklin and Stan Openshaw, originally called Geography of Pollution. I, the interloper, broadened it (necessarily as it became deluged with students who have, on the whole, been appreciative) and the crew now meet regularly to dine and for the 'crack' of environmental views. We have been much encouraged by Iain Stevenson at Belhaven who thought it would work. The teamwork continues because virtually everyone amongst the non-academic staff in Geography at Newcastle has been drawn in to get the book out, principally Lynne Martindale, Ann Rooke, Eric Quenet, Brian Allaker, Sheila Spence and Yvonne Lambord. Watts Stelling provided metal analyses, custard creams and wine gums. Photographs were rescued from the slides by Monochrome, St. Thomas'. Joe Satriani's guitar kept the editor going through the night when the promises of high-tech editing became the reality of losing files and dealing with printer crashes. Vanessa Harwood's team expertly cleaned up the contaminants which arrived, 'end-of-pipe', at Belhaven; any 'traces' remaining are the Editor's responsibility.

Malcolm Newson
Newcastle upon Tyne, 1992

Part 1 Principles

1

Introduction: managing the natural environment — why and how

Malcolm Newson

> The environmental crisis is an outward manifestation of a crisis of mind and spirit... the crisis is concerned with the kind of creatures we are and what we must become in order to survive.
>
> Lynton K. Caldwell

1.1 Background to environmental concern

Few would quarrel with Caldwell's views but perhaps we need to pause more often to enquire how it was that human society surrounded itself with environmental problems of its own making or to ask, 'Under what practical, spiritual or ethical principles will we bring about improvements and survival?' To attempt to answer these questions we of course need to know more about the human condition than about natural environments; there is no better way in these circumstances than to examine the lessons of history. Because our species has intelligence it has the power both to think in the abstract and in advance of observations and to learn and react to observations. We therefore need to examine the roles of both the abstract and the concrete in shaping our modern environmental concern.

1.1.1 The groundswell: culture, ethics and environment

Our hominid ancestors are said to have evolved under conditions of environmental stress: the savannah plains of Africa, habitat for our early evolution, exhibit highly irregular climate cycles and marked seasonal shortages of food, together with a need for shelter. Thus, it has been argued, the very origins of our communal behaviour, communication and technology occurred in what editors might now headline as 'environmental crisis'. We may therefore take a measure of optimism about the innate ability of our species to manage out of trouble, but only if certain lessons are learned about the basic dimensions and limits of human society; these lessons have been central to the 'deep green' movement (e.g. Naess and Rothenburg, 1989) and the 'small is beautiful' attitude to technology and settlement (Schumacher, 1974).

It is also clear from archaeological and anthropological studies, as well as from contemporary investigations of the behaviour of indigenous peoples, that some form of environmental spirituality — a reverence for the individual elements and totality of the natural environment — steered the sustainable lifestyles of early humans. The widespread occurrence of artefacts symbolizing an earth mother female fertility deity is evidence of this spirituality.

However, the guiding spiritual element for that contemporary civilization putting most pressure on environmental systems, i.e.

Western, Christian society, is less unanimously associated with sustainable use of the earth. The form of the 'right' relationship between 'Mankind and Mother Earth' (Toynbee, 1976) has been the subject of many deep philosophical reviews (e.g. Passmore, 1980; Thomas, 1984). Toynbee (p. 38) charts the progress of human impacts from prehistoric times, stressing that

> when a hominid chipped a stone with the intention of making it into a more serviceable tool, this historic act, performed perhaps two million years ago, made it certain that, one day, some species or genus of the hominid family of primate mammals would not merely affect and modify the biosphere, but would hold the biosphere at its mercy.

The reviews by Passmore and Thomas inevitably point to the role of the Christian faith in forming Western environmental behaviour patterns, patterns which were to be exported throughout the world by exploration, conquest and colonialism. Attention focuses on the Bible as text and on one particular passage: Christianity teaches that mankind has dominion 'over all the earth and over every creeping thing that creepeth upon the earth'. We are to be 'fruitful and multiply and replenish the earth and subdue it' (Genesis 1: 26 and 28).

Naess and Rothenburg (1989, p. 187) point out that the most commonly occurring biblical terms for mankind's relationship with nature are 'guardian', 'administrator' and 'steward' and they attack this notion as follows:

> The arrogance of stewardship consists in the idea of superiority which underlies the thought that we expect to watch over nature like a highly respected middleman between Creator and Creation. We know too little about what happens in nature to take up the task.

There is no doubt that a Christian ethic of stewardship (supported by Passmore, 1980, as representing the seed-corn for a new environmental ethic) received a setback from the development in the seventeenth Century of Baconian, mechanistic, science. To Francis Bacon the very essence of science was its enabling and emancipating power for humans in relation to wilderness and unharnessed natural potentials. The logical outcome of this Western ethic was the triumphant period of exploitation which we now call the Industrial Revolution. However, science had also presided over what Thomas (1984) labels 'the dethronement of man', for example, through theories of organic evolution, though even these tended to encourage a view from the 'pinnacle' achieved in the 'perfection' of *Homo sapiens*.

Outside the question of animal rights, into which Thomas enquires very closely, Christianity had few reasons to question an association with the 'civilizing' influence of resource exploitation until the effects of an Industrial Revolution became manifest. Air pollution had begun in London as early as 1273 and by the time John Evelyn was writing his *Fumifugium* in 1661 the situation in terms of ill-health and other damage was far beyond Evelyn's subtitle phrase of . . . *the Inconvenience of the Aer* (see the 1972 reprint)! Aside from blaming Newcastle coal for the problem, Evelyn was a practical environmental manager, suggesting that 'all those works be removed five or six miles distant from London'.

Others who recorded the plunder of resources and the many effects which would now be labelled 'externalities' included George Perkins Marsh (1864), who recorded mainly rural plunder and Friedrich Engels who, during the mid-nineteenth century in industrial northern England, witnessed the appalling toll of industrialization upon the home, work and natural environments of the new cities. However, urban politics became based upon arguments over the means of production rather than the ends or effects of production. Karl Marx, for whom Engels' observations formed field data, took a characteristically materialist view of nature and diverted attention away from more spiritual and ethical arguments concerning the use of the environment and of Nature. It was left to unexpected groups, for example the Lake District romantic poets, to reinvigorate a passionate love of wilderness for its own sake. John Ruskin was perhaps unique in criticizing both the political economy of resource development and the impacts it was clearly having upon landscape, flora and fauna (Landow, 1985).

In a recent global overview of the ethics of environment and development Engel and Engel (1990) assemble contributors from faiths other than Christianity: Islam, Buddhism, Hinduism, together with traditional Chinese and African perspectives. Clearly none is much 'better' nor 'worse' than Christianity and the variability is mainly accountable to the role of animal rights and of a dimension which might be labelled 'closeness to earth'. Another, growing axis to environmental ethics, though not in a spiritual form outside the 'Earth Mother' sects, centres on the role and rights of women in the environment. Women are closer to practical environmental management in many societies and have distinctive environmental perspectives, but both of these potential contributions are generally subordinated in patriarchal societies. Questionnaire surveys in the University of Newcastle in the last five years have consistently shown females to be more concerned about environmental issues than males.

Engel (1990) describes sustainable development as representative of the core of a 'new' environmental ethic. He claims (p. 5) that

our present political and economic arrangements are only retained because they are perceived to be legitimate, and their legitimacy rests ultimately on the perception that they are ethically justified. History teaches that once this ethical justification is challenged by new moral sensibilities, and legitimacy withdrawn, the arrangements are likely to change.

However, we should beware that, in the case of global environmental problems, ethical stances are likely to be yet more diverse than world religions; we cannot take Western, Christian, mechanistic science perspectives as in any way pervasive. Engel goes on to admit that 'It is difficult to make even the most accepted moral principles operational in a world of sovereign and antagonistic states' (p. 5). He states four ways in which an interest in spirituality, morality and ethics can have practical utility:

1. it helps explain the role of human values;
2. it explains human motivations;
3. it gives moral guidance to alternative courses of action;
4. it helps resolve conflicts.

Such an interest, therefore, is not merely a framework for analysing historical trends but a very practical tool in the crucial social context of environmental management. Arvill (1967) was bold enough to offer candidates for environmental ethics (Table 1.1) in an era where mankind is dethroned by biological science, depressed by the impacts of applied science/technology, and too far from primitive origins to follow the sustainable patterns of behaviour followed by indigenous peoples.

Table 1.1 Robert Arvill's candidate environmental ethics

1. INTEGRITY	to infuse ecological precepts and a true environmental ethos into your lifestyle
2. HUMANITY	to share the earth's resources more equitably with all life on earth today and tomorrow
3. DETERMINATION	to arrest pollution and squalor and to promote quality in your surroundings
4. JUDGEMENT	to choose wisely between competing and conflicting aims and values in order, with humility, to promote the trusteeship of society for the environment

1.1.2 Modern times: issue-attention cycles

Arvill's canidate ethics may win respect but not a following; Naess's more profound agendas penetrate very little into active environmental management (see below). The dominant resonance of 'Mankind and Mother Earth' in modern times has been, and continues to be an 'issue-attention cycle'.

Human beings, even when armed with a range of scientific methodologies championing prediction, are notoriously poor at anticipatory action unless the hazard signs are very clearly displayed. Humans approach a hazard with caution only if they have experienced that one in the past. For these reasons environmental events (often disasters — see Chapter 10) have tended to power the somewhat jerky progress of modern environmental management; ethics have had little influence despite their potential utility as part of coping strategies. In the twentieth century the rapid and all-pervasive influence of publicity through the media has meant that 'issues' and 'crises' are frequently the main channels through which the public are kept informed about the natural environment and their own impacts on it.

Downs (1972) identified five phases in the 'issue-attention cycle' followed by American public opinion:

1. pre-problem; it exists but is not recognized;
2. discovery and enthusiasm — often prompted by dramatic events;
3. appreciation of complexity and costs of the problem;
4. gradual decline of public interest;
5. post-problem — in the case of environmental issues a long phase with periodic resurgences of interest in the 'solution' (or its failure).

Sandbach (1980) uses press coverage of environmental issues to develop broader, composite versions of the issue-attention cycle, such as the rapid growth in interest in all environmental issues in the late 1960s and the plummeting of this concern in favour of the economy and work after the energy crises of the early 1970s. In the developed world it seems that this composite rise and fall has now smoothed out; the environment is regularly selected as important, despite the condition of the national or international economy. Public opinion, such as that recorded by opinion pollsters and questionnaire surveys in the developed world, regularly lists the environment and its component management problems as a matter of concern. However, within the 'motor' of progress towards environmental management the 'sparking plugs' still consist of individual issues and their attention cycles: dead seals, cancer clusters, fish kills and ozone thinning come and go by the week.

Certain issues have, through their widespread impact, had a more profound effect on the progress of management. Several reviews have been made of the landmark events (which can include action by leading scientists and politicians) of the 'New Environmental Age' (Nicholson, 1987). Nicholson himself selects, from personal, professional experience, the influence of such books as *The Earth as Modified by Human Action* (Marsh, 1864) and *Silent Spring* (Carson, 1962), events such as the wreck of the oil tanker *Torrey Canyon* off the Scilly Isles in 1967, and conferences such as the United Nations Conference on the Human Environment in Stockholm in 1972.

Other reviewers, such as McCormick (1989) have charted the build-up of pressure for positive management from environmentalists (i.e. active campaigners, rather than passive operators) in the 'global environmental movement'. Membership figures for pressure groups are obvious sources of data; the point in the late 1980s when these groups, in Britain, overtook membership of trade unions was clearly a landmark to our changing priorities. McCormick uses the device of tabulating the organizations and treaties which chart the jerky progress of the movement at an international level. However, his chapters themselves provide a broader classificatory guide to the growth of

Introduction

Table 1.2 Significant individuals and events in modern environmental management

Date	Category	Details
1864	Book	George Perkins Marsh's international evidence of land and water degradation. Raised the issue of longer-term implications of productive land use.
1956	Book	Thomas's edited compilation constituting an update on Marsh's warnings.
1952	Disaster	London smog, 5–10 December. 445 direct deaths and 4000 longer-term fatalities. Led to the Clean Air Act 1956 — UK first developed nation to act.
1962	Book	Rachel Carson's *Silent Spring* revealed the environmental effects of the early, crude pesticides such as DDT; established the wide geographical spread and persistence of chemical pollutants.
1967	Disaster	Wreck of the oil tanker *Torrey Canyon* — 117,000 tonnes of crude oil spilled and washed on to holiday beaches. Drew attention to the dangers of increasing scale of energy sources/trade and inadequacy of government response and dispersant technology. Led to formation of a UK government department (of environment).
1969	Picture	Apollo space programme — lunar landing/widespread 'spiritual experience' among astronauts. Earth-view established in environmental campaigns and advertising.
1972	Conference	Stockholm Conference on environmental issues affecting development. Led directly to environmental policy-making in the European Community.
1979	Book	James Lovelock's *Gaia* further developed the global biosphere concept and hypothesized the mutual dependence of organic material and its habitat. Transcendentalists take up Gaia as a spiritual guide; environmental science uses it to prioritize the global management task, e.g. controls on the loss of stratospheric ozone.
1986	Disaster	Chernobyl nuclear power plant USSR: 50 miles from Kiev, 32 deaths and 135,000 evacuated. Long-term, international health effects. Established, with 'acid rain', the international transport of pollution in the atmosphere. Discussion of risks in a world based upon nuclear energy given a case example.
1990	Book	Clark University's global review in the Marsh (1864), Thomas (1956) . . . series. First attempt to separate the influence of three centuries of development from natural trends and cycles, i.e. the true impact of mankind.
1992	Conference	United Nations Conference on Environment and Development — 'The Earth Summit' in Rio de Janeiro.*

* This is a hopeful entry!

environmental issues:

'The roots [largely Victorian]'
'Protection [conservation/wildlife] 1945–1961'
'The environmental revolution 1962–1970'
'Prophets of doom [resource exhaustion, pollution] 1968–1972'
'The Stockholm Conference 1970–1972'
'The United States Environment Programme 1972–1982'
'Northern Hemisphere politics 1969–1980'
'Southern Hemisphere — environment and development 1972–1982'

Table 1.2 provides a sample of the flesh of issues on this skeleton structure.

Every author is bound to produce a biased and partial view of the significant developments; the bias of Table 1.2 is towards events and individuals who have contributed towards academic and legislative progress in the United Kingdom in the modern era. The contents of the table illustrate both the issue-attention cycle on a large scale and the longevity of Downs's 'post-problem' stage.

Clearly scores of environmental issues have quickly faded from our attention in the

Table 1.3 Niches in the pattern of policy opportunities

	INDIVIDUAL	Scale CORPORATION*	STATE
Agent			
MORAL PERSUASION	'home ecology'	environmental audits	environmental assessment on policy
ECONOMICS	conversion to clean fuels	production of 'green' goods and technology	conservation-for-debt tax incentive
LEGISLATION	planning law smokeless zones tree preservation	litter laws labour and the environment	law of the sea catch limits

* We have concentrated upon commercial corporations here.

modern era because news editors have discarded them, because they were purely local or evidence was circumstantial or because in an unspecified way human society decided to continue facing the hazard.

We can detect in those events which have had a longer impact the popularity of command-and-control measures in policy-making (e.g. new legislation or new government departments). Such policies react to events; alternatives to command-and-control require anticipatory, precautionary action. The contribution of academic ideas seems mainly to have been to confirm at an understandable level the broader patterns, especially geographical patterns (see Calder, 1991), of human impacts rather than to predict them and to shape precautionary action.

1.2 People and environment: scales of interaction

The high content of geographical principles in this book makes the question of scale of measurement, management and policy imperative; put another way, section 1.1 has hinted at roles for 'actors' in environmental concern and remedies, and we need at least a cast list or order-of-appearance for them.

Johnston (1989) (p. 110) argues a division between individual and collective action in environmental management:

although individual initiative in the attack on environmental problems is desirable there are major limits to it, especially within competitive modes of production, and viable approaches to both solving existing environmental problems and preventing the creation on others must involve collective agreements.

Johnston's chosen geographical scale for analysis becomes the nation state but since he also identifies capitalism as the dominant mode of social production we also need to operate on a second collective scale for which the term 'corporation' is appropriate, although it must be given extraordinary breadth to include, for example, non-governmental organizations and commercial companies. We see the corporation as intervening between the state and the individual.

1.2.1 Management requires policy: how is policy formed?

There are said to be three ways in which policy on environmental management changes:

moral persuasion (or 'good practice');
legislation (command and control);
economic manipulation.

These could equally well be routes along which policy operates (see also Chapter 14), although Johnston (1989) prefers four types of action at state level:

goal-oriented: planning for the future;
problem-solving: amelioration of the present;
allocative: promoting certain trends;
exploitative: opportunity-seeking, precautionary.

Seeking symmetry we retain the first, triple division and merge it with our inherent triple of geographical scales in Table 1.3.

1.2.2 Environmental ideologies, politics and policy

This theme is close to that chosen by Sandbach (1980) in his far-sighted compilation of Marxist and pluralist alternatives in formulating environmental policy at the scale of the state, and his analysis of the roles of pressure groups and political parties. He separates the functionalist, pluralist environmental ideologies of the twentieth century from the Marxist, materialist standpoint. The former, according to Sandbach, is the more political since it often naïvely reinforces the existing power structures:

popular ecology, systems analysis, cybernetics, decision theory, technology assessment and cost-benefit analysis could be seen as agents of social control rather than liberation. (p 26)

There is no doubt that policy-making on the environment must not continue as an uncoupled muddle of the ethics and issues raised in section 1.1 above; a coherent political debate is badly needed and is inevitable as the decisions needed for long-term environmental management become more profound and hard-hitting. When human lives face profound changes there is bound to be a political dimension but it remains difficult to see quite what this will be in existing Western societies, even after the rise of their Green Parties.

In the transition from environmentalism to environmental politics it is impossible to neglect material interests; even the forerunners of the deep ecology movement (practitioners of what Bramwell, 1989, calls 'ecologism') represented at best a High Tory, landed, privileged axis in politics and at worst a Nazi connection (from which the German Green Party still seeks to distance itself).

It remains difficult, therefore, to see whether environmental politics will turn out to be (in terms of UK political colours) 'blue-green' or 'red-green'. Paehlke (1989, p. 273) suggests that the latter is more likely because of the stage of development which has been reached by environmentalists:

Environmentalism as an ideology is now at a stage of development comparable to that of socialism a century ago. Environmentalism may never obtain a mass base similar to that of conservatism, liberalism or socialism, but it has already transformed the way many people understand the political world. Environmentalists have produced a sociological, political, economic and philosophical literature of remarkable breadth, depth and variety that has significantly affected the politicial and administrative agendas of most nations of the world.

Paehlke lists ten priorities in an agenda for 'contemporary environmental progressivism'; he also notes a divergence of fundamental proportions from the socialist agenda because 'distributional issues take second place to choice of technology, design of technology and use of technology' (p. 189). He designs a simple axis diagram in which 'left–right' issues are swung by ninety degrees to 'environmentalism–antienvironmentalism'. If readers cannot identify an 'antienvironmentalist' view perhaps they should consider the extremely hostile reaction to Rachel Carson's *Silent Spring* (Hynes, 1989) or the speeches and articles written in 1991 by the UK Conservative MP Theresa Gorman on 'eco-terrorism'! Bramwell (1989) is a little more coherent in her rejection of the ecological movement which she accuses of advocating a return to primitivism and 'concomitant anarchy'.

What these analysts neglect by grouping

environmentalists as a political force is the inherent breadth and lack of cohesion in environmental ideology; 'holism' and 'pluralism' are semantically inadequate to cover the spectrum of pressure groups within which environmentalism now resides. The ecologists' theory of niche is perhaps the best guide; each individual has found a component of the movement with which they are comfortable and the lack of dialectic has removed any need for discipline in debates.

John McCormick (1991, p. 35) reviews British environmental politics and bravely attempts a typology of the pressure groups:

The words 'movement' and 'lobby' imply homogeneity, cooperation, singleness of purpose, unity and steady evolution. But this is misleading . . . There are sectional groups with demands which can often be conceded with minimal public controversy. There are economic and ideological groups. There are functional groups (mainly pursuing economic interest) and preference groups ('united by common tastes, attitudes of pastimes'). There are emphasis groups . . . and promotional groups.

The struggle of the UK Green Party to convert an increasing sympathy with sustainable environmental management to a political support takes place against this background. Meanwhile conventional political parties appropriate elements of the environmentalists much like a clothes purchaser in the fitting room. It seems inevitable that policy-making in environmental management may have to proceed for the foreseeable future against an incoherent political force field. The irony is, however, that it is unlikely that policies will succeed unless they are the subject of the widest public consultation (see Chapter 14), almost the equivalent of government by plebiscite.

1.3 Knowledge and institutions: defining the breadth of policy

As revealed in the foregoing sections, one of our yardsticks for recording progress in environmental management has been the connection between events, issues and their interpretation by academic authors *en route* to institutional incorporation. As we discuss in Chapter 14, however, the development of active environmental policies depends as much on a sound and unified knowledge-base as it does upon similar qualities of political pressure. If we regard current environmental politics as 'undisciplined' we must throw the same insult at environmental science (the irony being that it is the organization of knowledge into 'disciplines' which is much to blame!). A tentative wish-list for environmental futures might well include a sound knowledge-base ('sound' means both accurate and acceptable) and adaptive institutions of learning, finance and law to put policies into practice.

1.3.1 Academic disciplines: how do we teach and research the environment?

Because of ever-increasing specialization in education and research, an economist is expected to be ignorant about glaciers and a glaciologist about economics. So when experts discuss the state of the planet, it can be like hearing a chiropodist and a dentist arguing about a patient's heart condition. (Calder, 1991, p. 23)

The specialization of academic disciplines has both intrinsic and extrinsic origins; within disciplines progress is driven by very detailed research at locations near 'the cutting edge'. Research is gratifying intellectually and brings rewards for the individual, his/her discipline and the institution. Were we without academic institutions and were they not, currently, so poorly funded, it is unlikely that competition between disciplines would be so well-developed as it is in a modern Western university, for example. Many such institutions have gladly embraced environmental problems as a further opportunity for detailed 'reductionist' research (see Chapter 14) but, on discovering the inevitable multidisciplinary nature of the topic, have found severe problems of combining breadth and depth; even greater problems then emerge of peer-group acceptance, publication and further funding.

To geographers the quandary is not new; during the Victorian period geography struggled politically to join the growing range of specialisms in British universities. It succeeded mainly because of the serviceability of its knowledge-gathering about a far-flung empire. Halford Mackinder's vision of the subject stressed the interplay of the natural sciences and their joint relevance to an emerging political geography of resource exploitation.

Nevertheless, the holistic approach cannot be said to have triumphed in geography at all; lack of methodological rigour and an imitative specialization to find favour with other disciplines have dogged the subject. Even the integration of geology, geomorphology, biogeography and climatology under the Victorian heading of 'physiography' broke down (Stoddart, 1986) in the face of specialization, and the challenge of linking the physical and human environments was unanswered because both aspects of the subject became prey to the fashion of reductionism and of mechanistic, not organicist, science. If we regard geography as a talisman of holism in academic institutions we emerge with, at best, the impression of a field of knowledge which encourages an appreciation of environmental problems, rather than of a discipline which solves them. One danger is that this appreciation leads to a glib academic self-satisfaction (Newson, 1992a, in press).

Nevertheless, on a more optimistic note, higher education in the western world is now showing signs of a willingness to form interdisciplinary teams and geographers are often towards the core of such groupings. At school level UK education policy has recently seen fit to develop a National Curriculum in geography and has offered the subject a key role in educating young people about environmental problems and their management. There is no doubt that the effect of the high profile given to environmental education (including TV and magazine coverage) in the last decade has made children highly aware of the problems, if not the solutions, and this is bound to have political and institutional effects in future decades.

Meanwhile methodologists face the continuing problem of defining and making operational a rigorous treatment of both the ontological depth and epistemological flexibility demanded by environmental management; geographers have much to offer this debate.

1.3.2 Practical management: institutions

We cannot blame academic leadership alone for an inappropriately structured knowledge-base for the New Environmental Age. If knowledge is to have practical applications these normally emerge in partnership with the institutions of state regulation concerned, for example, with law, finance and technology. Professionals enter these institutions from education and, like researchers, they appear to be performing better if specializing. Governments design sectoral policies, e.g. for agriculture, energy, forestry and water, to cement this separation, perhaps in the hope that 'creative tension' will offer the best policy outcomes on interdisciplinary issues.

Whilst the United Kingdom's Department of the Environment and the United States' Environmental Protection Agency are examples of two kinds of government institution with deliberate interdisciplinary aims, the ideal situation seldom translates into the critical field of administration. Policy-making, when it comes up against what is practicable, is not merely faced by the constraints of the financial resource but by the units by which administration is conducted. These are geographical units of nation states or tiers of local government.

It is interesting, therefore, to consider the fundamental administrative units by which interdisciplinary institutional structures might act. For example, air-pollution management through either legal or fiscal means operates in regions where the air circulation can be mapped, e.g. Los Angeles.

Waste management has many regional components, as revealed here by Chapter 8. The easiest prescription for environmental management units comes, perhaps, from the author's own specialism of river basins.

The river basin unit is easy to define; most people who live in a basin know the river's name and have an identity with the river as a source of wealth, beauty, being (a particular feature of Newcastle *upon Tyne*!). As a result it has been possible in a number of nations to translate the many policy recommendations on the desirability of holism in administration into institutional and administrative structures which fit the bill (Newson, 1992b).

Gardiner, Thomson and Newson (1992, in press) review river-basin management institutions in Ontario and England, concluding that successful strategies include boundaries of administration corresponding with basin limits, specific data-gathering and multi-disciplinary teams, a national underpinning (important in dealing with the private sector), and, above all, a firm policy of public consultation and involvement. It is critically important to emphasize that environmental management is public property! As municipal planners have recorded throughout the world, trees planted by planners are torn up, whereas trees planted by communities stand a much better chance (if vandalism occurs it becomes internally regulated). Clearly, therefore, the principle of subsidiarity in decision-making, and therefore of consultation in policy-making are likely to feature strongly in environmental management schemes of the future. Whilst river basins have an immediate resonance in the minds of both regulators and regulated, it is not beyond the wit of researchers to start to define air basins, land basins or other fundamental units for management and administration around which institutions will put in place the knowledge and policies. Dryzek (1990) considers that, at present, environmental administrations use inappropriate (reductionist) frameworks for knowledge, whilst Torgerson and Paehlke (1990) suggest that environmental politics (even in its 'Not In My Back Yard' format) represents the failure of conventional administrative states to manage the environment. We may well see a progression of the philosophy of 'small is beautiful' (Schumacher, 1974) which gripped the imagination of the hippy generation through its emphasis on energy generation, food supply and appropriate technology in small communities, to a new form, without reference to the precise scale of communities. The new slogan might therefore be 'subsidiarity makes sustainable solutions'.

The daunting challenge of global environmental issues is, however, to translate this institutional evolution for planetary scales.

References

Arvill, R. (1967), *Man and Environment: Crisis and the Strategy of Choice*, Penguin Books, Harmondsworth, 376 pp.

Bramwell, A. (1989), *Ecology in the 20th Century: A History*, Yale University Press, New Haven, 292 pp.

Calder, N. (1991), *Spaceship Earth*, Viking, London, 208 pp.

Carson, R. 1962 (1987), *Silent Spring*, Houghton Mifflin, Boston, 368 pp.

Downs, A. (1972) 'Up and down with ecology: the "issue-attention" cycle', *The Public Interest*, 38–50.

Dryzek, J.S. (1990), 'Designs for environmental discourse: the greening of the administrative state?' in R. Paehlke and D. Torgerson (eds), *Managing Leviathan: Environmental Politics and the Administrative State*, Belhaven, London, 97–111.

Engel, J.R. (1990), 'The ethics of sustainable development', in J.R. Engel and J.G. Engel (eds), *Ethics of Environment and Development: Global Challenge and International Response*, Belhaven, London, 1–23.

Engel, J.R. and Engel, J.G. (1990), *Ethics of Environment and Development: Global Challenge and International Response*, Belhaven, London, 264 pp.

Evelyn, J. (1661), *Fumifugium Or the Inconvenience of the Aer and Smoake of London Dissipated* (reprinted 1961, 1972, National Society for Clean Air, Brighton, UK), 39 pp.

Gardiner, J., Thomson, K. and Newson, M. (in press), 'Integrated watershed/river catchment planning and management: a comparison of selected Canadian and United Kingdom experiences', Institution of Water and Environmental Management, IWEM '92 Conference, Birmingham, April 1992.

Hynes, H.P. (1989), *The Recurring Silent Spring*, Pergamon Press, New York, 225 pp.

Johnston, R.J. (1989), *Environmental Problems: Nature, Economy and State*, Belhaven, London, 211 pp.

Landow, G.P. (1985), *Ruskin*, Oxford University Press, Oxford, 99 pp.

Lovelock, J.E. (1979), *Gaia: A New Look at Life on Earth*, Oxford University Press, Oxford, 157 pp.

McCormick, J. (1989), *The Global Environmental Movement*, Belhaven, London, 259 pp.

McCormick, J. (1991), *British Politics and the Environment*, Earthscan, London, 201 pp.

Marsh, G.P. (1864), *Man and Nature or Physical Geography as Modified by Human Action* (reprinted 1965, Harvard University Press), 472 pp.

Naess, A. and Rothenberg, D. (1990), *Ecology, community and lifestyle. Outline of an ecosophy*, Cambridge University Press, Cambridge, 223 pp.

Newson, M.D. (1992a), 'Twenty years of systemic physical geography: issues for a "New Environmental Age"', *Progress in Physical Geography*, 16(2), 209–21.

Newson, M.D. (1992b), *Land, Water and Development*, Routledge, London, 340 pp.

Nicholson, M. (1987), *The New Environmental Age*, Cambridge University Press, Cambridge, 232 pp.

Paehlke, R.C. (1989), *Environmentalism and the Future of Progressive Politics*, Yale University Press, New Haven, 325 pp.

Passmore, J. (1980), *Man's Responsibility to Nature* (2nd edn), Duckworth, London, 227 pp.

Sandbach, F. (1980), *Environment, Ideology and Policy*, Basil Blackwell, Oxford, 254 pp.

Schumacher, E.F. (1974), *Small is Beautiful*, Sphere Books, London, 255 pp.

Stoddart, D.R. (1986), *On Geography*, Basil Blackwell, Oxford, 335 pp.

Thomas, K. (1956), *Man's Role in Changing the Face of the Earth*, University of Chicago Press, Chicago, 1193 pp.

Thomas, K. (1984), *Man and the Natural World: Changing Attitudes in England, 1500–1800*, Penguin Books, Harmondsworth, 246 pp.

Torgerson, D. and Paehlke, R. (1990), 'Environmental administration: revising the agenda of inquiry and practice', in R. Paehlke and D. Torgerson (eds), *Managing Leviathan: Environmental Politics and the Administrative State*, Belhaven, London, 7–16.

Toynbee, A. (1976), *Mankind and Mother Earth: A Narrative History of the World*, Oxford University Press, Oxford, 641 pp.

Turner, B.L., Clark, W.C., Kates, R.W., Richards, J.F., Matthews, J.T. and Meyer, W.B. (eds) (1991), *The Earth as Transformed by Human Action: Global and Regional Changes in the Biosphere over the Past 300 Years*, Cambridge University Press, Cambridge, with Clark University, 713 pp.

2

The geography of pollution

Malcolm Newson

2.1 Global patterns framing environmental management

Recent, well-publicized pollution events and their lingering effects have had a profound educational effect on our geographical perception of the natural environment and its artificial modification on a large scale. The revelation that depletion of the natural layer of ozone in the upper atmosphere over the Antarctic had a direct link to CFC production in the developed world of the Northern Hemisphere and the global influence of CO_2 production in creating an enhanced 'greenhouse effect' are clear examples of the extent to which we all now appreciate one aspect of our modified environment: the global geography of pollution.

2.1.1 Images of the planet

Many observers of the prodigious global spread of environmental information via 'the media' have pointed to the importance of global images from space in the late 1960s and 1970s as a precursor of our ability to conceptualize global problems (Table 2.1). The pervasiveness of the 'earthrise' photograph from the Moon, taken by the crew of Appollo II (20 July 1969) has led Wombell *et al.* (1989) to compile a guide to a large selection of recent graphic representations of our planet. As Warn says in this compilation.

If global empires can be built by advertising companies like Saatchi and Saatchi for world products like Coca-Cola, MacDonalds or Levi Jeans it cannot be beyond the realm of possibility to achieve full globalisation for ideas that will protect our threatened planet from the consequences of narrow nationalism. (p. 76)

Fred Hoyle presaged the impact of this visualization on the perceptions of both scientists and lay people as early as 1948. His revolutionizing theme is extended in the recent compilation of the plethora of space views now available, presented by Calder (1991).

The contemporary perspective on the burden of global environmental management can be exemplified by the words of D.J. McLaren (1990, p. xiii), President of the Royal Society of Canada.

We now see the Earth as a small planet in space that is inherently changeable. In the very recent past, the emergence of the human race began to cause change on the environmental flux more rapidly than, and in a different manner from, the established system.

A paradox of the 'globalization' of environmental concern has been that it represents a return to many of the perceptions and scientific preoccupations of the past. After all, it was James Hutton the geologist who in 1795 advocated a scientific remit of global study by making a parallel between earth and organism: 'The explanation which is given of the different phenomena of the Earth must be consistent with the actual constitution of this Earth as a living world' (p. 546).

Geographers, as Clayton (1991) complains, have been concerned to move away from the globe as a scale of research and as a teaching device (though there is now a brisk market in inflatable globes). With the exceptions of those working on remote sensing (mainly physical geographers) and those studying world political systems, much of the methodology in geography has moved in the direction of reductionism. The benefits have been greater insight into the processes of environmental and social science but the danger is now that geography as a discipline will be forced, on environmental matters, to adopt merely superficial treatments on a global scale. In this book we attempt to show how that danger can be averted by a policy-orientated, carefully scaled geographical treatment.

2.1.2 Spatial differentiation of the global environment

We may consider the following brief list as a summary of the many factors leading to geographical patterns of the natural environment, of the threats to it and the management systems applied to it as comprising:

A. environmental background (e.g. climatic demarcation of drylands);
B. cultural/political overlays (e.g. developed urban centres of US drylands);
C. developmental/land use overlays (e.g. traditional irrigation in Middle East/African drylands).

In the example quoted the natural semi-arid condition, at the hazardous margins of human occupation and hence vulnerable to anthropogenic abuse, is the result of contemporary world climate patterns but its vulnerability results from overlays of human occupancy — a basic tenet of geography. It is fair to consider climate and bio-energetic patterns as the most important agent of environmental differentiation during the history of man as a widespread inhabitant of the planet. Whilst we are often encouraged to consider geology, geomorphology or human agencies to be dominant controls, the Holocene period has undoubtedly seen the atmosphere/biosphere interaction as of profound influence on environmental patterns. The primary back-cloth pattern to the geography of both pollution and the conservation of biodiversity (Chapter 3) is therefore that of climate. It is through the workings of the planetary atmosphere heat engine that the global pollution problems of 'greenhouse' gas production and ozone depletion are activated. Any 'cure' or mitigation depends upon a precise understanding of processes at this scale. Similarly, with or without climate change, the conservation of globally important habitats such as tropical forests or savannah grasslands requires a knowledge of the climatic forcing functions for these habitats and their natural productivity via conversion of solar energy. Traditional climate structures of mean temperatures and rainfalls do not hold the major key to habitat — balances (e.g. between rainfall and evaporation), seasonalities (e.g. monsoons) and extremes (e.g. hurricanes/floods) are major components of the earth's environmental patterns even in human land use. In recent years it has become clear to researchers that a proper name for our planet would be 'Ocean', since the majority of its surface is composed of water. It is now considered that the oceans hold the key to storages and fluxes in energy, moisture and chemical cycles. Figure 2.1(a) illustrates another aspect of natural global variation and introduces a source of one of the 'greenhouse' gases, methane. Wetlands are an important source of methane but are also considered to be a habitat deserving of conservation. Immediately we draw into our argument the management of the very important global

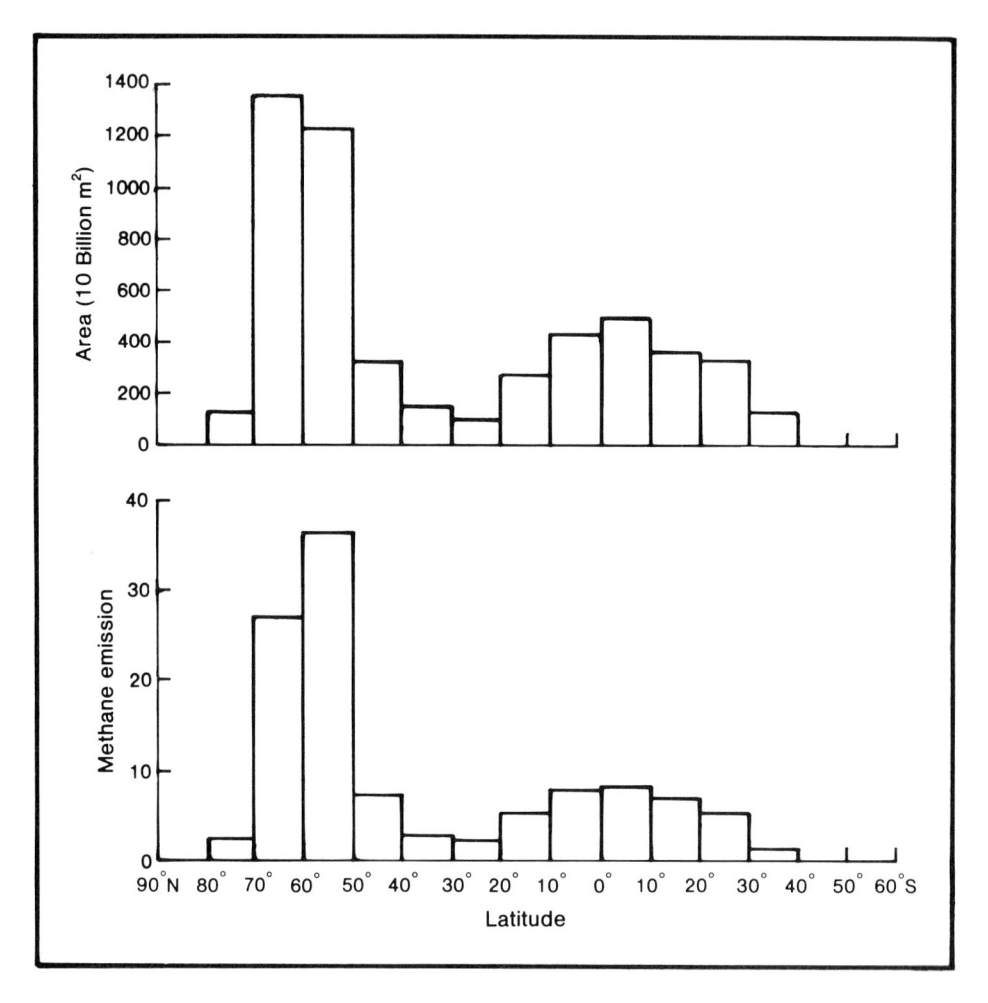

Figure 2.1a Global methane production by latitude, illustrating the importance of wetlands (above) as a source. (Matthews and Fung, 1987, *Global Biogeochemical Cycles*)

biogeochemical cycles whose fluxes of solids, solutes and gases form the background to our definitions of contamination and pollution (see below).

When pollution, such as chronic air pollution from urban and industrial sources, is considered on a global scale we can predict from a knowledge of physical, cultural and developmental overlays the likely spatial impact — for example, the patterns of rainfall acidification shown in Figure 2.1(b).

Before moving on from physical to cultural developmental aspects of world environment we must add that the most challenging way of considering the natural environment of the globe is one under the framework of environmental capacities (see also Chapter 13). Hitherto, the capacity of the Earth to sustain the activities of *Homo sapiens* has been considered mainly in terms of the exploitable, extractable resources available. We now need, however, much more information on the exploitable intrinsic resources represented by the capacities of oceans to receive pollutants, the atmosphere to circulate and disperse them, and the gene pool to rebound from grave losses of genetic stock. In terms of the spectrum of development,

The geography of pollution

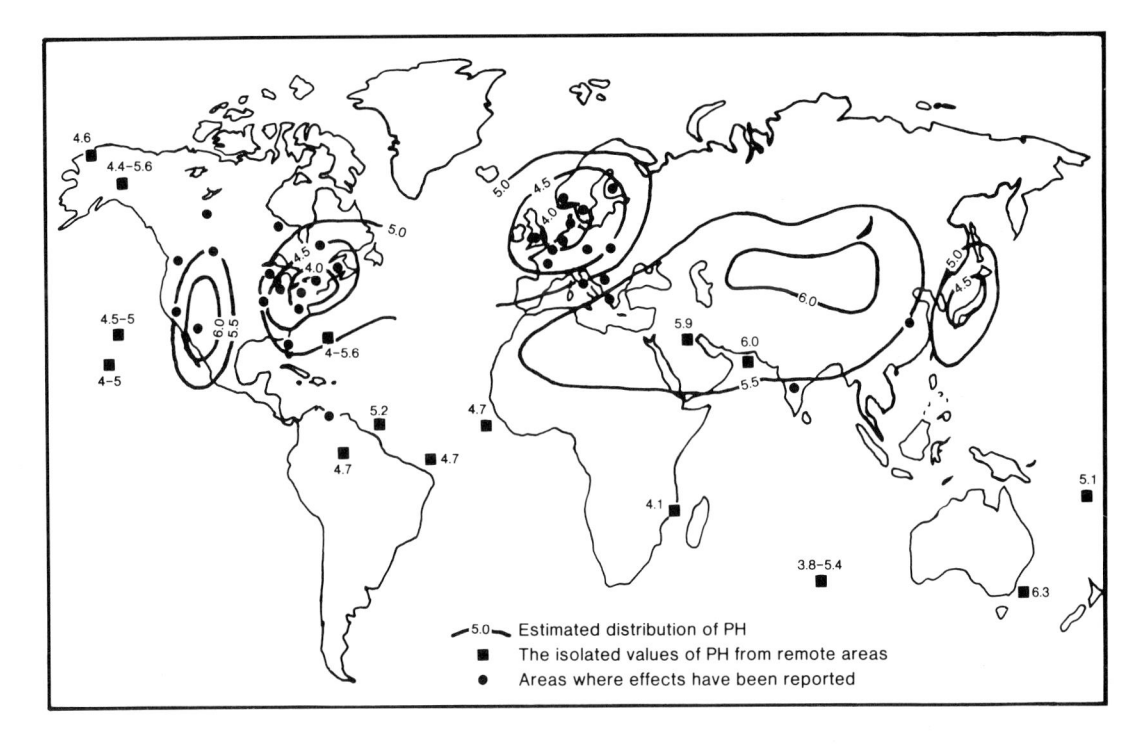

Figure 2.1b Global patterns of acidity in precipitation. (Park, 1987)

stages and strategies of which further account for spatial differentiation of the global environment, we may group as a caricature the priorities and preoccupations of the developed, and least-developed countries (LDCs).

In developed nations the most pressing environmental problems relate to the processes of production. Broadly these include safety (in relation to production processes, natural hazards in such nations being largely controlled), amenity and conservation.

Issues confronting the LDCs concern the use of natural resources in the exploitation of national assets. In turn control and ownership of these resources are politically paramount. Interest in environmental degradation may come first from researchers or aid organizations called in from outside the state concerned. For example, concern over soil erosion often results from site investigations for hydro-electric power schemes and their river catchments. Table 2.1 illustrates the range of attitudes in environmental management by sampling across the world development spectrum by national approaches.

Paehlke (1989, p. 156) distinguishes between the problems of the developed countries and the crises of the LDCs.

In the former the highly developed nature of civil society makes it possible for worker's and consumer's interests to be politically represented. Information on the state of the biosphere, hazards at work and so on reach governments in quantities unimaginable in most LDCs.

One may further caricature the overlay of development status and cultural identity by comparing attitudes to major aspects of environmental management prevalent in four nations or groups of nations (Table 2.1). The value of wilderness is more closely held in the United States compared with Europe. The attitude to primary resources and then exploitation differs profoundly between

Table 2.1 Attitudes in environmental management

Purpose of environmental management	Resource-population type			
	US	European	Brazilian	Egyptian
Exploitation of materials	Dominant aim but decreasing emphasis	Dominant but decreasing: imports always significant	Dominant as industrialization proceeds	Basic development often for export to USA and Europe
Life-supporting, i.e. food and ecological stability	Increasing realization of wider implications	Importance only recently realized: too densely populated to conserve wilderness	Food important, environment not much valued, outside aboriginals	Struggle for food often dominant, knowledge of other processes discounted
Beauty	Strong motivation for preservation as reaction to over-use	Always a feature of culture: getting stronger in face of increased impact	Residual or marginal land may thus be destroyed	Wildlife on residual land

Brazil and Egypt. These differences exemplify the three factors of nature, culture and development and then interdependence which this brief review seeks to elucidate.

Analysts of world patterns of environmental problems and their solution are increasingly turning to patterns of world trade further to structure and reinforce those produced by the development sequence. Both the Commissions headed by Willy Brandt (ICIDI, 1980; Brandt, 1986) and Gro Harlem Brundtland (WCED, 1987) identified connectivity between production and consumption of resources, hazards of resource depletion and damage from waste disposal. A simplified example will illustrate the point: oil from Saudi Arabia fuels the machines and makes the fertilizers and pesticides which allow marginal land in West Africa to grow a crop of cocoa for Switzerland to make into chocolate which it flies on American-made airplanes to Singapore for distribution in South East Asia; the profit made by the West African farmer allows him to purchase a Japanese motorcycle, Ethiopian coffee and Thai rice.

Perhaps the most interesting position in terms of development of nation states and the global environment is currently occupied by Japan. Japan's economy now accounts for 12 per cent of the world's revenues, a similar proportion to that of fast-growing economies of the past. Japan's need for raw materials has produced environmental degradation in supplier nations. Similarly, energy generation and the resulting atmospheric pollution now puts Japan into third place in the 'big league' of polluters. The island location of the Japanese nation has now had the effect of concentrating the politicians' minds on the 'greenhouse effect' and resulting rise of sea level. A white paper to the Japanese parliament in 1988 reported that:

Japan is a leading importer of resources in the world. *Japan is closely related to the global environment* due to the fact that she imports not only exhaustible resources such as metallic materials and petroleum but also renewable resources such as agricultural and marine products, wood and textiles. (Cross, 1988) (our italics)

Among the imports most concerning Japan's own environmental lobby are timber from tropical forests and shellfish from cleared mangrove coasts. Interestingly, however, only 26 per cent of the Japanese are concerned about the extinction of species, compared with 42 per cent in Western Europe (Cross, 1988) and there is not a

The geography of pollution

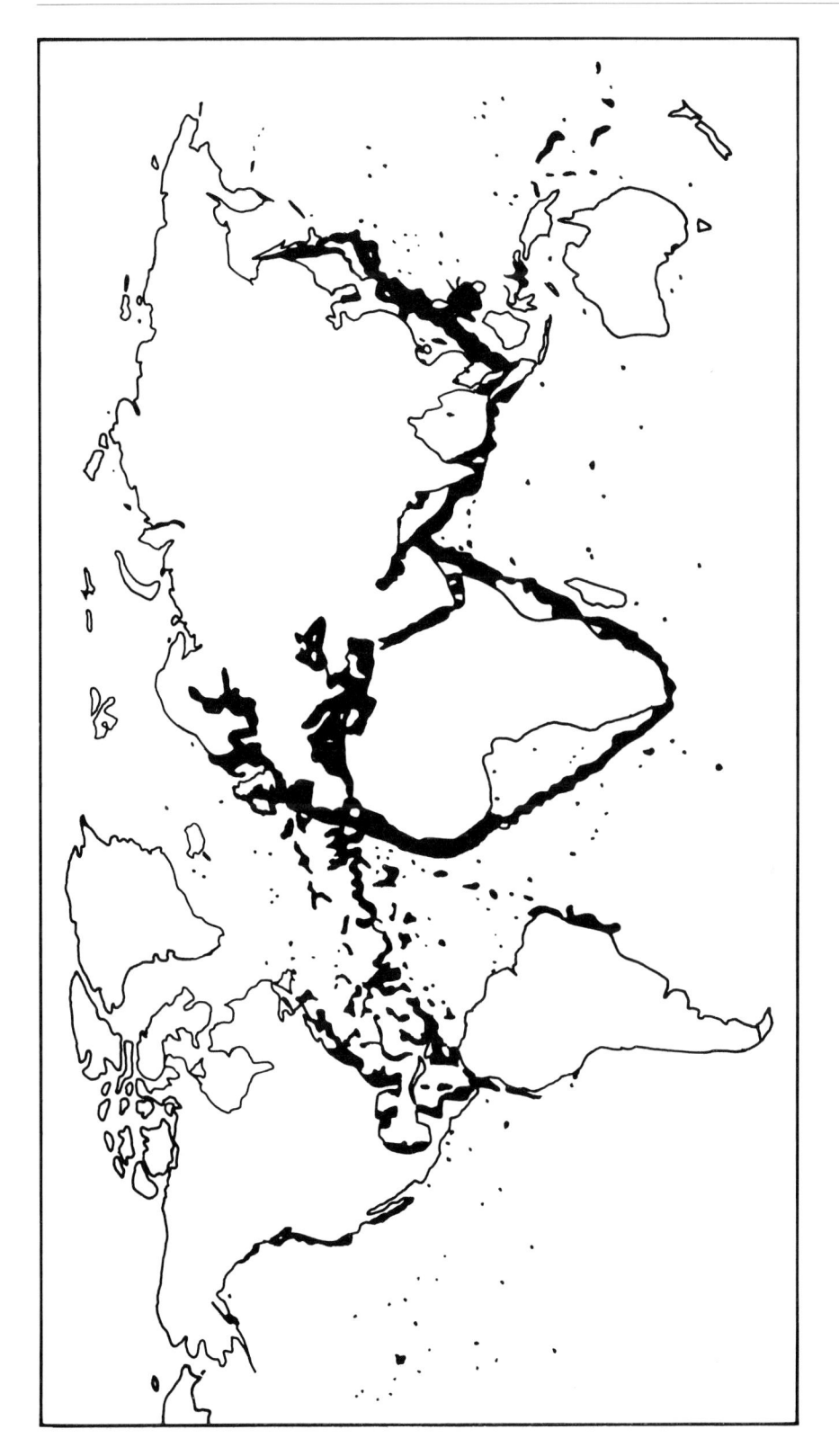

Figure 2.2 Oil contamination of the seas and oceans; note the major tanker routes. (OECD, 1985, 'The State of the Environment', Paris)

broad environmental research base from which might spring national concern.

Figure 2.2 illustrates a further component of global spatial differentiation brought about by resource location and world trade. The heavy dependence of development upon energy sources such as oil becomes manifested in the polluted quality of the major sea lanes, especially in the Northern Hemisphere. Another dimension to this basic resource drive is the relative lack of development in the Southern Hemisphere which raises the status and urgency of protection for biological reserves there (Chapter 3).

2.1.3 'Think globally, act locally': what does it mean?

This slogan has been widely adopted by environmentalists as a call to action. It suggests that in the realm of environmental management the individual is critical and that, either by consumer activity or protest action the local environment may be improved. However, there is a deeper meaning of the slogan for environmental science.

It is increasingly the view of historiographies of science that the utilization of science 'for the relief of Man's estate' from Francis Bacon (1561–1626) onwards has led scientific methodology down increasingly mechanistic and reductionist routes; the challenge of environmental issues, it is claimed, is for science to return to its holistic roots in order to consider the planet as James Hutton did.

However, if Baconian science became the servant of man's technological needs it also set out the controlled empirical experiment as a major means of confirming, refuting or extending the essential theoretical progress of the disciplines. Whilst environmental science has a special practical problem of verification (see section 2.3.4) it also depends on experimentation under controlled conditions. Consequently the phrase 'act locally' may also be taken to mean, in addition to social action, scientific action. The operation of the slogan in full to guide research therefore requires the geographers explicit use of scale; small-scale empirical findings require careful extrapolation using geographical variables to have any resonance with the practical and political needs of environmental management. In the case of pollution control for example, a result from toxicology may indicate the concentration of a pollutant which is lethal to birds or fish. However, to convert this to an emission control standard requires an understanding of patterns and processes in the transporting media of air and water.

For example, when in the 1970s Scandinavians recorded depletion of fish stocks in lakes which had become acidified, detailed experiments revealed that sulphur dioxide was a likely source (see Chapters 3, 6 and 12). They next used simple geographical principles — prevailing winds — to suggest with some vigour that the United Kingdom was the biggest source of this atmospherically transported pollution. However, investigation of the spatial pattern of airflows confirmed that much of the sulphur dioxide deposited in Scandinavia can originate in Central and Eastern Europe. The Chernobyl disaster provided an unfortunate 'tracer' material for this European 'pool' of air (see Figure 10.1).

2.2 Contamination and pollution

As James Lovelock, originator of the 'Gaia' hypothesis (Lovelock, 1979) consistently stresses, the majority of species which now dominate the world organic order are capable of survival only because of the emergence of an atmosphere containing oxygen.

Whilst Earth's atmosphere is now conductive to aerobic organisms it retains toxic gases (and gases which are toxic in some circumstances). The lithosphere and biosphere are also, of course, abundant in harmful substances, some released naturally (e.g. methane from wetlands) and others released, circulated or concentrated by human agency. Consequently we need to separate carefully phenomena via a definition

The geography of pollution

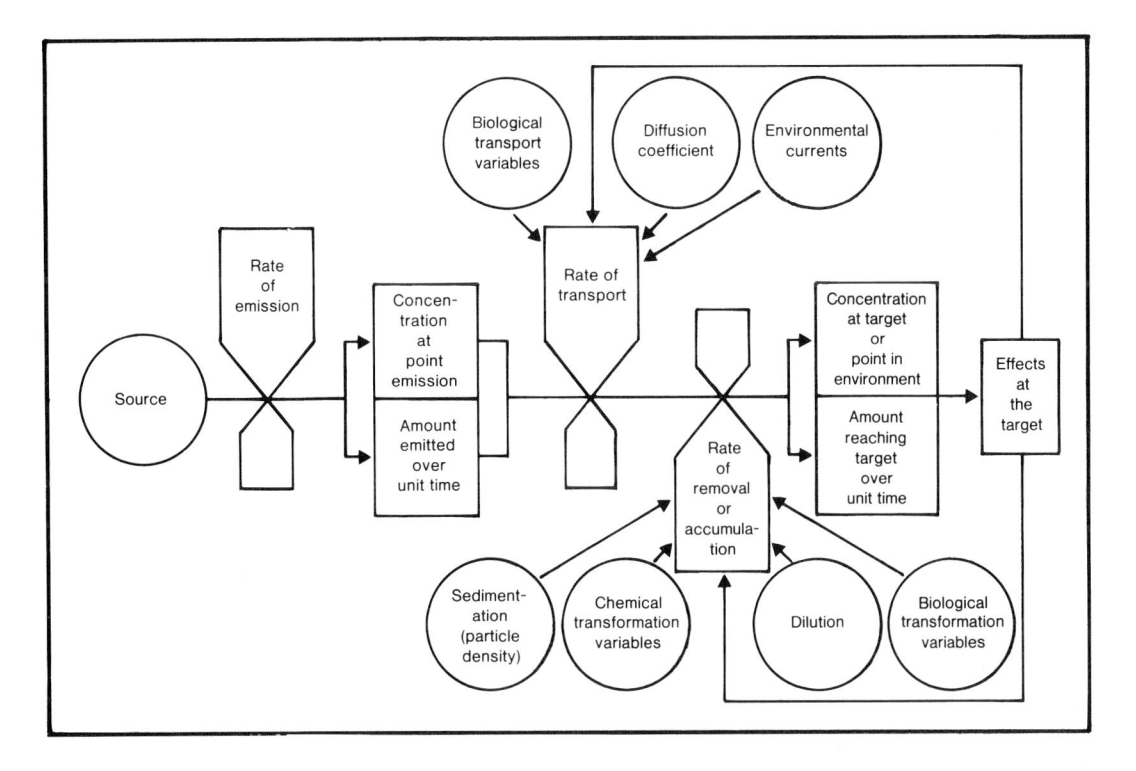

Figure 2.3 A systems diagram of pollution processes, stressing source–pathway–target linkage. (Holdgate, 1979)

which will stress either 'naturalness' or levels/combinations of substances in relation to their hazardousness to human society and biodiversity.

2.2.1 The importance of precision

In some senses the word 'pollution' has become overstated; it requires constant qualification and should be considered only as a counterpoint to 'contamination'. The latter word has also been used very imprecisely and is even considered by some campaigners to have 'more clout' than pollution. In their 10th report the UK Royal Commission on Environmental Pollution reserved the words contaminant/contamination for the introduction or presence in the environment of alien substances (or energy) without implying a judgement as to potential or actual damage or harm resulting. A recent pollution glossary (Murley and Stevens, 1991) has defined pollution as 'The introduction into the environment of substances that are potentially harmful to the health or well-being of human, animal or plant life, or to ecological systems' (p. 66).

A recent dictionary of environmental terms (Jones *et al.*, 1990) further distinguishes between primary pollution, resulting from any substance released into the environment in harmful concentrations and secondary pollution resulting from a chemical reaction between two or more sources (e.g. photochemical smog). The same source defines synergism as 'the combination of two harmless substances to produce impact greater than the sum of the separate effects' (p. 420).

2.2.2 Spatial distributions of contamination and pollution

Contaminants have sources from which they are dispersed; Holdgate (1979) has, however,

21

defined pollutants as having sources, pathways and targets (section 2.4). The notion of target is essential to our precise definition of pollution (section 2.2.1). Haigh (1989) reminds us that targets may be 'man, or animal or plant life or an inanimate structure (e.g. the stonework of a cathedral)'. Clearly if a potential pollutant becomes transformed into a harmless substance or becomes diluted along the pathway(s) no pollution occurs, though contamination has occurred.

Figure 2.3 is a modification of Holdgate's system diagram stylizing all pollution. Pathways through the environment involve not only the transporting media of air and water but the storage media of sediments and soils. Clearly society can consider intervention at any point between source and target to control or reduce pollution. There is therefore a geographical background to pollution management which may become explicit via the planning of source activity, the setting of standards for pathway media or the study of damage done to targets.

As we pinpoint in a number of chapters in this volume, argument abounds over the rival merits of source controls (favoured, for example, in the United States and Europe) and pathway management (the UK tradition). With this inherently geographical background to pollution control it is surprising that only recently has monitoring strategy become oriented to geographical presentations of contamination and pollution — for example, maps. The tradition has been to depict time series of critical pollutants measured at a number of index sites (often randomly located because of the foresight or interest of the observers).

However, as Figure 2.1(b) shows, it is now possible to compile maps of both contamination and pollution patterns. Environmental geochemistry has developed at a very fast pace in the last decade. Monitoring of the aftermath of pollution emergencies has helped focus our attention in the 'geography of pollution', whilst, for example, the public debate over water pollution or radiation damage has encouraged the presentation of 'background values', for example as contamination maps.

2.3 Definitions of pollution in terms of toxicity

Toxicity is the capacity of a pollutant to cause death or injury to an organism. Whilst pollutants impact on whole ecosystems, the science of toxicology inevitably is forced into controlled experiments, whose validity we examine below, which are carried out on single organisms (not necessarily the target organism).

Timbrell (1989) defines toxicology as 'The study of the harmful interactions between chemicals and biological systems' (p. 1). One of our contemporary problems is the rate of growth of industrial chemical production (e.g. Figure 2.4(a)). Timbrell estimates that around 65,000 chemical compounds are currently manufactured (US data) and that up to 1000 may be added each year. Whilst many new chemicals are fully tested for their environmental effects a degree of risk always attends these assessments and we do not by any means have an international law of impact assessment via toxicological tests.

On the other hand, farmers and manufacturers have complained bitterly of a prevailing 'chemophobia' amongst the public in the West. They have highlighted the complaints of gin drinkers about pesticides on the skin of the lemon (which they launch into their already toxic beverage!) as indicating imbalance in perception. A student class faced with anonymous coloured liquids (in fact, soft drinks) labelled only 'chemicals' will refuse to drink them. Water suppliers find objection to fluoridation in the interests of dental health because 'people don't like a chemical added to their water', and in 1989 the mineral-water suppliers Perrier withdrew a huge volume of stock to preserve the healthy image of their product when benzene contamination was proved in the filtration plant.

2.3.1 Toxicity testing: in vitro, in vivo

'The most commonly used assessment of toxicity is the measurement of short term

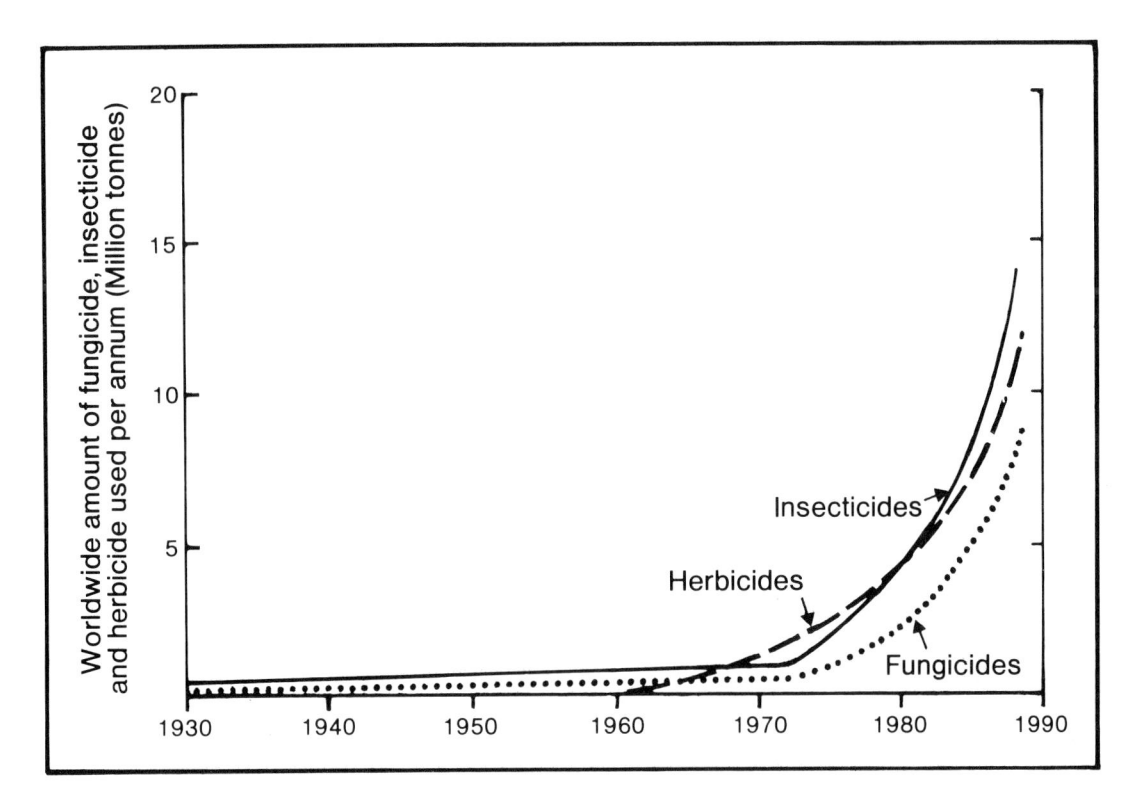

Figure 2.4a The growth of human use of manufactured chemicals as illustrated by pesticides. (Modified from *New Scientist*)

lethality' (Duffus, 1980). In laboratory tests on live subject organisms a median concentration is established which kills 50 per cent (usually) of the targets. There is generally a sigmoid curve, known as the dose-response curve (Figure 2.4(b)) which links the concentration of chemical with target mortality (though note in Chapter 9 that other relationships exist). The established median concentration is referred to as the lethal concentration (LC_{50}). If comparisons need to be made between toxic chemicals these should be related to the molecular weight of the chemicals concerned.

Variants of the text also include the dose required to kill just one organism (LD_1) and the 'maximum-dose-never-fatal' or 'minimum-dose-always-fatal'. It should be remembered, too, that many pollutant chemicals are in fact essential to life in small concentrations; the toxicological curve can be segregated in the general way from those of non-essential chemicals (Figure 2.4(c)). Toxicology is clearly a multidisciplinary activity; the uptake of toxic chemical involves knowledge of the target organism's physiology whilst the organs impacted are a matter for veterinary or medical science. But the main problems concerning toxicological testing, from a public viewpoint, stem from the selection of test organisms and the extrapolation of test results. On the latter topic, Conway (1986) employs the ludicrous case of Filov *et al.* (1973) in calculating the 'suitable dose of LSD to give to an elephant' (Table 2.2).

Tests in the laboratory seldom replicate the full breadth of environmental conditions involved with real pollution. *In vitro* tests are therefore very simple but tend to give a 'worst case' picture for this reason. However, clearly the target organism for LC_{50} tests is

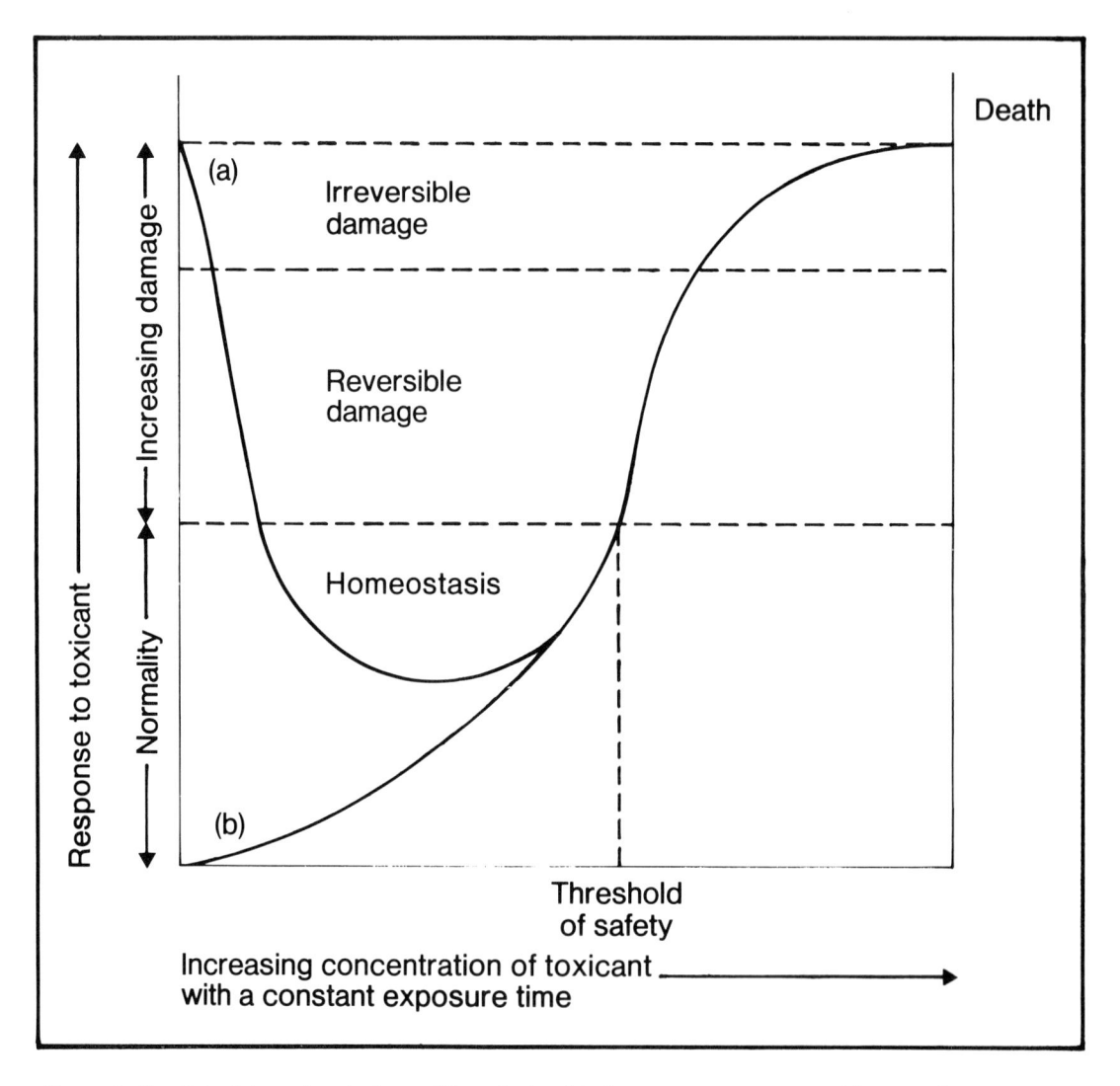

Figure 2.4b Principles of toxicology. The relationship between concentration (dose) and response. Chemical (a) is essential to life in small quantities; chemical (b) is toxic in some way at any dose. (Duffus, 1980, *Environmental Toxicology*)

Table 2.2 Possible calculations of a 'suitable' dose of LSD to give to an elephant

Based on body weight of elephant and dose effective in cats	297 kg
Based on metabolic rates of elephant and cat	80 mg
Based on body weight of elephant and dose effective in man	8 mg
Based on metabolic rates of elephant and man	3 mg
Based on brain sizes of elephant and man	0.4 mg

Source: Filov *et al.* (1973).

selected in a series of physiological and ethical steps. We do not test lethality on humans *in vitro*, though we may test cells or organs. The selection of test animals has aroused considerable and violent opposition from protest groups, as has their housing and care prior to (or after) testing.

In vivo testing represents the chemical, biological, medical and geographical study of chemicals or energy accidentally released to the environment or deliberately released on

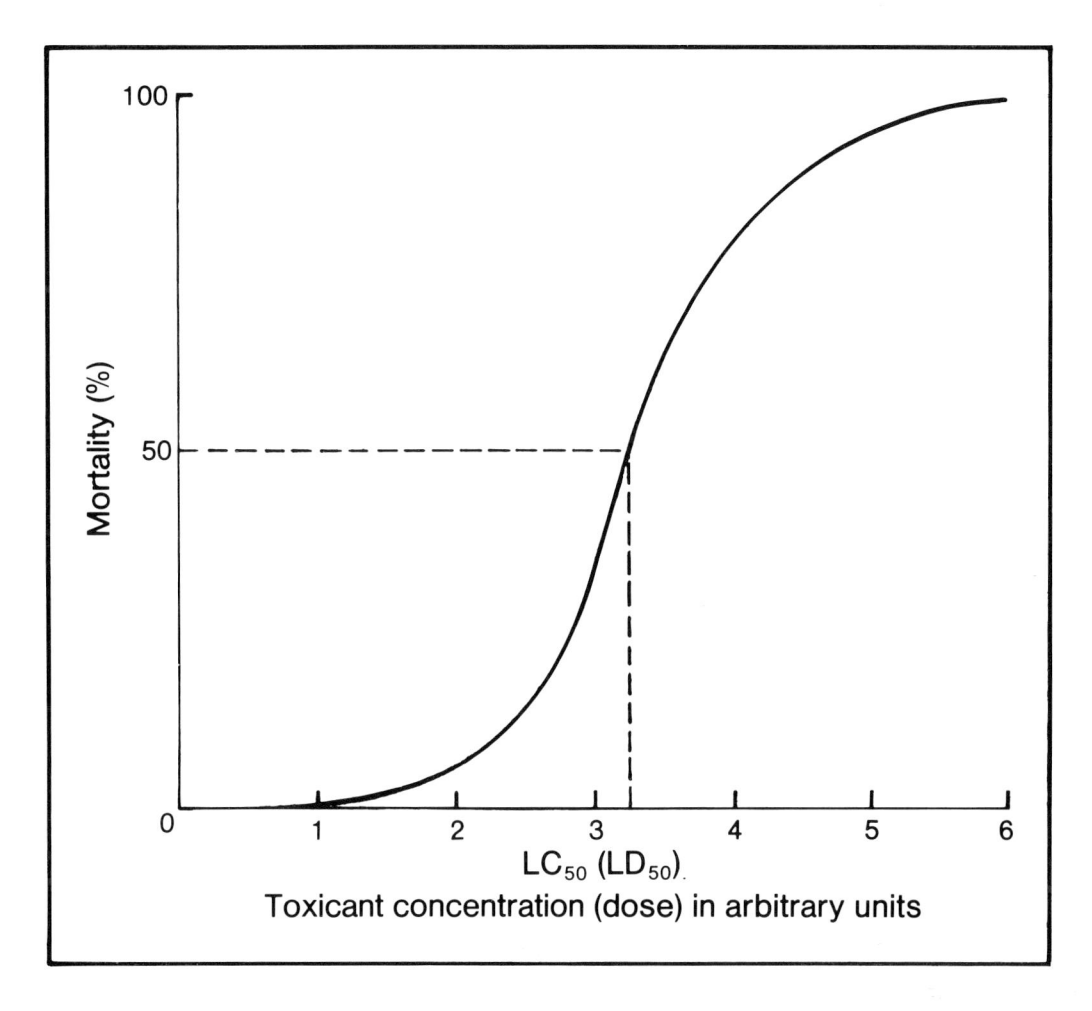

Figure 2.4c The result of a toxicity test, in vitro, on a chemical which is harmless at low concentrations. (Duffus, 1980, *Environmental Toxicology*)

the assumption that no harm would result. We may cite the growth of concern over pesticides following Rachel Carson's *Silent Spring* (1965) or the more recent identification of leukemia and other cancer 'clusters' (see below) as evidence of this unacceptable but highly necessary backward testing.

2.3.2 Medical geography and toxicity

Patterns often reveal processes, as we have already noted, and the spatial pattern of suspected pollution impact may reveal not only the sources and pathways but the exposure to and physiology of toxic materials in the environment. The earliest examples of this form of geographical enquiry come from disease epidemics. For example, in the nineteenth century cholera was the scourge of urban populations in the United Kingdom. In the celebrated case of the epidemic in Soho, London, Dr John Snow located the source of infected water from the spatial distribution of cases (see Chapter 7).

More recently Howe (1976) and Learmouth (1988) have mapped examples of

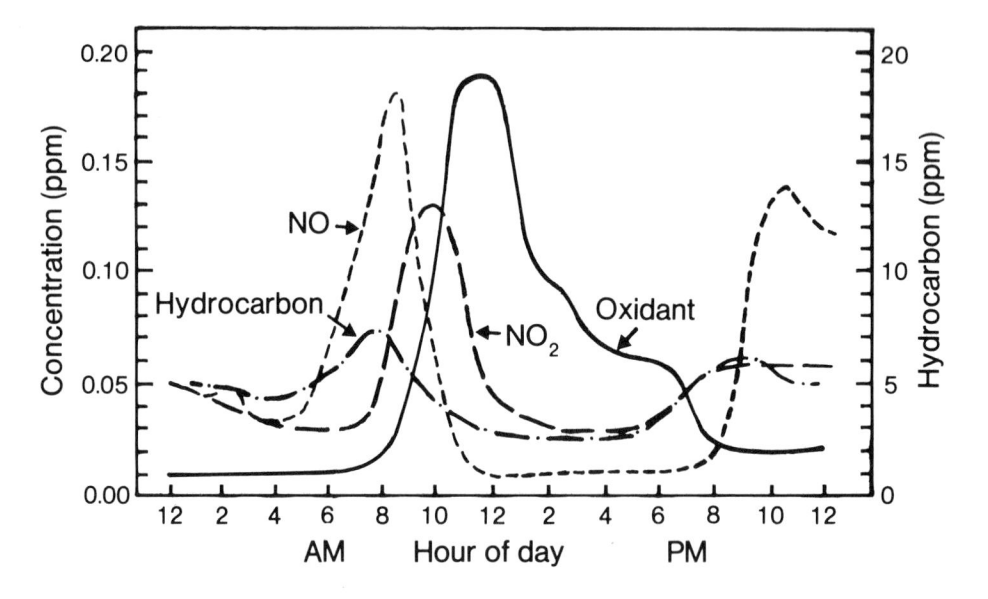

Figure 2.5a The pollution 'cocktail' of an urban atmosphere during the passage of a day, illustrating the synergistic development of oxidant (ozone). (US Dept. Health, 1965)

disease mortality and morbidity of human populations, making links with patterns of environmental pollution in the cases of, for example, bronchitis and stomach cancer. Bronchitis is clearly related to air pollutants which irritate or aggravate damage to bronchial organs; it is therefore an urban disease. By contrast, stomach cancer has both urban and rural clusters. At the time Howe wrote these appeared not to have a link to pollution and were thought to be linked to 'soft' water sources in the uplands; the spores of bracken growing in catchment areas have also been implicated.

One problem of modern medical geography which curtails its utility as a large-scale toxicity test is that the appropriate medical statistics are not gathered. The medical profession has often seemed unwilling to become involved with pollution-related health issues at the community scale, although Lloyd et al. (1985) provide a notable exception in relation to air pollution in the Scottish town of Armadale. Where special data-sets occur (e.g. for childhood leukemia in Northern England), sophisticated statistical analyses can identify links (e.g. Openshaw et al., 1987) through the assignment of significant levels to clusters of a particular type of morbidity (see Chapter 11).

2.3.3 Synergistic, delayed and cumulative effects

Synergism has become something of a buzz word in the realms of protest against pollution; the phenomenon may, however, be benign since it merely refers to the joint action of two or more substances to produce an impact greater than the sum of their individual action. For example, the combustion of fossil fuels releases sulphur which is relatively harmless. However, when sulphur oxidizes and then dissolves under atmospheric conditions an environmentally damaging pollution event of acidification may, in turn, release metals into solution which were previously present in contaminating, not polluting concentrations.

Synergistic reactions can be predicted or anticipated only when a complete analysis of pollution pathways is available (see section

The geography of pollution

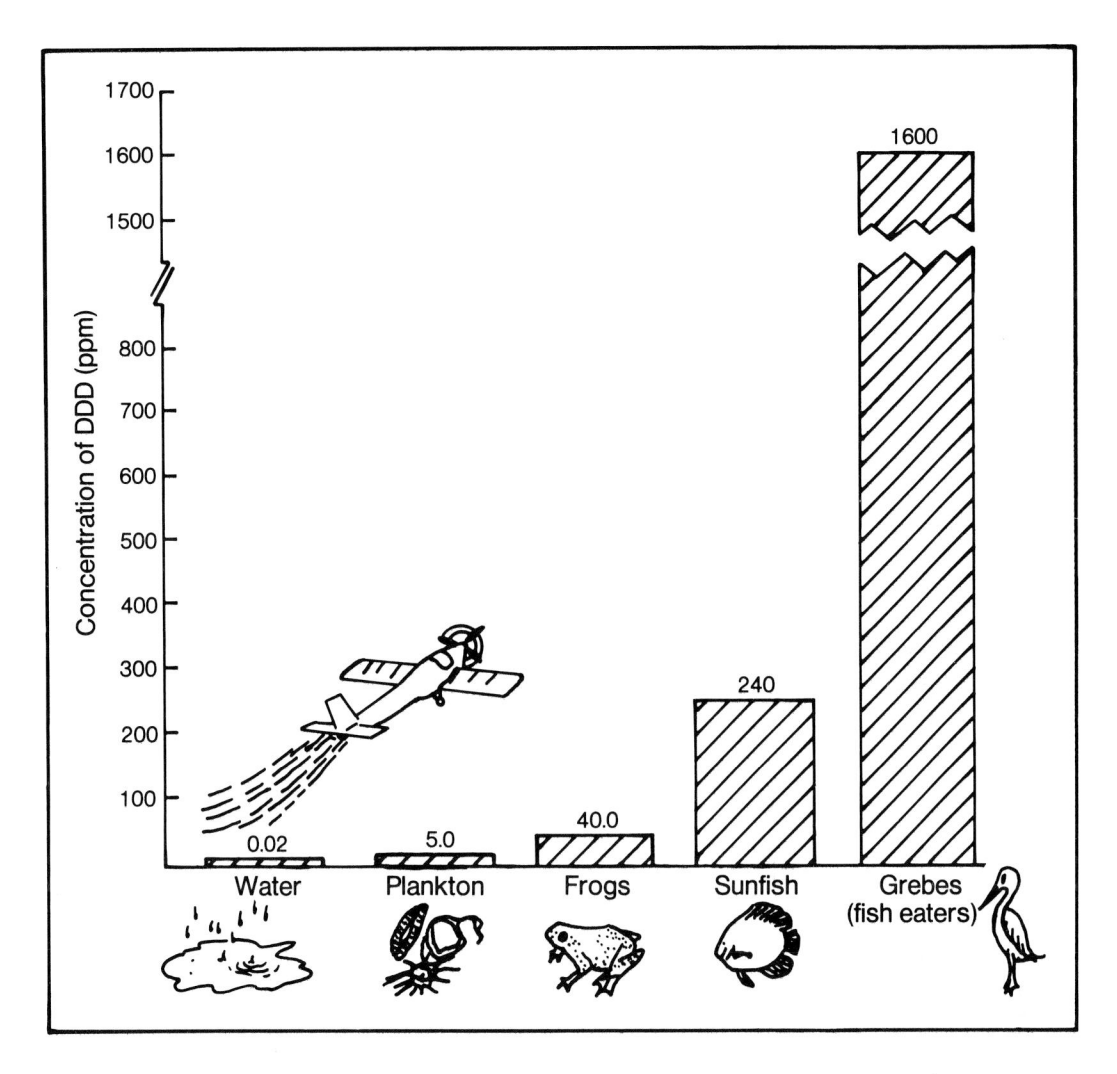

Figure 2.5b Concentration of the pesticide DDD in food chains from plankton to grebes. (Kupchella, C.E. and Hyland, M.C., 1989, *Environmental Science*, Allyn & Bacon, Boston)

2.4). In a similar way, delayed effects of pollution can arise when, for example, a pollutant becomes stored, perhaps physically or chemically bound to sediment particles but subsequently dissolved (e.g. phosphates released from intensively farmed land during episodes of soil erosion). In another case, shown in Figure 2.5(a), the release of nitrous oxide from car exhausts during an urban 'rush hour' leads, over a period of hours, to the formation of nitrogen dioxide; during the early afternoon sunlight produces a photochemical reaction leading to the formation of the direct pollutant gas ozone. Such a delayed action makes monitoring a problem, but once understood can lead to anticipatory monitoring and action. Because storage of pollutants occurs throughout natural circulation patterns there is inevitably in a variety of locations (both geographical and physiological) an accumulation of toxic materials or sources of harmful energy. A pristine location may therefore become contaminated and eventually polluted as the threshold of harm is exceeded.

The food chain is a means of accumulating pollutants in both organisms and organs for two reasons. As larger organisms feed upon quantities of smaller ones (or in vegetation) there is inevitably a concentration process (Figure 2.5(b)). Physiological processes of protection against toxicity such as excretion may store pollutants within individual organs (such as the human liver). The food chain also produces a further transport opportunity for pollution. It was this pervasiveness of pesticides which Rachel Carson's *Silent Spring* first drew to the attention of an international audience.

2.3.4 Uncertainty and precaution

One of the biggest crises of confidence faced by politicians in dealing with environmental problem is that of the credibility of the scientific guidance they receive. Because of the difficulty of extrapolation of scientific results to political spheres, for example, even apparently conclusive results are attended by uncertainty, both spatial and temporal. A result 'here' may not be serious enough to demand action 'there' and an impact 'now' does not rule out recovery in the future. As Santos (1990) admits:

It is a solemn thought that scientific uncertainty is so complex and pervasive in environmental decision making. Science is hardly ever sure about how pollutants affect humans, biota or the ecosphere as a whole. The relationship between the concentration of a pollutant and its effects may be linear, exponential or even polynomial. Also there are the additional problems of time lag and synergistic effects of interacting factors. Consequently, scientists can provide little detail on how to monitor the earth's life support system. (p. x)

In the light of this uncertainty it is instructive to consider an element of (West) German public policy, the *Vorsorgeprinzip*. The 'precautionary principle' has been advocated for general application by the UK Royal Commission on Environmental Pollution (RCEP, 1988, p. 59):

Faced with a complex and very large ecosystem, and consequently with tenuous cause and effect relationships to link sources of pollution to identifiable environmental effects the Vorsorgeprinzip became a key element in developing an argument for measures before damage occurs.

We return to this theme in Chapter 14.

2.4 Sources, pathways, targets and control

Clearly sources, pathways and targets have geographical definitions and distributions but their use as a framework is not merely a transparent attempt to gain pollution as a central concept for geography. In a number of ways geographers were dealing with pollution long before the 'pure' sciences (e.g. in biogeographical studies and in medical geography). The source — pathway — target chain is of principal value in terms of environmental management. It offers, for example, a simple agenda for decisions about monitoring: How many sources? What pathways? Where are the targets? More importantly it raises the question, What do we control — the source, the pathway or the target?

2.4.1 Sources, point and diffuse

The traditional depiction of pollution is a smoking chimney or a belching pipe: these are point sources. A point source of pollution is not always easy to identify but once located can be controlled more directly than can diffuse sources. Diffuse or 'non-point' pollution is that which occurs from an extensive land-use activity such as agriculture (fertilizers, pesticides, etc.) or forestry (concentration of aerosols, sedimentation). Geographically the two forms of pollution may be likened to the contrast between a rifle wound and shot-gun injuries.

Point sources are represented by a discharge or emission of some type into one of the transporting media or direct to storage — as in the case of landfill of untreated wastes.

Clearly we can, if worried by point sources fix emission standards or introduce pollution costs (e.g. the UK National Rivers Authority now practises 'charging for discharging'). However, point sources pose a difficulty, too; by their very nature they represent prior location decision for a plant or process. Thus, controls frequently need to be introduced retrospectively and there are obvious political problems of bringing this about efficiently and equitably.

If point sources become the focus of some form of *allocation* policy (e.g. of a site, a standard, a cost or a licence) diffuse sources can often be controlled by an *accommodation* between land-users and the targets of pollution. For example, an amelioration of soil erosion problems can often be achieved by timely cultivation or by choice of crop. Certain diffuse pollution sources have, however, been considered serious enough to warrant relocation or change of use.

2.4.2 Pathways in space and time

Pathways of pollution are most obvious as plumes — the plume of smoke from a chimney or of discoloured waste water from a pipe. Generally plumes illustrate very visibly the two main transformations available in transporting media: dispersion and dilution. Dispersion occurs as the particles or molecules of pollutants spread through the receiving medium; dilution inevitably occurs as a result but can also occur by the addition of quantities of unpolluted air or water (as with a change of wind direction or at a tributary junction in a river system).

The transportation media for pollutants are by no means chemically or biologically inactive; as a consequence transformations occur (both ameliorative and synergistic) as a result of mixing between pollutants and reactants in the medium concerned. There may also be storages or 'sinks' in the system; the former term is to be preferred because seldom is there a long-term loss of a potential pollutant from a pathway. Diversions into storage may be very short term as in the case of an entry to the food chain via micro-organisms. On the other hand, storage in sediments may involve a delay in reaching a target until geological events have occurred or there is an environmental change which re-releases the pollutant.

Pathways also involve real-time dimensions; this becomes evident when dealing with chemical emergencies (Chapter 10). Point sources may produce pollution episodes which pass through the environmental pathway at a rate modified by dispersion and dilution. The receiving medium itself is often modified through time, for example, it is a principle of sewerage pipe design that overflow to a river can occur when that river is likely to be in flood and therefore capable of transporting, dispersing and diluting the polluting load. Theoretical difference between load (as a precise term) and concentration of a pollutant in a pathway is returned to in section 2.5.

In control terms, the temporal variability of transport processes means that standards are hard to fix; control is often, therefore, approached as a series of desirable objectives which can nowadays be met in systems of sensor-based, real-time strategies responding themselves to temporal changes such as flood and drought, gale and calm.

2.4.3 Targets: sitting ducks?

There is a measure of inevitability about the term target, introduced by our strict definition linking pollution to damage. Targets are, however, seldom static if they are organic. Whilst a cathedral cannot 'dodge' acidic deposition, organic targets may either move or metamorphose both into and out of pollution danger. The young fish cannot easily escape from a thoroughly acidified river system and is much more susceptible than the adult. Adults can move into unpolluted tributaries if available but have no means of obtaining warnings of a pollution episode.

Human beings are clear examples of

Table 2.3 Some critical pathways and critical groups identified in the UK

Site	Pathway	Nuclide	Group
Dounreay	Contaminated fishing gear by particulate material	Cerium-144	Fishermen's hands
	Suspended matter carried into rocky clifts on coastline	Various	All those who use the beaches
Springfield	Grazing animals	Iodine-131	Those consuming local milk — principally young children
Sellafield	Harvesting locally grown seaweed (*Porphyra*) to produce lava bread	Ruthenium-106 Cerium-144	Local lava bread consumers, critical organ is the G.I. tract
	Contamination of foreshore	Ruthenium-106 Zirconium-95 Niobium-95	All those who use the beaches
	Sea to land transfer	Plutonium 231	Increase in ambient dose rates in coastal locations
	Locally caught fish	Caesium-137 Caesium-134	Local fish consumers
Bradwell	Oysters caught locally in the Blackwaste estuary	Zinc-65 Silver-110m	Local consumers of oysters
Trawsfynydd	Trout and perch	Caesium-137 Caesium-134	Local consumers of fish and fishermen

Source: Newson 1992a (p. 149) and SI 1982 (p. 210).

targets with both physical and cultural mobility into and out of pollution danger. A middle-class family move house because of industrial smoke; they then let their children walk to school through traffic fumes. Some members of the family may smoke cigarettes despite their hatred of ambient industrial smoke.

It is therefore increasingly fashionable to group pollution with other risks taken by human beings (see Johnston, 1989, p. 174). It is also important to investigate the relationship between target 'status' and lifestyle including, principally, workplace environment, home environment and other sets of controlled conditions which influence target exposure. These principles are to some extent recognized in the UK radiation monitoring (see Chapter 11). Radiation workers are monitored directly but sampling programmes also consider certain critical groups involved at key points in food production (see Table 2.3). A certain amount of pollution control is feasible by target management not only by evacuation in emergencies but by planning residential development and by encouraging safety in the workplace. Target monitoring has long been practised at the point of consumption, for example, in water-supply testing where a random sample of domestic taps is tested. Individuals may choose to fit water filters, wear masks, etc., as a means of control but generally this is considered an unjust shifting of costs away from sources.

2.5 Pollution monitoring: a measure of concern

In general, science has not elevated monitoring as an activity, preferring the controlled experiment and the extrapolation of its results as the definitive good practice. However, environmental science has a greater regard for monitoring and, whilst it remains difficult to raise funds for monitoring, environmental protection (e.g. of biodiversity) and environmental control (e.g. pollution control) demand effective monitoring systems.

2.5.1 The aims of monitoring

First we may contrast monitoring with survey or surveillance (Hellawell, 1991); surveys have no preconception about their findings and are generally of short duration or even 'one-off'. Surveillance refers to an extended programme of surveys but monitoring is 'Intermittent (regular or irregular) surveillance carried out in order to ascertain the extent of compliance with the predetermined standard or the degree of deviation from an expected norm' (Hellawell, p. 2). Attitudes to the detection of change have changed profoundly in the last ten years. The successful use of data from the longest-running sites, be they climatic, chemical or biological, to point to the direction and rate of environmental changes (global warming, acidification, loss of species), has gained a new respectability for monitoring. Monitoring is often an expensive chore to carry out and its auditing or related functions tend to appear more obviously beneficial. Nevertheless, national and local organizations concerned with environmental management all carry out monitoring programmes; legislation concerned with environmental control normally prescribe the compliance monitoring appropriate to the problem from which relief is sought.

Geographical patterns, if understood and accepted, are a powerful guide to monitoring networks. As a banal example one does not monitor for water pollution determinants *upstream* of a suspected point source! However, one would by the same token monitor for the unpolluted state variables upstream of the source. Geographical patterns can be used, for example, to rationalize a network of sampling sites on the basic of critical groups or targets. Once some basic knowledge of a source–pathway–target chain is available it is possible to allocate monitoring to locations in the chain. Impact monitoring chooses the target end but of course compliance monitoring favours the source end.

The monitoring of pollutant pathways involves the following problems:

A. pathways are often multiple and not every route can be identified or covered;
B. movement along pathways may be episodic; for much of the time monitoring may be redundant;
C. synergistic or storage effects may negate or invalidate the results of simple pathway monitoring.

2.5.2 Monitoring networks in practice

Given the purposeful nature of monitoring one must be careful in allocating the plethora of recent global measuring systems, for example, satellite imaging, to this category. Much of the expansion of high-technology 'monitoring' is in fact surveillance. The general shortage and insecurity of international environmental legislation results in a relatively slim body of true global monitoring. Large improvements in harmonizing scientific techniques are made during the 'International Year' of this or 'International Decade' for that science, but much of the true monitoring remains organized on the nation-state scale. For reasons of confidentiality or to allay public fears the details of monitoring networks may be difficult to uncover. In the United Kingdom part of the reason for apparent secrecy over monitoring until recent years appears to have been the absence of co-ordination of the monitoring networks. However, since 1974 the Department of the Environment has had an interdepartmental committee to ensure co-ordination on monitoring in the freshwater, marine and atmosphere environments (DOE, 1986).

Nevertheless, the Department's own description of the integrated system admits that 'monitoring is largely a decentralised

Table 2.4 UK environmental monitoring register: planned coverage

Air:	Emissions of pollutants from factories, etc.
	Urban and rural networks for smoke and sulphur dioxide; sulphur dioxide baseline network
	Nitrogen oxide networks
	Particulates monitoring schemes
	Lead in air
	Acid deposition network
Fresh water:	Industrial discharges
	Surface and groundwater quality
	Discharges from sewage treatment works
Sea water:	Discharges into estuaries and coastal water
	Bathing water quality
	Estuarial water quality
	Water disposal at sea
	Chemical and biological monitoring of the sea
Land:	Use
	Disposal of wastes, etc.
	Quality
Waste:	Disposal sites
	Special and hazardous wastes
	Recycling
	Leaching
Fauna and Flora:	Habitat quality
	Endangered species and extinctions
	Health
	Radioactivity and heavy metal levels
Lead:	In humans, food, etc.

Source: DoE, 1986.

activity in the UK' (p. 63). 'The actual measurements are made by local authorities, water authorities, individual factories, Ministry of Agriculture, Fisheries and Food, the Department of the Environment and other government departments and so on'.

The shortlist of topics for a centralized register of monitoring data planned for the United Kingdom includes the basic forms of environmental monitoring, both of impact and compliance, together with 'good housekeeping' monitoring such as that of land use and land quality (see Table 2.4).

It can be observed that, despite the difficulties of pathway monitoring it is often the case that the land, air and water pathway media are chosen for reasons of convenience, objectivity and as a hybrid with surveillance of ambient values of environmental parameters. The monitoring tends to be concentrated on:

A. the most significant sources;
B. the critical groups at risk as targets;
C. 'hot-spots' and sinks or storages;
D. important sites or periods of change.

As for reporting and presentation of monitoring data, recent legislation has opened up compliance data to public inspection via registers. Impact data tend to be confidential where issues of personalized data are concerned. However, pathway data, whilst available, are often difficult to assemble to a coherent whole. For this reason the Department of the Environment's water division, for example, established a subset of water quality monitoring (surveillance) stations as the 'harmonized network' (see Figure 2.6(a)). Results from these stations can be assembled for academic or predictive purposes (see, for example, Rodda and Jones, 1979).

2.5.3 Intensity of monitoring

The harmonized network of river water quality stations in the United Kingdom to some extent 'designs itself' in terms of spatial intensity. If the objective of a scheme is to monitor all major pathways in a medium the river network is merely subdivided to a level at which major nodes are equipped with sampling apparatus; there is a further deterministic location factor — the need for the river's flow to be monitored also, so as to allow calculation of loadings as well as concentrations of natural solutes, contaminants and pollutants.

Spatial intensities of monitoring sites for other polluted media are more difficult to fix. However, purely practical exigencies such as availability of observers, existence of

The geography of pollution

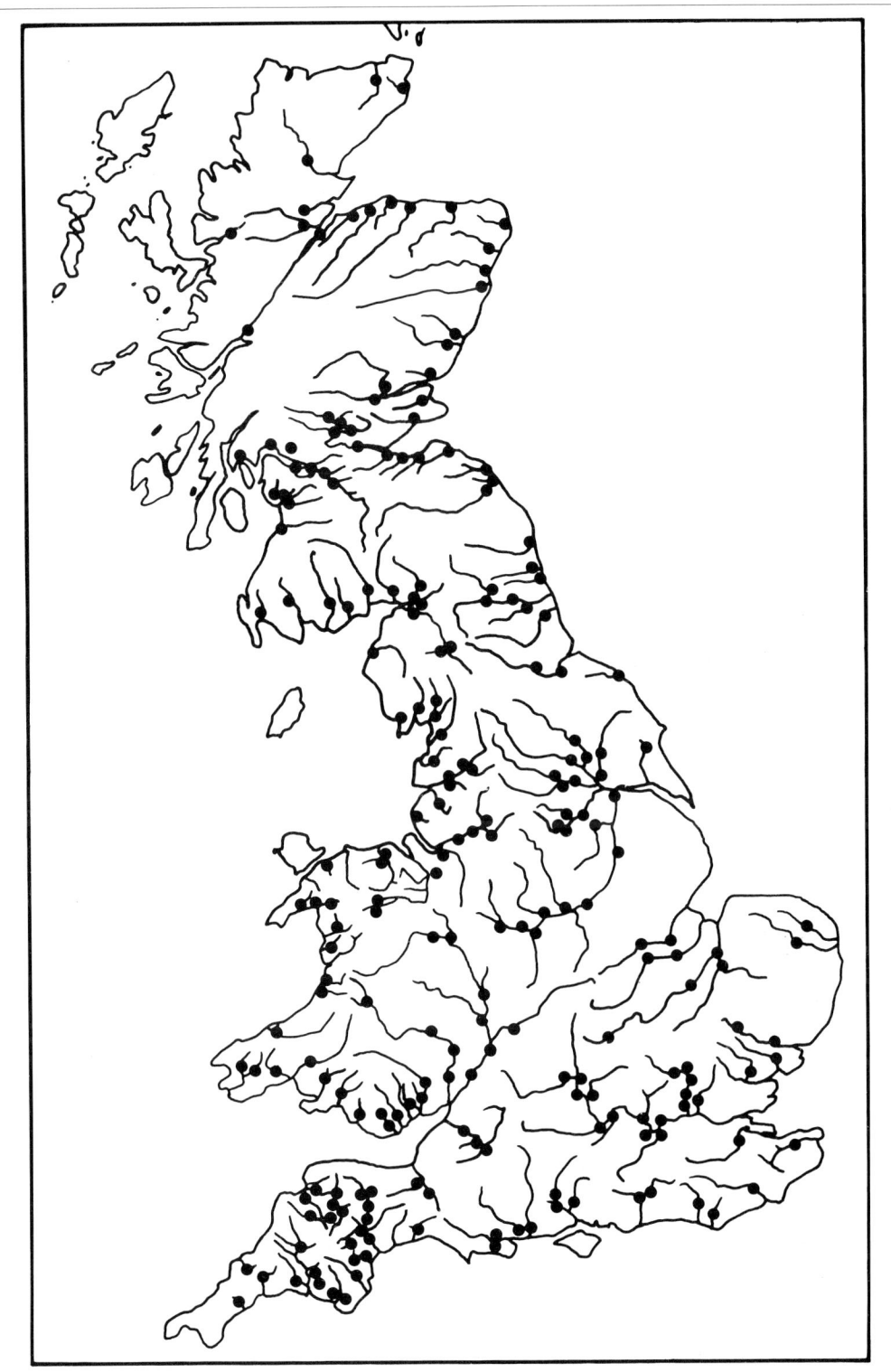

Figure 2.6a The UK harmonized monitoring network for river water quality. (Rodda, J.C. and Jones, G.N., 1983, *Journal of the Institution of Water Engineers & Scientists*, 37(6), 529–39.)

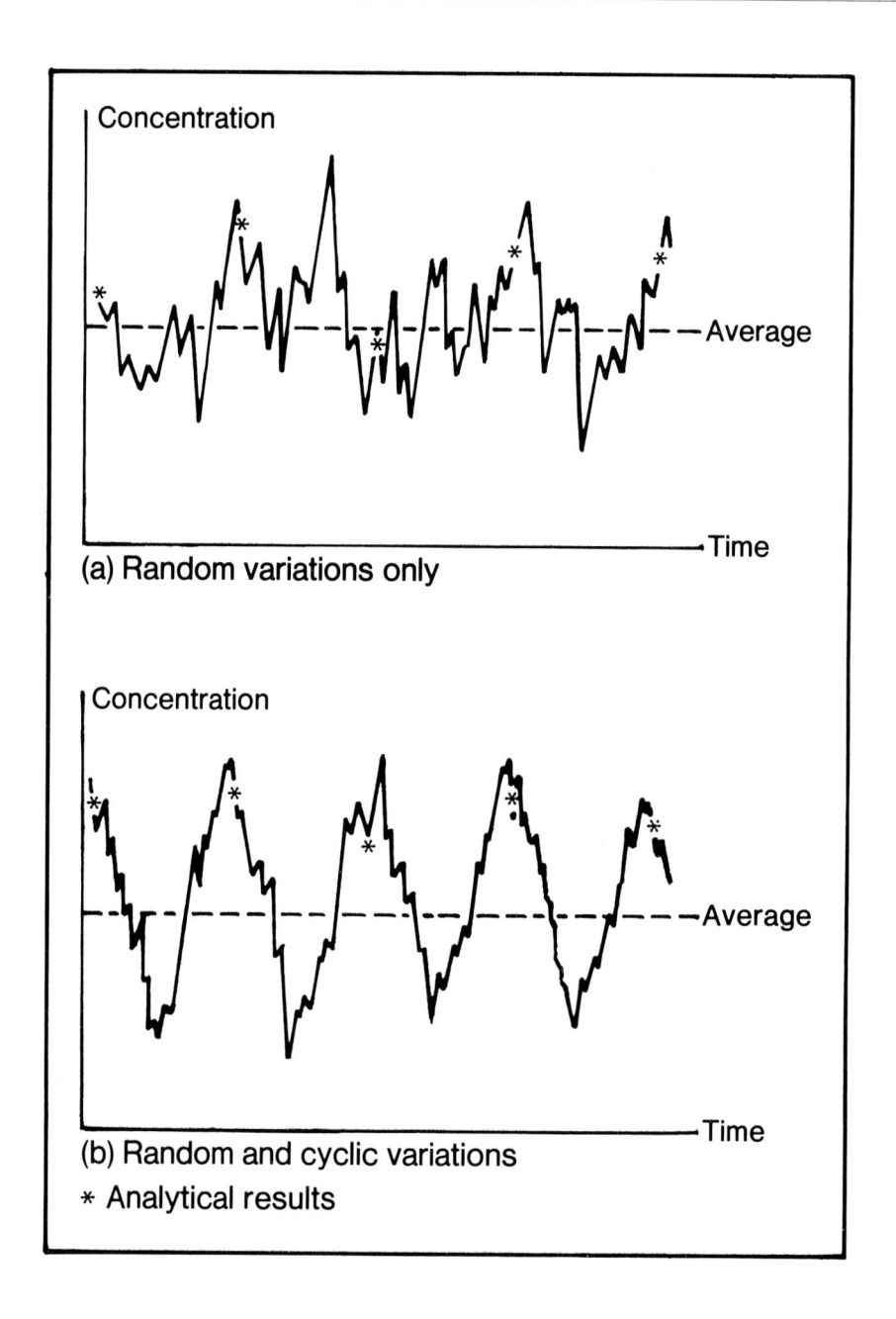

Figure 2.6b Problems of sampling river water quality as it varies in time, either (a) randomly or (b) with cyclic additions. In neither case will regular 'spot' samples (asterisks) reveal the true pattern

other relevant measurements (e.g. weather conditions for air pollution) and cost (see below) are again important. It must be faced that perfect monitoring systems are impossible and that a 'best-choice' system is often the one which makes the best trade-off between spatial coverage and continuity. Chemical analyses are often time-consuming and expensive, with the result that laboratory procedures, not field equipment/observers, set the contrasts for a monitoring programme as a whole.

In some cases a monitoring programme for a particular kind of pollution may be implemented before the physical behaviour of the polluted medium is fully understood at the sampling site. Sampling for coastal pollution very close to the shore, as advocated by European Community procedures, ignores the known cellular pattern of circulation which may invalidate the representativeness of each sample.

In many cases a spatial design may seem adequate until a particular incident; for example, radiation monitoring until the Chernobyl accident in 1986 concentrated with respect to national, but not international, sources of pollution. There are continuing doubts about the ability of chemical analysis to isolate new toxic substances; for this reason water supply authorities often pass a sample of their treated product through a tank of live fish as a final toxicological test.

Pollution is by nature episodic; even the major continuous sources tend to become episodic if control is only partial (e.g. sewer overflows to rivers during floods). The temporal intensity of monitoring is therefore of great concern in environmental management (see Figure 2.6(b)). There are also difficulties of reconciling the environmental, manpower, equipment, analytical or economic need for regularity in sampling with the background and induced variability of concentration to be expected with most pollutants.

In recent years the use of (non-human) bio-indicators of environmental pollution has become popular, for example, lichens for air pollution and stream invertebrates for water pollution. These have the advantage of bypassing the need for laboratory analysis; they also integrate the effects of pollution over a period of time — there is little risk of 'missing' a pollution episode (see Chapter 7). They are used also by educational surveys and by low-financed pressure groups.

2.5.4 Costs, compliance and control

One of the overriding issues of environmental management is the cost of bringing it about (see Chapter 5); monitoring costs can become astronomical very quickly and this tendency often guides law-makers in their judgement of the forms of limit to pollution which are practicable (see Chapter 4). Even in a pluralistic society in which environmental bodies are often carrying out surrogate forms of monitoring there must be a system upon which the legislature can rely for its accuracy or at least for its broad acceptance to the relevant scientific community. In some cases this may mean that the precision of an analysis and a degree of 'ceremony' is necessary to emphasize the legal importance of this aspect of monitoring (Hawkins, 1984).

For compliance with pollution control legislation it is essential that monitoring is adequate, even if a certain amount of pure guile is necessary on the past of those policing compliance. It is, as Hawkins (1984) reveals in his penetrating social analysis of water pollution control, essential to discriminate between legislation and regulation. In the latter category, appropriate to pollution control, those 'policing' the law tend to do so in a conciliatory way in order, eventually, to achieve 'cheap' compliance rather than set up 'expensive' adversarial positions.

References

Brandt, W. (1986), *World Armament and World Hunger: A Call for Action*, Victor Gollancz, London, 208 pp.

Calder, N. (1991), *Spaceship Earth*, Viking, 208 pp.

Clayton, K. (1991), 'Scaling environmental problems', *Geography*, 76(1), 2–15.

Cross, M. (1988), 'Japan wakes up to the environment', *New Scientist*, 23 June, 38–39.

Department of the Environment (1986), *Digest of Environmental Protection and Water Statistics*, HMSO, London, 71 pp.

Duffus, J.H. (1980), Environmental Toxicology, Arnold, London, 164 pp.

Filov, V.A., Golubev, A.A., Liublina, E.I. and Tolokontsev, N.A. (1973), *Quantitative Toxicology*, John Wiley & Sons, New York.

Haigh, N. (1989), *EEC Environmental Policy and Britain*, Longman, Harlow, Essex, 382 pp.

Hawkins, K., (1984), *Environment and Enforcement Regulation and the Social Definition of Pollution*, Oxford Univerity Press, Oxford.

Hellawell, J.M. (1991), 'Development of a rationale for monitoring', in F.B. Goldsmith (ed.), *Monitoring for Conservation and Ecology*, Chapman and Hall, London, 1–14.

Holdgate, M.W. (1979), *A perspective of environmental pollution*, Cambridge University Press, Cambridge.

Howe, G.M. (1976), *Man, Environment and Disease in Britain*, Penguin, Harmondsworth, 302 pp.

Hutton, J. (1795), *Theory of the Earth with Proofs and Illustrations*, William Creech, Edinburgh (2 vols).

Independent Commission in International Development Issues (1980), *North–South: A Programme for Survival*, Pan Books, London, 304 pp.

Johnston, R.J. (1989), *Environmental Problems: Nature, Economy and State*, Belhaven Press, London, 211 pp.

Jones, G., Robertson, A., Forbes, J. and Hollier, G. (1990), *Environmental Science, Collins Reference Dictionary*, Collins, London/Glasgow, 473 pp.

Lloyd, O.L., Barclay, R. and Lloyd, R.M. (1985) 'Lung cancer and other health problems in a Scottish industrial town', *Ambio*, 14(6), 322–28.

Lovelock, J.E. (1979), *Gaia: A New Look at Life on Earth*, Oxford University Press, Oxford, 157 pp.

McLaren, D.J. (1990), 'Preface', in C. Mungall and D.J. McLaren (eds), *Planet under Stress*, Oxford University Press, Ontario.

Murley, L. and Stevens, M. (eds) (1991), *NSCA Pollution Glossary*, National Society for Clean Air and Environmental Protection, Brighton, 92 pp.

Openshaw, S., Charlton, M., Wymer, C. and Craft, A. (1987), 'A Mark 1 Geographical Analysis Machine for automated analysis of point data sets', *Int. J. Geogr. Info. Systems*, 1(4), 335–58.

Rodda, J.C. and Jones, G.N. (1983), 'Preliminary estimates of loads carried by rivers to estuaries and coastal waters around Great Britain derived from the Harmonized Monitoring Scheme', *Journal of the Institution of Water Engineers & Scientists*, 37(6), 529–39.

Royal Commission on Environmental Pollution (1988), *Twelfth Report: Best Practicable Environmental Option*, HMSO, London, 70 pp.

Santos, M.A. (1990), *Managing Planet Earth: Perspectives on Population, Ecology and the Law*, Bergin and Garvey, New York, 172 pp.

Timbrell, J.A. (1989), *Introduction to Toxicology*, Taylor and Francis, London, 155 pp.

Warn, P.R. (1989), 'The benefits of globalisation', in P. Wombwell (ed.), *The Globe: Representing the World*, Impressions, York.

Wombwell, P. (ed.) (1989), *The Globe: Representing the World*, Impressions, York, 80 pp.

World Commission on Environment and Development (1987), *Our Common Future*, Oxford University Press, 400 pp.

3
The geography of conservation
Anthony Stevenson

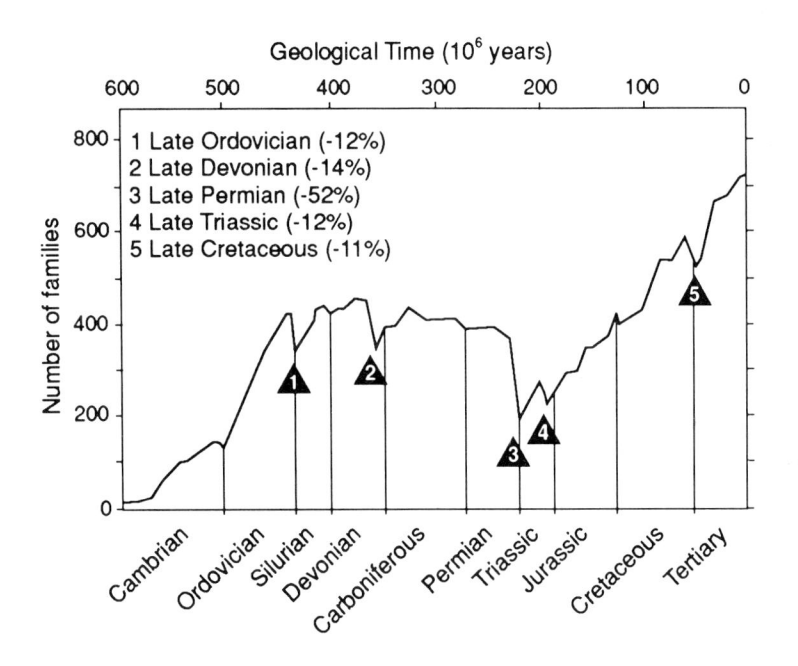

Figure 3.1 The five major extinction periods of families of marine vertebrates and invertebrates during geological time. (Redrawn from E.O. Wilson, 1988)

3.1 Biodiversity and environmental change

The world's present million or so species are the modern-day survivors of an estimated several billion species that have ever existed. While species extinctions have occurred periodically in the geological record (Figure 3.1) the present rate of loss of the world's biological diversity (Table 3.1) is at its greatest since life began (Wilson, 1988), especially in the tropical megadiversity centres (see Box). Up to 25 per cent of the earth's total biodiversity is at serious risk of

> *Biodiversity*: biodiversity is an umbrella term that includes not only the world's total genetic and species diversity of animals, plants and microorganisms but also the total diversity of the world's ecosystems. Although only 1.5 million species have so far been described some estimates reckon that there are another 5–10 million more species, chiefly invertebrates, as yet undescribed (May, 1988). NB: within this definition diversity is used synonymously with species richness and includes (not, as is commonly found in ecological textbooks) the concept of 'evenness'.
>
> *Megadiversity centres*: these are regions of the world where species diversity is great because of the presence of a large number of 'endemic' taxa (about 13.8 per cent of the world's total number of described endemic species).

Table 3.1 Current rates of loss of wildlife habitat in selected regions of the world

Region	Original wildlife habitat ($\times 10^3$ha)	Amount remaining habitat ($\times 10^3$ha)	Percentage loss
Tropical Asia	815,816	248,765	67
Africa (south of the Sahara)	2,079,641	773,774	65

Source: Adapted from McNeely *et al.* (1990, pp. 46–7).

extinction during the next 20–30 years (Raven, 1988).

While all past extinctions had natural causes, today humans are overwhelmingly responsible. Major threats to biodiversity include habitat alteration, over-harvesting, chemical pollution, climate change, introduced species and population increases. Of these, habitat alteration, especially tropical deforestation, and consequent habitat loss, fragmentation and degradation is primarily responsible (IUCN/UNEP, 1986a,b; McNeely *et al.*, 1990, see Table 3.1). However, even if deforestation rates are arrested, pollution of the land, sea and air on regional and global scales will maintain pressure on the world's biota, especially given current concerns about the greenhouse effect.

3.2 Species conservation and pollution: relationships

The effects of pollution on the world's fauna and flora have been recognized since the United Kingdom's alkali inspectorate was created in 1862 in response to widespread devastation caused by hydrochloric acid emissions from soda factories in the north west of England (Rose, 1990). Within a geographical framework, pollution from both point and diffuse sources can affect target populations at both point and diffuse scales. For example, seed dressed with the pesticides aldrin and dieldrin caused direct mortality in seed-eating birds and indirect mortality and reproductive failure, by biomagnification in birds of prey at local levels (Carson, 1988; Sheail, 1985). In marked contrast, more recent pollution problems are globalized and affect species populations diffusely, often via poorly understood, intermediate pathways, for example, CO_2, 'acid rain', nutrients and heavy metals. This globalization of pollution, together with the mediation of its effects by poorly understood and often lengthy pathways, results in dilemmas for control policies. What should the control approach be? Source, pathway or target? With the movement of many pollution problems away from easily controlled point sources, the diffuse nature of many current pollution problems results in control of the pollutant at the target without tackling the problem at source, e.g. 'acid rain'.

3.2.1 Regional problems: acid deposition

'Acid rain' (defined here as rain, mist, snow or aerosols which contain acid compounds of sulphur and nitrogen) is a regional phenomenon affecting the industrialized countries of northern Europe, North America and Asia. Figure 2.1(b) in Chapter 2 demonstrates the scale of the problem in Europe where lower-than-normal rainfall pH is found across most of the continent, with the lowest pH (most acid) levels recorded over the industrial heartland of Germany in the Rhine valley. Geographically, the greatest effects of acid deposition are found in those areas with high rainfall (where more acidity is transferred to the ground) and with base-poor geologies (i.e. they cannot neutralize the deposited acidity). The effect that acid deposition has had on plant and animal communities is now becoming clear although its contribution to the widespread forest declines observed over most of eastern and western Europe is still unclear (cf. Blank *et al.*, 1988).

The earliest link between acid deposition and adverse environmental effects was demonstrated by palaeo-ecological studies of lake sediments in Europe and America which showed that lake and stream acidification was the result of industrial sulphate (SO_4) emissions (Battarbee *et al.*, 1990). Regionally, 'acid rain' has caused great damage to fish (Muniz, 1984), bird and mammalian populations in the affected areas, and once-thriving and important sports fisheries, in geologically sensitive areas, in Scandinavia and Scotland have now disappeared, representing a significant financial loss. Stream acidification has also reduced the invertebrate food supply of the dipper, causing an 80–90 per cent reduction in populations in the River Irfon, Wales (Ormerod *et al.*, 1985), and dippers are now absent from over 60–70 per cent of suitable streams in Wales. Moreover, the lack of fish in acidified streams has prevented otter (*Lutra lutra*) populations from recovering from the hunting and pesticide problems of the early 1960s (Mason and Macdonald, 1987). Meanwhile in southern England, acidification of the breeding sites of natterjack toads (*Bufo calamita*) has reduced populations to very low levels (Beebee *et al.*, 1990).

Various approaches have been developed to ameliorate the effects of acid deposition. The retro-fitting of power stations with flue-gas desulphurization plants (FGD), fuel-switching policies such as burning of imported, low-sulphur coal, and the future development of combined gas turbine plants (CGT) will all reduce sulphur emissions. However, it is predicted that these reductions will only account for 25 per cent of the 70 per cent reduction in sulphur emissions that is required immediately to restore sulphate deposition back to pre-industrial levels. Moreover, the accumulated sulphate store in many soil types will result in a long lag period before the effects of sulphur reductions are seen in drainage waters (Wright *et al.*, 1988).

Until sulphur reductions have occurred, either through more efficient fuel burning or fuel substitution, control must occur at the pathway or target by liming. This approach has been used extensively in Sweden where thousands of acidified lakes have been treated with lime to restore successfully the fishery but at a cost dependent on lake volume and turnover rate. In Sweden, lake turnover time is slow and bi- or triennial liming is sufficient. However, in the United Kingdom a significant cost increase occurs because of the faster turnover times of UK lakes (Underwood *et al.*, 1987). Alternatively, treating the pathway by catchment liming reduces costs because only one application every five years is required, as calcium is released slowly from the catchment (Howells and Brown, 1987). While restoring the fishery, catchment liming causes significant deleterious effects on both the catchment vegetation, with up to 90 per cent of the *Sphagnum papillosum* within the

Table 3.2 Heavy metal inputs (tonnes per annum), excluding atmospheric inputs, by the UK to the North Sea, 1985–86

Substance	River inputs min max	Direct discharges	Dredgings	Sewage sludge	Industrial waste liquid	Industrial waste solid	Incineration at sea	Total input to the North Sea from all countries
Cadmium	8 14	16	5	3	<1	<1	<1	90
Mercury	4 5	3	5	1	<1	<1	<1	40
Copper	240 283	215	262	103	1	160	<1	2700
Lead	245 303	133	338	99	1	206	<1	3500
Zinc	1440 1451	986	970	219	5	396	<1	17000
Chromium	98 141	439	453	39	<1	17	<1	4000
Nickel	233 265	88	117	14	<1	51	<1	1150
Arsenic	28 65	220	—	<1	38	4	<1	800

Source: DoE (1990).

limed area killed (Loch Fleet, 1989), and aesthetic qualities.

Problems arise because the success of these studies is judged purely by the re-establishment of the fishery and little attention is paid to assess whether the whole ecosystem has been restored to its original non-polluted state. The palaeoenvironmental evidence suggests that lake-water chemistries created by liming may have never been recorded in the lake's post-glacial history (e.g. Loch Fleet, Galloway, S.W. Scotland — pH 7.0 after liming, but never above pH 5.8 during the post-glacial — Anderson *et al.*, 1986). In addition, the diatom, macrophyte and invertebrate populations created by liming do not represent the previous non-polluted fauna and flora. One priority area of future research must be in the derivation of suitable criteria for determining the success or otherwise of lake restoration programmes.

3.2.2 Regional/global effects: sewage disposal, eutrophication and the seas

Increasing worries are now being raised about the indiscriminate dumping by the United Kingdom of sewage effluent, heavy metals and power station fly ash in the North Sea Basin (Rose, 1990). The biological effects of this disposal are poorly known but it is becoming increasingly obvious that major problems are now occurring with many aspects of our coastal and marine ecosystems. West German scientists observed that in the German Bight, where dumping of sewage sludge took place from 1961 to 1981, species diversity dropped dramatically, anoxia occurred as a result of eutrophication and a large proportion of marine life became contaminated by viruses, posing problems to human health. Even after dumping ceased, sites 8–9 km from the dumping sites still suffered from lower species diversity than comparable areas far away from the dumping (Dethlefsen, 1986). Elsewhere others have shown that there has been a dramatic increase in the level of pollutants recorded in marine life (Goerke *et al.*, 1979) especially shellfish and seals where PCBs have reached toxic levels.

In the Adriatic, the River Po discharges large quantities of phosphate and nitrate, derived from intensive farming and sewage disposal, promoting enormous blooms of diatoms, algae and dinoflagellates (Marchetti *et al.*, 1989). The death of these organisms in the shallow, semi-enclosed, coastal waters of the northern Adriatic Basin leads to rapid oxygen depletion (anoxia) of the water body and very large kills of marine life (in 1990 some 7000 tonnes of dead fish, mollusca and crustaceans were washed up on beaches in the area). Moreover, a large amount of money is having to be spent on cleaning up the huge quantities of mucilage that these algal blooms produce which is despoiling vast areas of local coastline and reducing the recreational values of these areas (Degobbis, 1989).

Control policies here are far more difficult to implement because not only are the pollution sources mostly diffuse in nature, but also whole river basins are involved which often cross international boundaries. It is clear that while the United Kingdom has committed itself to abandoning its current practice of sewage sludge dumping at sea by 1998 this will only reduce heavy metal inputs by 10–20 per cent and pressure will still be maintained on the marine ecosystem because of continued riverine inputs of lead, zinc, nitrate and phosphate, dredging residues, other direct discharges and industrial-waste dumping (DoE, 1990, Table 3.2). Since so little is still known about the effects of these residues on the marine ecosystem, it is inevitable that ameliorative measures must involve reduction of these pollutants into the drainage water system at source.

3.2.3 Global effects: CO_2 and climate change

While 'acid rain' and sewage-sludge dumping pose problematic regional problems, in the future it is the enhanced 'greenhouse effect' that will have the greatest set of global consequences on terrestrial, freshwater and marine ecosystems. Section 13.1.3 in Chapter 13 explores this subject in detail.

3.3 Biogeography and habitat conservation

The history of conservation has seen it move from a non-interventionist, preservationist outlook (according to Margalef, 1968) through a reductionist scientific viewpoint to (recently) a more holistic approach. The first aspect of change in emphasis was prompted by the realization that a large number of our most valued ecosystems, chalk grasslands, wetlands (broads) and heathlands, are the creation of man's activities and need active management to conserve them. Dissatisfaction with the reductionist, scientific approach and its inability to address and integrate the local populace, especially in the developing world, prompted the latter move. Today, conservation is viewed as synonymous with 'the wise use of the country's resource of land, water and wildlife' (IUCN, 1980).

The question 'why conserve?' raises many issues about the criteria by which the need for conservation is justified. Broadly speaking, these can be divided into those based on the environmental character of the site (including both abiotic, biotic and cultural variables); a set of socio-economic variables; and a response based on potential threats to a site (Table 3.3). In reality, all approaches are equally valid and subsets of these criteria have been used extensively. In practice, for a large number of cases it is the potential threat criterion that triggers off the conservation process and then various combinations of criteria are then deployed, the choice being constrained by the socio-political systems in which the various interest groups operate.

3.3.1 The scientific approach

The former UK (Great Britain) Nature Conservancy Council (now split into three regional conservation bodies — English Nature, Countryside Council for Wales, and Scottish Natural Heritage) evaluated sites using a combination of mainly biotic and abiotic variables: size (area), diversity, naturalness, rarity, typicalness, fragility, recorded history, position in ecological/geographical unit, potential value, intrinsic appeal (Ratcliffe, 1977). It is not intended here to go into a detailed discussion about the

Table 3.3 A classification of criteria used to evaluate and justify the conservation of a site

---

A. *Environmental characteristics*

Biotic and abiotic factors
    Conservation of representative and/or unique ecosystems and habitats
    Biological productivity (maintenance, sustainability and reconstitution of the stocks
    Maintenance of genetic diversity
    Physico-chemical parameters (climatology, geomorphology, hydrology, sedimentation, geochemistry
    Evolution and dynamics of the environment

Cultural elements, representative of civilizations and traditions

    Ethnographic considerations
    Archaeological and historical heritage
    Architectural and/or landscape elements
    Potential for research and education

B. *Socio-economic benefits of conservation*

    Direct values: (i) consumptive; (ii) productive
    Indirect values: (i) non-consumptive use; (ii) option value; (iii) existence value

C. *Potential threats*

Source: Glineur (1990).

use of these criteria as much has already been written on this subject (Usher, 1986). It is clear that while some of the NCC's criteria can be assessed precisely (size, diversity), others are vague (naturalness) or reflect threat (fragility), what could be there (potential value), science (rarity) or human activity and emotion (intrinsic appeal, recorded history). Some of these criteria are difficult to define (typicalness, naturalness) and some are mutually exclusive (rarity and typicalness). For much site selection the first six or so of these criteria (the so called 'scientific' criteria) tend to be used.

3.3.2 Towards an objective evaluation?

This pathway down the scientific evaluation of conservation reaches its most logical conclusion in the development of indices that attempt to measure objectively the conservation potential of sites. For limestone pavements at Malham, north Yorkshire, Ward and Evans (1976) developed an index based on species rarity. Species were allocated to four different rarity groups (from very rare to common) and the index calculated by summing the number of plants in each group (weighted by abundance) and differentially weighting the rare groups. They found that large values of the index were caused by pavements with high species richness and large numbers of rare species. Usher (1980) showed that pavement area had a major effect on the index since large pavements tended to have a high index. When the species richness for each pavement was plotted against the cartographic area he found that objective decisions could be made, depending on the criterion used, about the types of pavement that should be conserved (Figure 3.2). If the most typical pavement were to be conserved then pavements T_1, T_2 and T_3 are obvious candidates since they lie close to the average species/area line. If rarity is your criterion then R_{max}, the pavement with the highest number of rare species, should be conserved. If diversity were the criterion then D_{max}, the pavement that deviates most from the average species/area line, should be conserved. If area were considered important, A_{max}, the pavement with the largest area, should be selected.

Others (e.g. Helliwell, 1985) have produced indices that seek to combine different criteria. These have been the subject of numerous objections (Goldsmith, 1991a) since the summation or multiplication of different parameters, which are not equivalent to one another, is often meaningless. This approach only works if the habitats under study are similar.

Overall, most conservation evaluations in the 1970s and early 1980s depended on the formal 'scientific' criteria. This is in contrast to site selection in urban areas where the needs of the general public have an

The geography of conservation

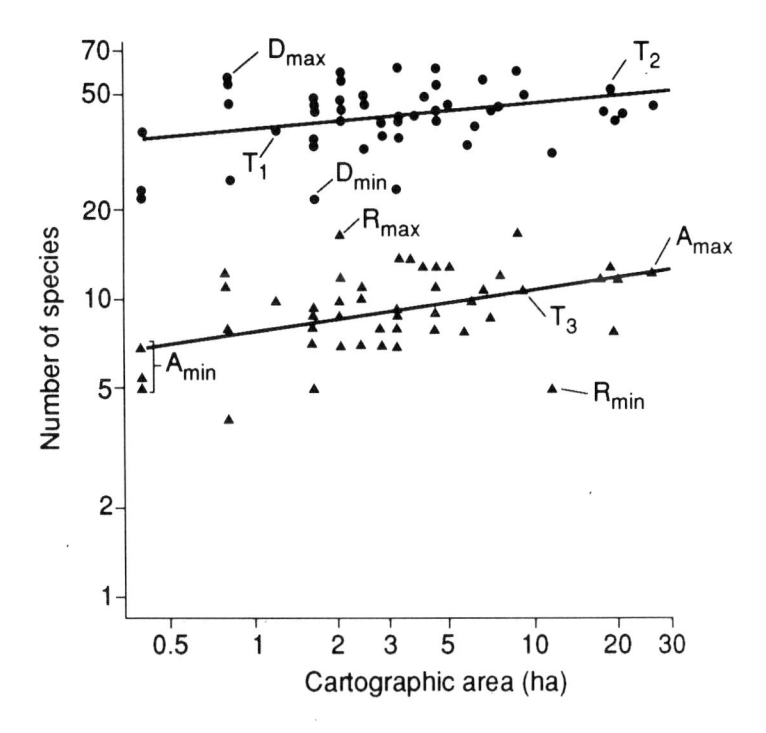

Figure 3.2 Species–area relationships for higher plants on 49 limestone pavements on the Malham-Arncliffe plateau. Triangles represent two groups of rare species. Circles represent all plant species. For an explanation of codes, see text. (Redrawn from Usher, 1986)

important role and criteria like amenity value, accessibility and aesthetic appeal are important (Goldsmith, 1991a). Some have argued that this élitist 'scientific' approach should be broadened to include other criteria, especially the inspirational and emotional demands of the general populace, although few studies have actually addressed this problem in a rational way.

3.3.3 Population size and the rise of socio-economics

Wetland ecology, because of the ornithological interest, was initially dominated by population size criteria. This was aided by the early formulation in 1971 of an international convention to protect wetlands of international ornithological importance known as the Ramsar convention. The Ramsar convention initially used a set of population size criteria and sites were included if they satisfied various percentage rules based on the known total population of the species in the western Palaearctic region (Table 3.4). Notably, these rules only applied to waterfowl; criteria such as other species, especially other vertebrates, the site's scientific value or amenity value played an ancillary but minor role. In the early 1980s the convention had a major impact in the western developed world but few countries from the less developed world became signatories. One of the main reasons, apart from the fact that all documentation was initially only in English, was the impractability of applying the numeric criteria in the tropics. Not only were bird numbers utilizing these wetlands poorly known but their world population sizes were unknown. Moreover, the criteria were seen as not providing any role for the local populations who have been sustainably utilizing these systems for long periods of time.

Table 3.4 Heilinghenhafen criteria used to select wetlands of international importance as used by the Ramsar convention, 1971–87

1. A wetland should be considered internationally important for waterfowl if it regularly supports one or more of the following:

    (a) either 10,000 ducks, geese or swans, 10,000 coots, or 20,000 waders;
    (b) 1 per cent of the individuals in a population of one species or subspecies of waterfowl;
    (c) 1 per cent of the breeding pairs in a population of one species or subspecies of waterfowl.

2. A wetland should be considered internationally important for plants and animals if it:

    (a) supports an appreciable number of a rare, vulnerable or endangered species or subspecies of plant or animal;
    (b) is of special value for maintaining the genetic and ecological diversity of a region because of the quality or peculiarities of its plants and animals;
    (c) is of special value as the habitat of plants or animals at a critical stage of their biological cycles;
    (d) is of special value for its endemic plant or animal species or communities.

3. A wetland should also be considered internationally important if it is a particularly good example of a specific type of wetland characteristic of the region

Source: International Waterfowl Research Bureau (1981).

Maltby (1986) demonstrated that wetland degradation often had a substantial financial cost in addition to the loss of ornithological interest, since many of the goods and services that wetlands have traditionally provided (such as flood prevention, aquifer recharge, flood-plain agriculture and sewage treatment) to the country/region have traditionally been valued as free (Hollis *et al.*, 1988). Consequently, at the regular four yearly conferences that review the Ramsar convention the selection criteria have changed to reflect this increasing emphasis on the socio-economic potential of wetland sites. At Regina, Canada in 1987 the numeric criteria were downgraded and other criteria based on 'wise use' and 'sustainability' rose to become the primary criteria for site selection (Table 3.5).

3.3.4 Socio-economic criteria

These trends culminated in the growth and application of the field of environmental economics to provide a socio-economic evaluation of the conservation of sites and known in the wetland literature as 'wise use' (Pearce, 1976, 1987; Pearce *et al.*, 1989; Turner, 1987; McNeely *et al.*, 1990; Chapter 5 here).

For many countries, especially the less-developed ones, a socio-economic approach is probably the only way sites can be conserved. Here, any amount of scientific rationale has limited effect not only because the societies perhaps operate a different value system, but also because it is common to find that the local populace have never been involved in the conservation scheme, let alone have they been asked their opinions. It is only if planners and conservationists go beyond a narrowly based scientific evaluation of a site's conservation potential and begin to take a more holistic perspective, by identifying the site's functional units and placing economic values on them, that sites may be actively conserved. While there are many difficulties involved in placing economic values on biological properties (cf. Adams, 1991), unless an attempt is made to tackle the government planners in their own terms, sites will inevitably be lost.

One example of a case in which an initial socio-economic appraisal could have preserved a site (instead of its eventual implementation as an *ex post* implementation), can be found in the Ramsar site of Garaet el Ichkeul which lies on the Mediterranean coast in north-west Tunisia (Figure 3.3). The lake (87 km^2) and its surrounding marshes

Table 3.5 Revised criteria used to identify wetlands of international importance as adopted by the Ramsar conference in Regina, Canada, 1987

1. Criteria concerned with the socio-economic value of wetland products, including fish. A wetland or wetland system should be considered internationally important if it:
   (a) yields or supports fish, forest, agricultural or wildlife products, which make a significant contribution to the country's international trade in these products,
   (b) supports activities, including tourism, whose role is of outstanding importance to the country's gross national product;
   (c) borders an international river that now has or historically did have anadromous fish runs.

2. Criteria concerned with ecological or hydrological processes. A wetland or wetland system should be considered internationally important if it:
   (a) has a substantial hydrological or ecological role in the functioning of an international river basin or coastal system;
   (b) recharges water resources in an aquifer that underlies an international boundary;
   (c) forms a significant proportion of the total wetland surface within an international river basin;
   (d) exceeds 100,000 ha in total surface area or is a significant proportion of that wetland type remaining.

3. Criteria concerned with the selection of representative or unique wetlands. A wetland or wetland system should be considered internationally important if it:
   (a) is a representative example of a wetland community characteristic of, or rare and endangered in, its biogeographical region;
   (b) exemplifies a critical stage or extreme in ecological or hydrological processes;
   (c) is an integral part of a peculiar or unique feature.

4. Criteria pertaining to a wetland's importance to populations and species. A wetland or wetland system should be considered internationally important if it:
   (a) regularly supports a substantial proportion of the flyway or biogeographical proportion of one species of waterfowl;
   (b) supports an appreciable number of an endangered species of plant or animal;
   (c) is of special value for maintaining genetic and ecological diversity because of the quality and peculiarities of its flora and fauna.

5. Criteria concerned with the research or educational value of wetlands. A wetland or wetland system should be considered internationally important if it:
   (a) is outstandingly well suited or well equipped for scientific research and education, or is regularly visited by substantial numbers of foreign visitors, researchers or students;
   (b) is well studied and documented over many years and with a continuing programme of research of high value, regularly published and contributed to by the scientific community;
   (c) offers special opportunities for promoting public understanding and appreciation of wetlands, open to people from several countries;
   (d) preserves important archaeological remains or evidence of past environments.

Source: Ramsar (1987). Revised criteria used to identify wetlands of international importance as accepted by the Ramsar conference in Regina, Canada.

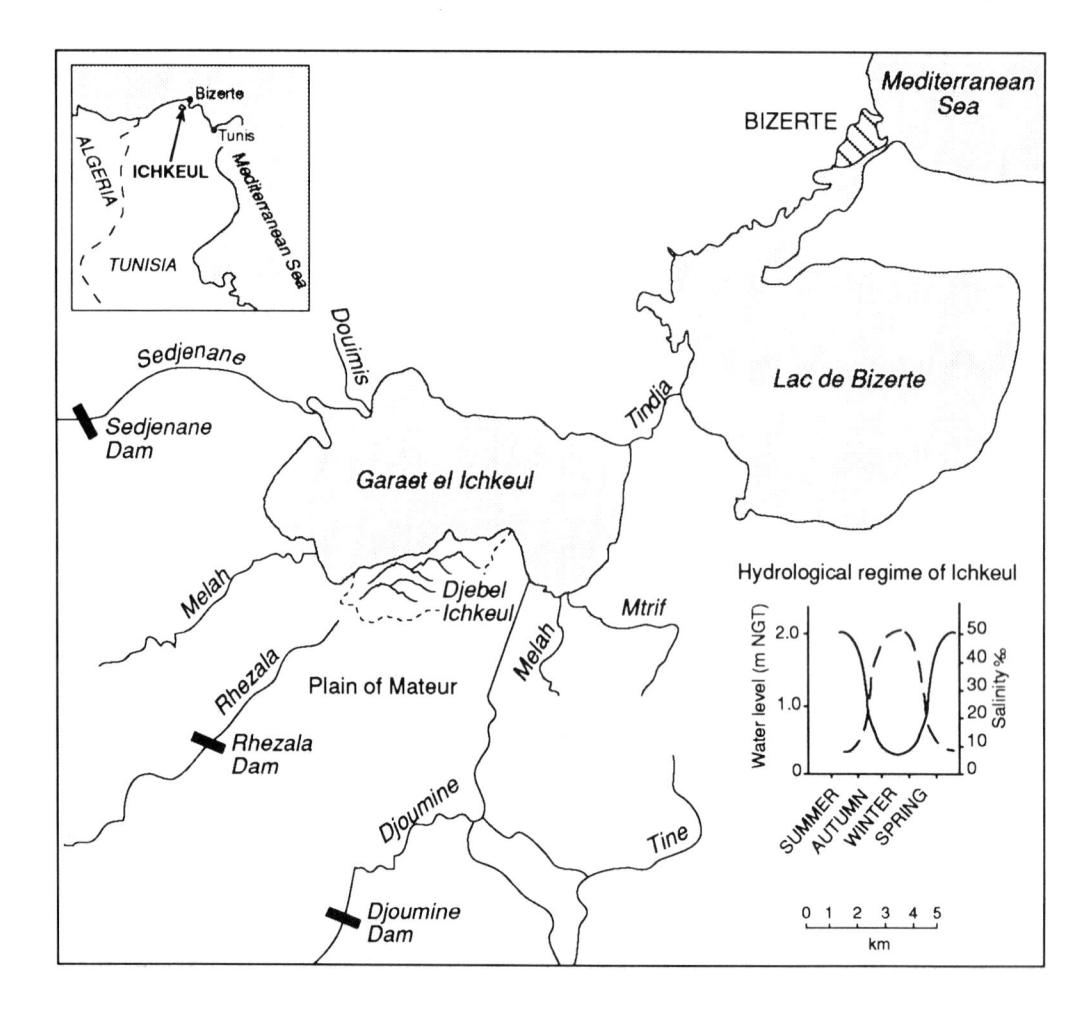

Figure 3.3 Location map for Garaet el Ichkeul showing the major rivers, dams and hydrological regime (inset). Inset key: solid line = salinity, dashed line = lake level

(22 km^2) were internationally important for their wintering waterfowl populations (240,000 assorted widgeon, pochard, coot and greylag geese — Hollis, 1986; Scott, 1980), but are currently threatened by a water resources scheme which includes damming the major inflowing rivers (Djoumine and Sedjenane for potable water supply) and an agricultural improvement scheme (irrigated from a dam on the Rhezala River) on the Plain of Mateur.

The lake is fed by six rivers and drains to Lac de Bizerte, a tidal marine bay of the Mediterranean. The outflow channel is unusual since its flow reverses in the summer when the level of Ichkeul falls below that of Lac de Bizerte as a result of cessation of flows in the major rivers and high evaporation rates from the lake. As a consequence sea water enters Ichkeul and the lake undergoes a variable level and salinity regime (Hollis, 1986; Figure 3.3). Hollis (1986) demonstrated that without any ameliorative measures such as a sluice to control the flow of fresh water out of, and sea water into, the lake together with a system of compensatory water releases from the dams, Ichkeul National Park will rapidly become a permanently saline sebkha. It will have lost its current status as an international bird

Table 3.6 Comparison of economic benefits of water releases into Ichkeul National Park compared to the use of water on the agricultural scheme

	Theoretical optimum of agricultural scheme	Fisheries and grazing in Ichkeul National Park
Average annual water requirement million m^3	5.533	1.8
Gross annual income (TD)	1,733,389	909,846
Gross annual income per m^3 of water (TD)	0.239	0.505
Net annual revenue (TD)	775,589	828,745
Net annual revenue per m^3 of water (TD)	0.141	0.460
Annual costs per m^3 of water (TD)	0.204	0.0
Net profit per m^3 of water (TD)	−0.063	0.460

TD = Tunisian Dinars.
Source: Adapted from Thomas (1989).

reserve, an important economic fishery and valuable grazing lands.

When an *ex post* economic evaluation of the present 'wise use' of the Park's fishery and grazing compared to the agricultural improvement scheme was made it was discovered that the economic benefits of the improvement scheme were illusory and the result of market and intervention failures according to Turner and Jones (1991) (Thomas, 1989). Table 3.6 demonstrates that the Park generates more income from its eel fishery and associated grazing of the marshes than would the agricultural scheme under its theoretical maximum potential. Every m^3 of water released from the Ghezala dam would actually produce a profit for the eel fishery and grazing while it is producing a deficit for the agricultural scheme. Moreover, this profit is generated without any additional indirect economic benefits being included such as tourism, willingness to pay to conserve the system, or more rational grazing of the marshes.

In spite of this economic analysis the Tunisian government is still pushing ahead with the full implementation of the agricultural scheme. Moreover, the ameliorative measures that were identified as being needed as the various dams were completed still have not been enacted. Furthermore, drainage channels have been dug across some of the river-fed marshes, preventing inundation, and have caused a further loss of grazing land for the local graziers and food supply for the waterfowl (Stevenson, unpublished data). It is interesting to speculate that if a proper economic evaluation had taken place earlier Ichkeul may well have been saved.

The Coto de Doñana is a protected area of 50,720 ha located at the mouth of the Guadalquivir River in south-west Spain, containing an extensive marsh habitat which regularly supports upwards of 200,000 wintering waterfowl. Fifty per cent of this marshland has been lost to drainage and rice cultivation since the 1930s, but it is in the planned developments on the right bank of the river where the greatest threats now lie. The Almonte-Marismas plan was developed to exploit the confined aquifers that lie under the National Park for irrigation purposes (Montes and Bifani, 1991). This, twinned with a proposed expansion of the nearby holiday resort of Matalascañas along the coastline of the National Park, would involve an appreciable additional demand in water resources and result in the depletion of these aquifers. Since the vegetation of Doñana depends on high ground-water tables for

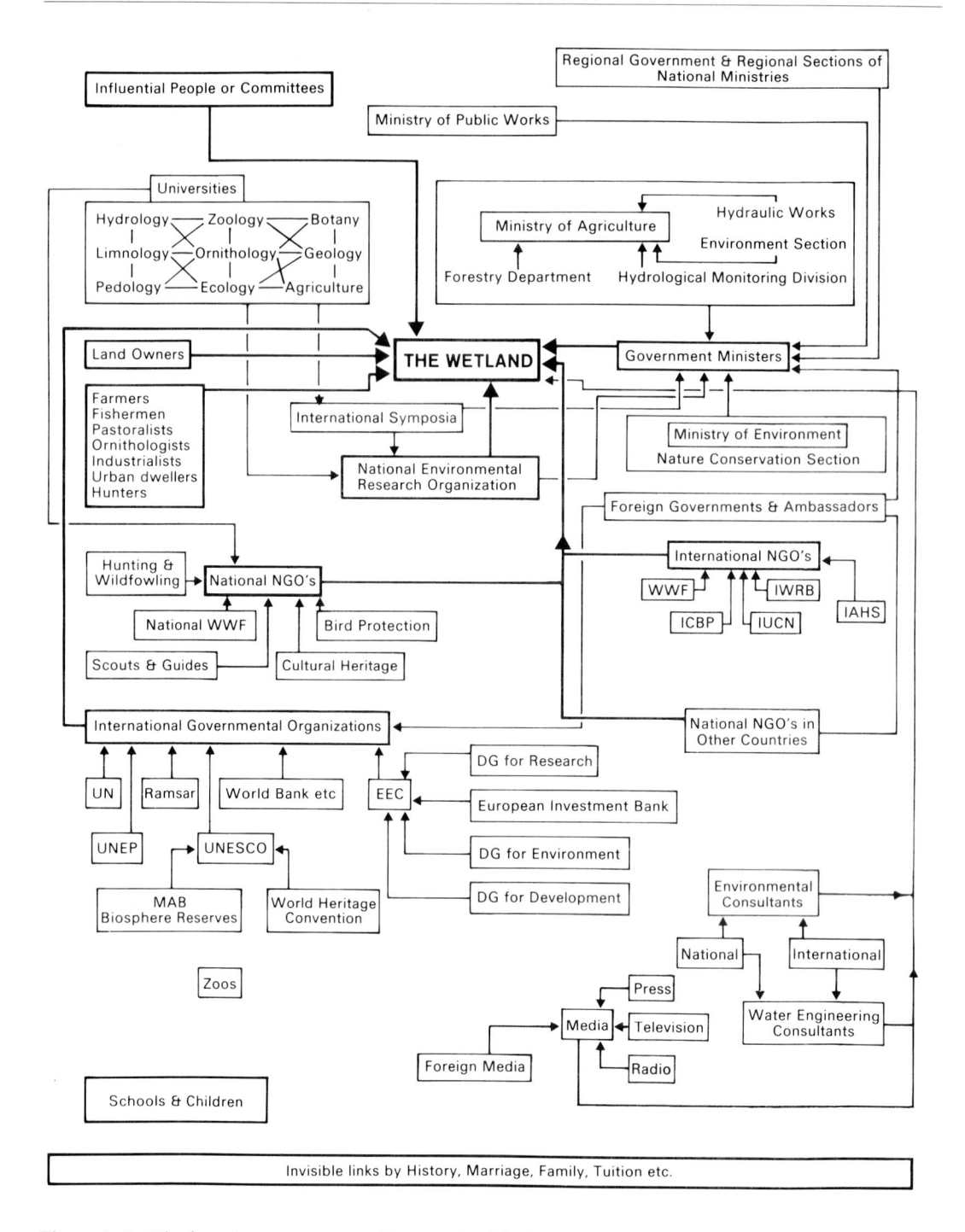

Figure 3.4 The broader system controlling wetland hydrology, development and conservation. (Redrawn from Hollis, 1990)

survival it is clear that these two programmes pose a real threat to the Park (García Novo, 1979). It is interesting to note that it took a study by three outside wetland specialists to identify that most problems stemmed from the lack of integration and co-operation between various state, regional and local administrations, each of which had some interest or involvement in the National Park (Hollis et al., 1989). Furthermore, in spite of the great scientific esteem in which Doñana is held and the large amount of scientific work being carried out there, it was clear that very little attention had been paid to surveillance or to monitoring and that there was a complete lack of knowledge of trends in ground-water levels and water quality within the National Park, a fact all the more surprising considering the importance the ground-water tables play in maintaining the various ecosystems.

Both the Ichkeul and Doñana studies illustrate that conservation of sites cannot be carried out within a scientific vacuum but that a holistic view of how the site is affected by outside interests is essential for proper management. Figure 3.4 from Hollis (1990) generalizes these interrelationships and illustrates the complex web of interrelationships that impinge on wetlands. These 'organigrammes' can be drawn for most sites and illustrate that for most conservationists, wetland managers and planners the problem is not in understanding the wetland's science base but it is the understanding of the overall socio-political system, and eventual targeting of the key links under which a site operates, that holds the key to successful conservation.

Nevertheless, a large number of objections have been raised to the application of economic values to biological entities, with the accompanying observation that this corrupts the conservationist because they are using the very value judgement system they despise (cf. Adams, 1991). In reality, planners and developers must be confronted head on, and while in the case of the developed western world attempts can be made to appeal to aesthetics and science, in the developing world where economic planners hold sway, the only way forward is to question their economic analysis and force them to accept that wetlands have traditional values and economic benefits that are normally omitted from their analysis. Even in the case of the conservation of the limestone pavements in Malham, while it would be difficult to place an economic value on the direct values of the site, it would be the indirect values (option value — genetic diversity, rarity arguments) and the existence value (amount people would be prepared to spend to preserve the site) that should be examined in order to place an economic value on the site.

3.4 Biogeography and nature reserve design

When the theory of 'island biogeography' was proposed by MacArthur and Wilson in 1967, many conservationists noted that the theory could predict the minimum sizes of nature reserves, their shape and connectivity. The theory therefore has potential to offer a rationale analysis and planning of a country's nature resource (e.g. Diamond, 1975; Diamond and May, 1976; Higgs, 1981; Higgs and Usher, 1980). The theory predicts that the species richness of an island, be it geographic or habitat, is the result of the balance between the immigration rate of new species and the extinction rate of existing species. This produces a dynamic species equilibrium which is more or less fixed, depending on the size of island. Therefore, large islands may be expected to have a higher species-richness than smaller islands, and islands of the same size, but located at different distances from an immigration source, will result in the farther island containing lower numbers of species than nearer islands. These conclusions stimulated a large amount of debate, especially concerning whether it is better to have one single large reserve or lots of small reserves (the so-called SLOSS debate), and what shape should reserves be — long and narrow or circular.

While it has been theoretically shown that several small reserves adding up to the same area as one large one would be better because they reduce the probability of disease epidemics breaking out, small reserves result in reduced population sizes and if a species can no longer maintain a certain degree of outbreeding this can quickly lead to problems of extinction as a consequence of genetic drift and loss of genetic vigour. While the theory has been put to good use on real geographic islands, its applicability to insular 'islands' of land has become the subject of much criticism, especially since it only deals with species number and not actual species; it is ahistorical and difficulties arise in definition of the species pool available for immigration (Gilbert, 1980; Spellerberg, 1991). Although the theory allows some general points to be made, for example, general species/area relationships (see section 13.1.3, Chapter 13), these should not be a substitute for a thorough understanding of the biological and ecological needs of the organisms concerned. Since most conservation is directed at large predators/herbivores with requirements for large territories, in practice one large reserve is better than several smaller ones. Furthermore, with much conservation practice still *ad hoc*, mostly in response to imminent threats, the luxury of being able slowly to plan your nature resource is often not available and there is rarely any choice as to the shape, size and location of nature reserves. In conclusion, although the theory has prompted much debate, little of any practical use has been derived from it to enhance the conservation of the world's biota.

3.5 Historical perspective

In order for wetland managers to understand fully the dynamics of their wetlands so that the consequences of various external forcing mechanisms can be predicted, it is essential that the timing, trends and causes of ecological change are established. While the impacts of some current and potential threats to wetlands are easy to quantify (e.g. drainage), other impacts may have been operational in the past that are outside human memory (e.g. eutrophication, acidification) and subtle in their effects. While documentary evidence may be useful in resolving the issues the record is often ambiguous, incomplete or in many cases absent and does not have the necessary temporal resolution. Routine monitoring, while providing high temporal resolution has only offered this, if at all, for the last 20–30 years.

One way of resolving some of these problems is to elucidate the timing and trends of environmental change by palaeoecological analyses of sediment cores from wetlands and lakes. Such sediments act as repositories of environmental information over time and their analysis has been vital in elucidating the timing, trends and causes of ecological change in wetlands all over the world (e.g. Battarbee *et al.*, 1990; Stevenson and Battarbee, 1991; Stevenson and Flower, 1991; Huntley, 1991; Jacobsen *et al.*, 1991). Two examples are presented here to exemplify the approach and outline the contribution that such analyses can make to conservation.

Lake Mikri Prespa, located in mountainous countryside in north-west Greece, is of international importance for dalmatian and white pelicans and pygmy cormorants. Pyrovetsi *et al.* (1984) and Koussouris *et al.* (1989) have claimed that Mikri Prespa is, and has become, eutrophic by means of runoff of agricultural fertilizers and untreated raw sewage entering the lake. Stevenson and Flower (1991) conducted diatom analyses on a dated sediment core from the lake to assess the magnitude of recent change resulting from catchment disturbance and water enrichment. It was found that the diatom populations of the lake had not changed substantially over the last 90 years, suggesting that phosphorus enrichment over the period spanned by the core has been minor.

Peat erosion is widely recognized in Britain and Ireland and is causing many problems

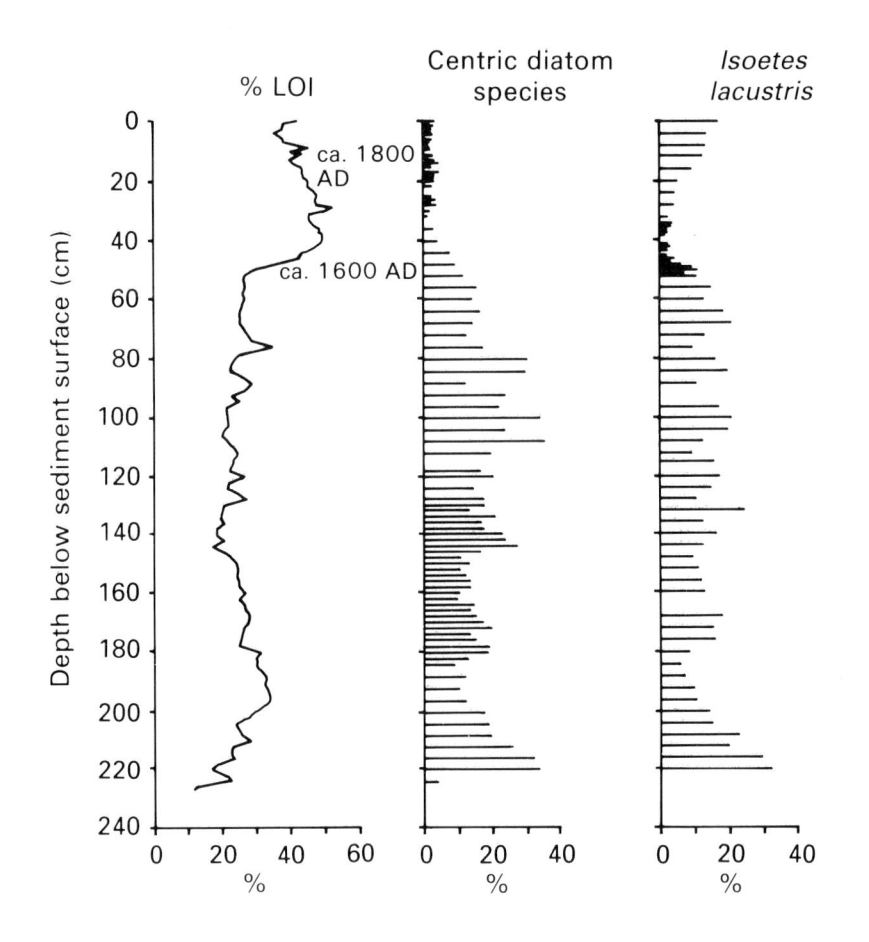

Figure 3.5 Variations in the organic matter content (% LO1), diatom plankton (centric species) and *Isoetes* spores for a sediment core from the Round Loch of Glenhead, Galloway, S.W. Scotland. Depressed levels of diatoms and spores from ca 1600 AD result from reduced water transparency caused by peat erosion. (Redrawn from Stevenson, Jones & Battarbee, 1990)

for upland managers, for example, reservoir siltation, grazing reduction, habitat loss and aesthetic problems. Many hypotheses have been advanced to explain the initiation of erosion and range from climate, inherent instability of peat masses, anthropogenic burning, grazing and trampling, air pollution to a combination of these factors. Accurate dating is critical to evaluating the various competing hypotheses, while the identification of the causes of peat erosion on a catchment basis is crucial to the long-term management of peatland systems.

Studies of the lake sediment from the Round Loch of Glenhead, south-west Scotland showed that peat erosion is recorded in the sediments by changes in the loss on ignition, diatom and aquatic spore records (Stevenson et al., 1990). Although precise dating of these changes is problematical at present, the data do allow some of the competing hypotheses to be evaluated (Figure 3.5). The data suggest that the peats were stable until a few hundred years ago indicating that the 'inherent instability' hypothesis is invalid at this site. The main erosion is associated with high charcoal values in the catchment peats which suggests that burning is important, but a climatic influence cannot yet be ruled out since the erosion period probably began during the Little Ice Age. Acid deposition is of little significance at this site since severe erosion began at least 100 years before the first evidence of significant air pollution in the lake sediment core (Jones et al., 1989).

It is clear from these two examples that such studies can offer enormous benefits in resolving some critical problems about the nature, timing and trends of environmental change. Moreover, it allows the wetland managers to see in perspective the current functioning of their systems.

3.6 Monitoring success/failure

The successful conservation of a site is not in itself a guarantee that the site will remain unchanged in the future. Careful consideration must be given to ways of following changes in the ecosystem through time. Where the direction and/or cause of change in an ecosystem are unknown then this is known as surveillance; if these are known then this is known as monitoring.

Monitoring can be performed on a variety of temporal and spatial scales, the scale being determined by the underlying aims of the monitoring exercise (cf. Goldsmith, 1991b). On a small scale the effect of a grazing regime on a grassland is often followed by means of permanent quadrats/transects. However, the monitoring of water quality and definition of critical loads (see Chapters 6 and 7) in relation to regional and/or global air pollution occurs across a wide spatial scale. Thus the critical loads advisory group (CLAG), which is determining the critical loads for sulphate and nitrate on the terrestrial and aquatic biota, has undertaken a monitoring programme to determine the vulnerability of surface waters to acidification (DoE, 1991). This has involved sampling lakes and streams in every 10 km grid square throughout the country and has resulted in the production of critical loads maps for the United Kingdom which show those areas most at risk from increases in SO_2 and NO_x emissions. These maps can also be used to evaluate various policy options for reducing acid emissions by quantifying the extent to which emissions have to be cut to reduce the number of exceedances of a given percentage of grid squares above their critical loads and thereby result in an improvement in that area's fresh-water biota (see Chapter 6 for more detail).

In other cases, monitoring can occur remotely. Satellite data over the last 20 years for the Ichkeul National Park have been analysed to examine the relationship between the area of *Potamogeton pectinatus* (the chief food resource for the wintering waterfowl) and various lake hydrological parameters (Stevenson, unpublished). At ground level it is impossible precisely to define the area of lake occupied by *Potamogeton* and would involve at least 15 man-days to survey the lake accurately. With satellite data, cloud-free images are normally available at least 10–15 times during the growing season and a reasonable estimation of the extent of *Potamogeton* can be obtained from the images in about 30 minutes. A model can subsequently be derived between the area of *Potamogeton* and various lake hydrological variables. This allows not only the lake to be monitored in almost real time but also permits the development of predictive models that forecast how the food resource may change under future hydrological scenarios.

Some of the most sophisticated techniques of monitoring change in water quality in lakes can be achieved by using the lake-sediment record. High resolution diatom analysis of the most recent sediment deposited in acidified lakes in Scotland (involving examining the core at every 2 mm) has demonstrated that the decline in sulphate emissions that occurred from the mid-1970s onwards in the United Kingdom has resulted in a return of diatom communities that were present in the early 1920s prior to the major onset of acid deposition (Battarbee et al., 1988). Since the lake sediments provide a historical record of past environmental conditions they could be used to erect target water quality and biotic conditions that would result in the return of the lake to its pre-perturbation state.

References

Adams, J.G.U. (1991), 'On being economical with the environment', *Global Biogeography Letters*, 1, 161–3.
Anderson, N.J., Battarbee, R.W., Appleby, P.G., Stevenson, A.C., Oldfield, F., Darley, J. and Glover, G. (1986), 'Palaeolimnological evidence for the recent acidification of Loch Fleet, Galloway', research paper no. 17, Palaeoecology Research Unit, Department of Geography, University College London.
Battarbee, R.W., Renberg, I., Talling, J.F. and Mason, B.J. (eds) (1990), *Palaeolimnology and Lake Acidification*, Royal Society, London.
Battarbee, R.W., Flower, R.J., Stevenson, A.C., Jones, V.J., Harriman, R. and Appleby, P.G. (1988), 'Diatom and chemical evidence for reversibility of acidification of Scottish lakes', *Nature*, 332, 530–2.
Beebee, T.J.C., Flower, R.J., Stevenson, A.C., Patrick, S.T., Appleby, P.G., Fletcher, C., Marsh, C., Natkanski, J., Rippey, B. and Battarbee, R.W. (1990), 'Decline of the Natterjack Toad (*Bufo calamita*) in Britain: palaeoecological, documentary and experimental evidence for breeding site acidification', *Biological Conservation*, 53, 1–20.
Blank, L.W., Roberts, T.M. and Skeffington, R.A. (1988), 'New perspectives on forest decline', *Nature*, 336, 27–30.
Carson, R. (1988), *Silent Spring*, Houghton Mifflin, Boston.
Degobbis, D. (1989), 'Increased eutrophication of the northern Adriatic Sea: second act', *Marine Pollution Bulletin*, 20, 452–7.
Dethlefsen, V. (1986), 'Marine pollution mismanagement: towards a precautionary concept', *Marine Pollution Bulletin*, 17, 54–7.
Diamond, J.M. (1975), 'The island dilemma: lessons of modern biogeographic studies for the design of natural reserves', *Biological Conservation*, 7, 129–46.
Diamond, J.M. and May, R.M. (1976), 'Island biogeography and the design of nature reserves', in R.M. May (ed.), *Theoretical Ecology: Principles and Applications*, Blackwell, Oxford, pp. 163–86.
DoE (1990) *Digest of Environmental Protection and Water Statistics*, HMSO, London.
DoE (1991) *'Acid Rain': Critical and Target Loads Maps for the United Kingdom*, Air Quality Division, Department of Environment.
García Novo, F. (1979), 'The ecology and vegetation of the dunes in Doñana National Park (South West Spain)', in R.L. Jeffries and A.J. Davy (eds), *Ecological Processes in Coastal Environments*, Blackwell, Oxford.
Gilbert, F.S. (1980), 'The equilibrium theory of island biogeography: fact or fiction', *Journal of Biogeography*, 7, 206–36.
Glineur, N. (1990), *METAP Biodiversity Component*, draft report from the World Bank on the Environmental Programme for the Mediterranean, Washington, DC.
Goerke, H., Eder, G., Weber, K. and Ernst, W. (1979), 'Patterns of organochlorine residues in animals of different trophic levels from the Weser estuary', *Marine Pollution Bulletin*, 10, 127–32.
Goldsmith, F.B. (1991a), 'The selection of protected areas', in I.F. Spellerberg, F.B. Goldsmith and M.G. Morris (eds), *The Scientific Management of Temperate Communities for Conservation*, Blackwell, Oxford.
Goldsmith, F.B. (ed.) (1991b), *Monitoring for Conservation and Ecology*, Chapman and Hall, London.
Helliwell, D.R. (1985), *Planning for Nature Conservation*, Packard, Chichester.
Higgs, A.J. (1981), 'Island biogeographic theory and nature reserve design', *Journal of Biogeography*, 8, 117–24.
Higgs, A.J. and Usher, M.B. (1980), 'Should nature reserves be large or small?' *Nature*, 285, 568–9.
Hollis, G.E. (1986), *The Modelling and Management of the Internationally Important Wetland at Garaet el Ichkeul, Tunisia*, IWRB Special Publication No. 4, IWRB, Slimbridge.
Hollis, G.E. (1990), 'Environmental impacts of development on wetlands in arid and semi-arid areas', *Hydrological Sciences Journal*, 35, 411–28.
Hollis, G.E., Holland, M.M., Maltby, E. and Larson, J.S. (1988b), 'Wise use of wetlands', *Nature and Resources*, 19, 2–13.
Hollis, G.E., Heurteaux, P. and Mercer, J. (1989), 'The implications of groundwater extraction for the long term future of the Doñana National Park', report to WWF-International, Gland, Switzerland.
Howells, G. and Brown, D.J.A. (1987), 'The Loch Fleet project, SW Scotland', *Transactions of the Royal Society of Edinburgh: Earth Sciences*, 78, 241–8.
Huntley, B.H. (1991), 'Historical lessons for the future', in I.F. Spellerberg, F.B. Goldsmith and

M.G. Morris (eds), *The Scientific Management of Temperate Communities for Conservation*, Blackwell, Oxford.

IUCN (1980), *World Conservation Strategy*, IUCN, Gland, Switzerland.

IUCN/UNEP (1986a), *Review of the Protected Areas System in the Indo-Malayan Realm*, IUCN, Gland, Switzerland.

IUCN/UNEP (1986b), *Review of the Protected Areas System in the Afrotropical Realm*, IUCN, Gland, Switzerland, 259 pp.

IWRB (1981), *Conference on the Conservation of Wetlands of International Importance Especially as Waterfowl Habitat*, Cagliari, 1980, IWRB, Slimbridge.

Jacobsen, G.L., Almquist-Jacobsen, H. and Winne, C. (1991), 'Conservation of rare plant habitat: insights from the recent history of vegetation and fire at Crystal Fen, Northern Maine, USA', *Biological Conservation*, 57, 287-314.

Jones, V.J., Stevenson, A.C. and Battarbee, R.W. (1989), 'Acidification of lakes in Galloway, south-west Scotland: a diatom and pollen study of the post-glacial history of the Round Loch of Glenhead', *Journal of Ecology*, 77, 1-23.

Koussouris, Th., Diapoulis, A. and Balopoulus, E. (1989), 'Assessing the trophic status of Lake Mikri Prespa, Greece', *Annls. Limnol.*, 25, 17-24.

Loch Fleet (1989), 'Vegetation changes in the Loch Fleet catchment following liming', Loch Fleet Information Sheet No. 4, Loch Fleet Project, CEGB, Leatherhead.

MacArthur, R.H. and Wilson, E.O. (1967), *The Theory of Island Biogeography*, Princeton University Press, Princeton.

Marchetti, R., Provini, A. and Crosa, G. (1989), 'Nutrient load carried by the River Po into the Adriatic Sea, 1968-1987', *Marine Pollution Bulletin*, 20, 168-72.

McNeely, J.A., Miller, K.R., Reid, W.V., Mittermeir, R.A. and Werner, T.B. (1990), *Conserving the World's Biological Diversity*, IUCN, Gland, Switzerland, WRI, CI, WWF-US, and the World Bank, Washington, DC.

Maltby, E. (1986), *Waterlogged Wealth*, Earthscan, London, 200 pp.

Margalef, R. (1968), *Perspectives in Ecological Theory*, University of Chicago Press, Chicago.

Mason, C.F. and Macdonald, S.M. (1987), 'Acidification and otter (*Lutra lutra*) distribution on a British river', *Mammalia*, 51, 81-7.

May, R.M. (1988), 'How many species are there on earth?' *Science*, 241, 1441-9.

Montes, C.M. and Bifani, P. (1991), 'Spanish wetlands: an overview and historical perspective', in R.K. Turner and T. Jones (eds), *Wetlands: Market and Intervention Failures*, Earthscan, London.

Muniz, I.P. (1984), 'The effects of acidification on Scandinavian freshwater fish fauna', *Philosophical Transactions of the Royal Society of London*, 305B, 517-28.

Myers, N. (1988), 'Threatened biotas: "Hotspots" in tropical forests', *Environmentalist*, 8, 1-20.

Ormerod, S.J., Tyler, S.J. and Lewis, J.M.S. (1985), 'Is the breeding distribution of Dippers influenced by stream acidity?' *Bird Study*, 32, 32-9.

Pearce, D.W. (1976), *Environmental Economics*, Longman, London.

Pearce, D.W. (1987), 'The sustainable use of natural resources in developing countries', in R.K. Turner (ed.), *Sustainable Environmental Management: Principles and Practice*, Belhaven, London.

Pearce, D., Markandya, A. and Barbier, E.B. (1989), *Blueprint for a Green Economy*, Earthscan Publications, London.

Pyrovetsi, M., Crivelli, A.J., Gerakis, P.A., Karteris, M.A., Kastro, E.P. and Kouminus, N. (1984), 'Integrated environmental study of the Prespa National Park, Greece', final report to the EEC DG XI PM1/183/83.

Ratcliffe, D.A. (ed.) (1977), *A Nature Conservation Review*, Vol. 1. Nature Conservancy Council, Peterborough.

Raven, P.H. (1988), 'Biological resources and global stability', in S. Kawano, J.H. Connell and T. Hidaka (eds), *Evolution and Coadaptation in Biotic Communities*, University of Tokyo Press, Tokyo, pp. 3-27.

Rose, C. (1990), *The Dirty Old Man of Europe: The Great British Pollution Scandal*, Simon and Schuster, London, 366 pp.

Scott, D.A. (1980), 'A preliminary inventory of wetlands of international importance for waterfowl in W. Europe and N. Africa', *IWRB Spec. Publ.*, 2, Slimbridge, UK, 127 pp.

Sheail, J. (1985), *Pesticides and Nature Conservation: The British Experience*, Clarendon Press, Oxford.

Spellerberg, I.F. (1991), 'Biogeographical basis of conservation', in I.F. Spellerberg, F.B. Goldsmith, and M.G. Morris (eds), *The Scientific Management of Temperate Communities for Conservation*, Blackwell, Oxford.

Stevenson, A.C., Jones, V.J. and Battarbee, R.W. (1990), 'The cause of peat erosion: a palaeolimnological approach', *New Phytologist*, 114, 727-35.

Stevenson, A.C. and Battarbee, R.W. (1991), 'Palaeoecological and documentary records of recent changes in Garaet el Ichkeul: a seasonally saline lake in N.W. Tunisia', *Biological Conservation*, 58, 275–95.

Stevenson, A.C. and Flower, R.J. (1991), 'A palaeoecological evaluation of environmental degradation in Lake Mikri Prespa, N.W. Greece', *Biological Conservation*, 57, 89–109.

Thomas, D.S. (1989), 'Irrigation and drainage development adjacent to the Ichkeul National Park, Tunisia', report to the CEE (DG X11), Department of Geography, University College London.

Turner, R.K. (ed.) (1987), *Sustainable Environmental Management: Principles and Practice*, Belhaven, London.

Turner, K. and Jones, T. (1991), *Wetlands: Market and Intervention Failures*, Earthscan, London.

Underwood, J., Donald, A.P. and Stoner, J. (1987), 'Investigations into the use of limestone to combat acidification in two lakes in West Wales', *Journal of Environmental Management*, 32, 29–40.

Usher, M.B. (1980), 'An assessment of conservation values within a large Site of Special Scientific Interest in North Yorkshire', *Field Studies*, 5, 323–48.

Usher, M.B. (ed.) (1986), *Wildlife Conservation Evaluation*, Chapman and Hall, London.

Ward, S.D. and Evans, D.F. (1976), 'Conservation assessment of British limestone pavements based on floristic criteria', *Biological Conservation*, 9, 217–33.

Wilson, E.O. (1988), 'The diversity of life', in H.J. de Blij (ed.), *Earth '88: Changing Geographic Perspectives*, National Geographic Society, Washington, DC, pp. 68–81, 392 pp.

Wright, R.F., Lotse, E. and Semb, A. (1988), 'Reversibility of acidification shown by whole-catchment experiments', *Nature*, 334, 670–5.

4

Environmental law

David Harte

4.1 The relationship of law to other disciplines and its role as a tool for managing the natural environment

Law provides an indispensable framework for regulating and developing the natural environment. However, it is used to achieve policies which are mainly determined by other disciplines. Thus, the study of human society is fundamental to an understanding of environmental change. The manner in which such factors operate and the ways in which they may be taken into account are studied by a wide range of disciplines, including geography, sociology and economics. Indeed, the law itself influences the way people behave and for this reason its study is often classed as a social science. Thus, the very fact that a certain activity such as depositing litter is made a criminal offence is likely to have social consequences. These may not always be what is expected! Such a law may reduce the amount of litter or it may increase the problems of those in authority by making litter-dropping a new form of entertainment.

Lawyers are often blamed for defects in the law and such criticism is often partly justified because it is usually lawyers who draft the legislation which is passed by national parliaments and other law-making bodies. Lawyers also interpret the law when they advise people who are affected by it and when a dispute is brought to the courts for resolution. However, the decision to regulate a particular topic by law in the first place is a political one. Comprehensive laws for pollution control or for conserving wildlife are now being devised because these have become popular political issues throughout the world. To this extent it is non-lawyers, politicians and the public, who are likely to be at fault if the law is used where it is unsuitable or for unpopular purposes. The wide aims to be achieved in environmental policy must be worked out through theoretical disciplines other than law. Such disciplines may have self-contained philosophies, as do some schools of politics or economics, or they may look for more fundamental guidance through moral philosophy or theology.

The forms of practical control chosen to protect the environment depend to a large extent on what technologies those in scientific disciplines have been able to devise. Typical are the methods evolved by chemists and engineers to detect and cope with pollution. The siting of major roads or pipelines will be determined initially by social needs and economic factors but then by engineering considerations of what is safe and practical. Again, laws on the abstraction or impounding of water will depend partly on social considerations but equally on the insights of hydrology, such as the understanding of

Practical, administrative and legal decisions affecting water abstraction from inland waters

Archer, a farmer, needs to increase his water supply if he is to enlarge a pedigree herd of cattle and expand his agri-business by setting up a large new riding stable. His existing water comes from the river Wren which flows through his farm. He currently abstracts less than 20 cubic metres in any period of 24 hours. His estimate is that he may need up to 50 cubic metres in a day. He wishes to construct a small reservoir to ensure a regular supply. He wants this to have a capacity of 1500 cubic metres and to replenish it from the Wren.

The following legislation is relevant to Archer's decision: Reservoirs Act 1975 imposes restrictions on the construction of 'large reservoirs' (ap. s 1(1)(a)); 'a raised reservoir is a 'large raised reservoir' if it is designed to hold, or capable of holding, more than 25,000 cubic metres of water above [the natural level of the land adjoining the reservoir].'

Water Resources Act 1991, s 24(1)(a) forbids any person from abstracting 'water from any source of supply' save under a licence from the National Rivers Authority. There are exceptions where licences are not required. In particular under s 27(3) and (4) the restriction does not apply where water (a) is abstracted for use on a holding consisting of the contiguous land with or without other land held with that land; and (b) is abstracted for use on that holding for either or both of the following purposes, that is to say (i) the domestic purposes of the occupier's household; (ii) agricultural purposes other then spray irrigation, unless the abstraction exceeds 20 cubic metres, in aggregate, in any period of 24 hours.

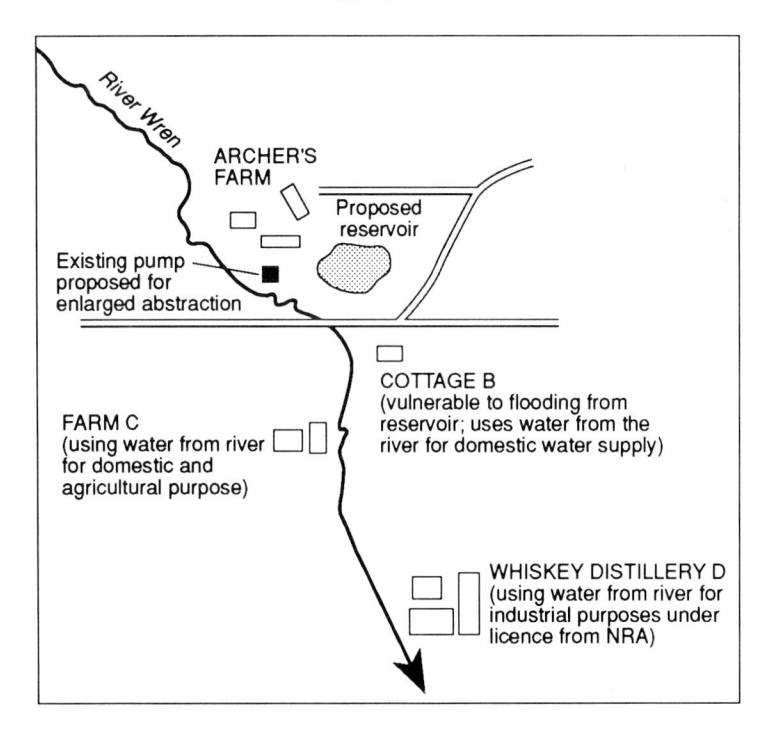

Figure 4.1a A sketch map of the Archers Farm area.

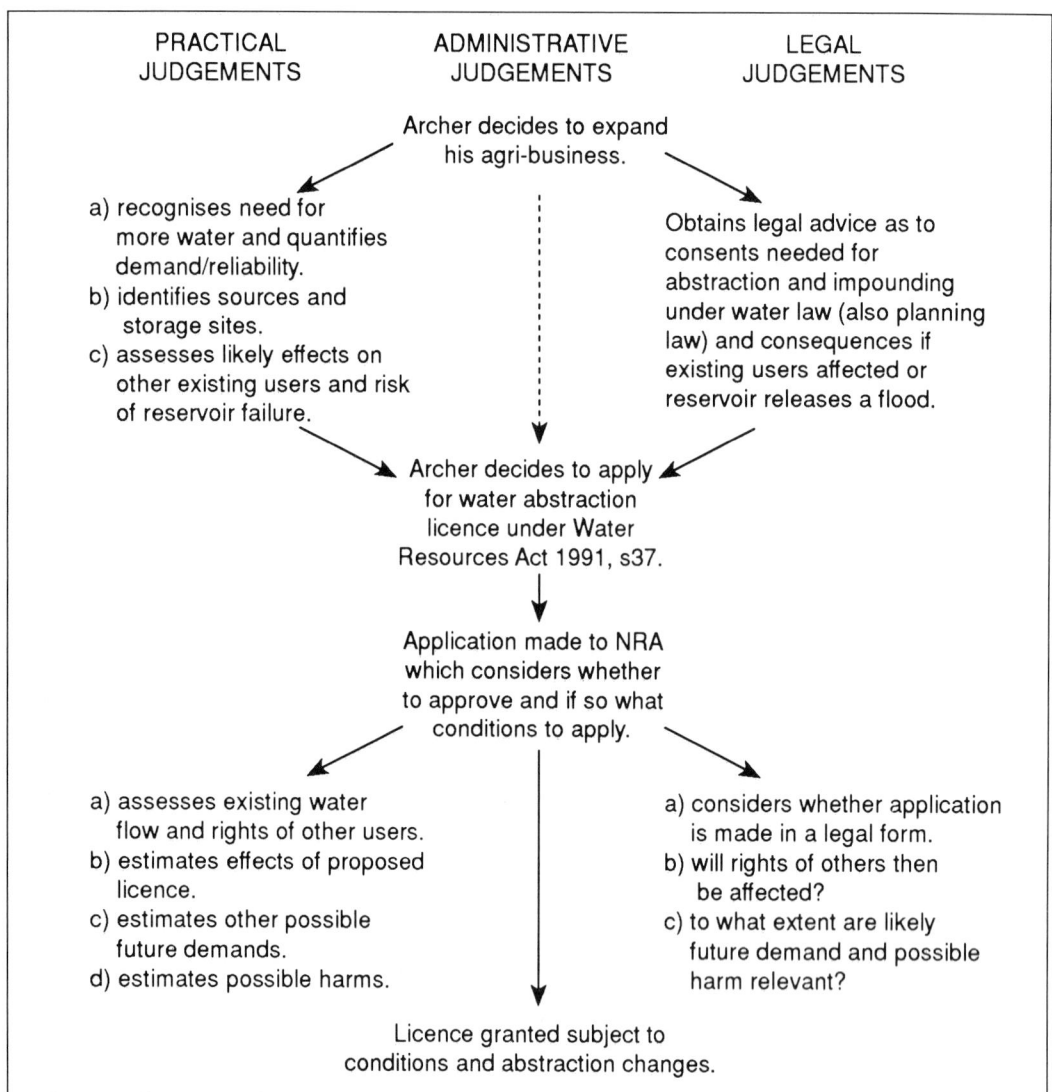

```
                PRACTICAL              ADMINISTRATIVE              LEGAL
               JUDGEMENTS                JUDGEMENTS             JUDGEMENTS

                              Archer decides to expand
                                 his agri-business.

    a) recognises need for                           Obtains legal advice as to
       more water and quantifies                     consents needed for
       demand/reliability.                           abstraction and impounding
    b) identifies sources and                        under water law (also planning
       storage sites.                                law) and consequences if
    c) assesses likely effects on                    existing users affected or
       other existing users and risk                 reservoir releases a flood.
       of reservoir failure.
                              Archer decides to apply
                              for water abstraction
                              licence under Water
                              Resources Act 1991, s37.

                              Application made to NRA
                              which considers whether
                              to approve and if so what
                              conditions to apply.

    a) assesses existing water                       a) considers whether application
       flow and rights of other users.                  is made in a legal form.
    b) estimates effects of proposed                  b) will rights of others then
       licence.                                         be affected?
    c) estimates other possible                       c) to what extent are likely
       future demands.                                  future demand and possible
    d) estimates possible harms.                        harm relevant?

                              Licence granted subject to
                              conditions and abstraction changes.
```

Figure 4.1b Practical, administrative and legal judgements involved in decision-making over the Archers Farm application

s 38 provides for the consideration of licensing applications. s 46 empowers the Secretary of State to stipulate the form of abstraction licences. In particular these are to make provisions as to the quantity of water which may be abstracted in a given period from a particular source of supply and how this is to be measured; also the methods to be used for abstraction and the place or places of abstraction.

s 38 requires the authority to take account of users with existing abstraction rights who may be affected by a licence, and not to derogate from their rights without their consent.

s 48 makes an abstraction licence a good defence against any legal action by an existing user whose rights are affected but s 60 gives such a person an action for breach of statutory duty against the National Rivers Authority instead.

processes like flooding, erosion and soil acidifification.

The law also provides the mechanism for allowing economic and other controls to operate. Thus licence fees, fines and taxation, or, more positively, grants and tax concessions, may make certain decisions more or less economically attractive. Such economic incentives depend upon legal rules for their implementation. A major economic principle which was embedded in the Treaties of the European Community by the Single European Act 1986 is the requirement that the polluter shall pay for dealing with the problems of pollution. This is being worked out in many different ways, but a clearly economic control which is in the process of being developed is a carbon tax, whereby activities generating carbon into the atmosphere may be taxed in accordance with the amount of carbon they produce. The level of tax and how it is applied to different producers will depend on economic and scientific analyses, but who has to pay and in exactly what circumstances will involve detailed legislation.

4.1.1 The scope and operation of law

Law is inseparable from any human activity, because it provides the rules by which each activity is carried out. In its most obvious form, law includes regulations made by the state to forbid certain actions, such as destroying or damaging natural resources. However the law of a state will also oversee voluntary arrangements. It will provide machinery for settling private disputes and may enforce private contracts, such as agreements between adjacent landowners to look after their land with special care. Such law is not normally dealt with by lawyers, but it is still law.

Thus, standards for controlling the deposit of waste or for allowing water abstraction are supervised by inspectors with scientific training and usually with no legal qualification. However, on the rare occasions when their interpretation of the standards is challenged in the courts the problem will clearly be seen as a legal one. Applying standards of this sort depends on scientific judgement. It will also often involve an administrative exercise of discretion, but both these aspects of any decision must be made within the terms laid down by the law. For example, if permission is sought to abstract and impound water from a river for a new factory or some other use, the law will lay down what permission is needed and the procedures for obtaining it (see box and Figures 4.1(a) and (b)). To some extent also the law will probably lay down what matters should be taken into account in deciding whether or not permission should be given and what weight should be attached to each factor.

4.2 Legal systems and legal sources

Law is very much a national phenomenon. This chapter, therefore, is mainly concerned with British, and, in particular, English law. However, as different legal systems have to contend with similar problems, the means by which English law is used to manage various forms of pollution give a fair indication of the methods applied in other countries. English law is closely related to other common law systems which historically evolved from it during the colonial period. Systems which split off at an early stage such as US law still have much in common. Those which have become independent more recently, notably those of Australia and New Zealand, remain more similar. Other states as diverse as India, Malaysia, South Africa and Canada have legal systems which partly derive from Britain but also, at least as much, from other sources, varying from local customary law to Islamic law or to the laws of other major European nations.

The legal systems of continental Europe can be traced back to the law of the Roman Empire but were recast in the form of codes by Napoleon and the other creators of the modern states of Europe which emerged in the eighteenth and nineteenth centuries. The codified systems of continental Europe have been incorporated by other major nations, notably Japan. During the second half of the twentieth century countries subject to communist doctrines developed distinctive regimes of socialist law. They largely eliminated the idea of private property in land, but in practice did not enable individuals to challenge the way state agencies managed the environment. The communist regimes of Eastern Europe relied upon written codes which were not essentially different in form from those of Western Europe. As the Eastern European states dismantle their previous regimes, their legal systems are likely to resemble increasingly those of Western Europe, whatever formal links may be established with the European Community.

4.2.1 EC environmental law

The European Community originated in the Treaty of Rome in 1957 between six European States. The United Kingdom joined in 1972. In 1992 the Community comprises 12 full members and the association has merged into a progressively closer union. In particular, the Single European Act Treaty between the Member States in 1986 has made the protection and the improvement of the environment a major new field for the development of an emerging pan-European legal system. The content of environmental law in Britain at the end of the twentieth century increasingly will be set at a European level, especially in Directives produced by the European Commission in Brussels. For this law to be enforced at the moment, it must normally be applied by the individual legislative processes of the Member States — in the United Kingdom, this means by Acts of Parliament, with detailed provisions left to delegated legislation and cases decided in the higher courts. However, as time goes on, European Community law is likely to be applied directly in Member States like the United Kingdom, by-passing national parliaments. Environmental law is a key area where the different legal systems of the United Kingdom and continental Europe will have to be harmonized by changes in legislation and by the work of the judges (see box).

The problems of pollution are worldwide, so that even a unified approach within the European Community will have limited value in, for example, reversing the 'greenhouse effect', halting the progressive poisoning of the oceans by toxic waste and oil or in removing the danger to whole hemispheres of radioactive contamination from nuclear power stations, demonstrated by Chernobyl. However, an integrated approach at a European level does offer a real prospect of stimulating the improvement and the integration of law by the individual industrialized Member States of the Community. At a continental level it offers a real prospect of effectively tackling certain forms of pollution, particularly to the atmosphere and river systems.

The commitment of the European Community to effective laws for the regulation of large-scale pollution may provide both a model and powerful political impetus for expanding the fragile framework of international environmental law. This already exists in the structures of the United Nations organizations and in a range of treaties relating to such matters as marine pollution and the protection of Antarctica. International law differs from the law of individual nation states in that it depends on the voluntary submission of governments. To a large extent international rules on such matters as pollution control only apply to states which have formally accepted treaties in which the rules were agreed. There is no permanent system of enforcement for such rules even against states which have agreed to observe them.

4.2.2 British environmental law

Before the accession of the United Kingdom to the European Communities, British law was independent and self-contained. Parliament was supreme and the primary source of

The outlawing of chlorofluorocarbons

A major global danger from pollution is the depletion of the ozone layer in the atmosphere which acts as a filter for harmful cancer-inducing sunlight. The ozone layer is affected by chemical reactions with chlorofluorocarbons, which are produced, notably, by aerosols and refrigerators. An EC Council Decision of 26 March 1980 spells out how a strategy for limiting these chemicals emerged. This Decision was clarified by a later decision in 1982 (82/795/EEC). At a world level, the Vienna Convention for the Protection of the Ozone Layer in 1985 and the Montreal Protocol on Substances which deplete the Ozone Layer in 1987 agreed the phasing out of a wider range of chemicals. In particular account was taken of the difficulty for Third world countries in extending effective control to refrigerators. These international agreements were dealt with in the EC by a further decision resricting import quotas for chlorofluorocarbons and by successive EEC Regulations.

The present Regulations (91/594/EEC) restrict import, production and consumption of products containing specified controlled substances. The Regulations include short articles prohibiting import and much much more detailed provisions to enable the phasing out of production over a period of years. The following extracts illustrate different elements of EC legislation; the explanation for such law, based on the historical background, specific restrictive rules and more rules creating more complex systems for control. Although Regulations are directly enforceable in Member States, it will still be necessary for individual governments to operate their own systems for imposing quotas on individual manufacturers and for monitoring and enforcing the controls.

Council Decision of 26 March 1980 concerning chlorofluorocarbons in the environment (80/372/EEC) OJ L90. p. 45

THE COUNCIL OF THE EUROPEAN COMMUNITIES
Having regard to the Treaty establishing the European Economic Community, and in particular Article 235 thereof . . .
Whereas, as stated in the resolution of the council of the European Communities and of the representatives of the Governments of the Member States, meeting within the Council, of 17 May 1977 on the continuation and implementation of a European Community policy and action programme on the environment, it is necessary to review continuously at Community level the impact of chemicals on the environment;
Whereas the Council resoultion of 30 May 1978 on fluorocarbons in the environment states that the problem of the effects of chlorofluorocarbons on the ozone layer and of ultraviolet radiation on health cannot be ignored;
Whereas the Member States, in accordance with the terms of the resolution of 30 May 1978, adopted a common position on 6 December 1978 concerning chlorofluorocarbons in the environment, to be put to the International Conference on chlorofluorocarbons held in Munich from 6 to 8 December 1978; whereas that conference adopted certain recommendations, in particular recommendation III . . .
HAS ADOPTED THIS DECISION

> Article 1
> 1. Member States shall take all appropriate measures to ensure that industry situated in their territories does not increase its chlorofluorocarbon production capacity F-11 (CC13F) and F-12 (CC2f2)
> 2. Member States shall take all appropriate measures to ensure that not later than 31 December 1981 industry situated in their territories achieves a reduction of at least 30% compared with 1976 levels in the use of those chloroflurocarbons in the filling of aerosol cans.
>
> COUNCIL REGULATION
> of 4 March 1991
> on Substances That Deplete the Ozone Layer (91/594/EEC)
> Article 5
> Importation of controlled substances from non-parties
> 1. The release into free circulation in the Community of chlorofluorocarbons or halons imported from non-parties shall be prohibited.
> 2. With effect from 1 January 1993, the release into free circulation in the Community of other fully halogenated chlorofluorocarbons, carbon tetrachloride or 1,1,1-trichloroethane imported from non-parties shall be prohibited.
> Article 9
> Exportation of controlled substances to non-parties
> With effect from 1 January 1993, the exportation from the community of virgin, recycled or used controlled substances to any non-party shall be prohibited.
> Article 10
> Control of production
> 1. Subject to the provisions of paragraphs 6 to 9, each producer shall ensure that:
> — the calculated level of its production of chlorofluorocarbons in the period 1 July 1991 to 31 December 1992 does not exceed the calculated level of its production in 1986. However, for those Member States whose calculated level of chlorofluorocarbons was less than 15,000 tonnes in 1986, the calculated level of their production of chlorofluorocarbons in the period 1 July 1991 to 31 December 1992 shall not exceeed 150% of the calculated level of their production in 1986.
> ... [A progressive reduction is provided for over a further 3 stages] ...
> — the calculated level of its production of chlorofluorocarbons in the period 1 January to 30 June 1997 does not exceed 7.5% the calculated level of its production in 1986.
> — there is no production of chlorofluorocarbons after 30 June 1997.
> [Similar phasing out is laid down for other chemicals; fully halogenated chlorofluorocarbons, halons, carbon tetrachloride and trichloroethane.]

law was an Act of Parliament. Generally, Acts of Parliament remain the principle source of law in the United Kingdom. Parliament legislates throughout the United Kingdom but Scotland retains its own separate courts, and the law produced by Parliament for England and Wales on the one hand and for Scotland on the other does vary somewhat. This is particularly true of land law and criminal law which form the background to modern environmental legislation.

In addition to Acts of Parliament, legislation includes delegated legislation. Typically, in the United Kingdom, this consists of Statutory Instruments produced by the Secretary of State for the Environment and other government ministers. For example, detailed rules as to minor changes of use which are permitted without planning permission are given in a Use Classes Order and a General Development Order. Detailed procedures for applying for planning permission and other forms of licence and for processing such applications are set out in Statutory Instruments. Another form of delegated legislation consists of local by-laws. These are made by bodies such as local authorities and may be used, for example, to set out detailed rules to protect sensitive or heavily used pieces of land. An example is by-laws prohibiting lighting of fires or damaging flora or fauna on open land which is used for public recreation.

A distinctive feature of the British legal system is that decisions of the higher courts contain binding authoritative interpretations of the law which must generally be followed on future occasions by all except another court further up the hierarchy. In England the most important courts are the House of Lords, the Court of Appeal, and Divisional Courts of the High Court. For Scotland the highest civil appeal courts are the House of Lords and the Inner House of the Court of Session, but final criminal appeals are heard in the Scottish High Court. There are now rules requiring disputed issues of European law to be referred by the national courts for a definitive ruling from the European Court in Luxembourg. In continental Europe, to reach consistent decisions, greater reliance was placed on academic analysis of the purpose and meaning of particular laws. The courts did not generally refer to previous cases. The European Court follows an essentially continental approach.

Common law, generally found in English-speaking countries, depends to a large extent upon the reported decisions of judges in past cases. Much English law concerned with balancing the competing interests of neighbouring landowners is still found in this form. This law was developed in the nineteenth century at a time when the emphasis was on expanding industrial production with the innovative techniques of the Industrial Revolution.

4.3 From the protection of private property to an integrated public environmental law

In certain areas, particularly the private law for protecting land, UK law has been left largely untouched by Parliament and is still based essentially on rules which must be gleaned from an accumulation of cases decided in the past by the judges. Traditionally, English land law was developed by centralized courts to avoid conflicts between individual landowners and between landowners and others whose actions might affect particular pieces of land, for example, by enabling claims to be made for rights of way. It is still a basic tenet of English law that rights in private property should be protected wherever possible, so that any new restriction on what a landowner can do with his land, or any provision allowing a public body to take it from him, must be very clearly spelled out by Parliament if it is to be upheld in the courts.

Private land law provides the background for land management and is usually judged on the one hand by its success in maximizing production, especially in agriculture, and on the other hand by the extent to which it enables the benefits of the land to be widely shared, particularly by ensuring security for people's homes and livelihoods. From early days this has meant providing a measure of protection for individual landowners from pollution and other harmful activities on neighbouring land. Such protection still constitutes a piecemeal, but nevertheless important, line of defence for the environment in the public interest.

4.3.1 The law of nuisance

The law of private nuisance was developed

Plate 4.1a Development in Exeter: (a) the luxury housing

R. v. Exeter City Council ex parte J.L. Thomas & Co Ltd. [1990] 1 All E.R. 413.

On 28 April 1989, Simon Brown J., a High Court Judge in the Queen's Bench Division, dismissed an application for Judicial review to overturn a grant of planning permission for residential development in Exeter. The case illustrates the relationship between public planning law and the private law of nuisance in England.

The planning consent was granted on 13 July 1988 to allow developers, called Tutorhome, to build 87 flats and maisonettes (Plate 4.1(a)) with a shop, on 7 hectares at the site of the old Texaco Depot in the most heavily industrialized part of Exeter. Although there was already some terrace housing nearby, the area was dominated by two firms. One was J.L. Thomas & Co Ltd which ran one of the only two animal waste processing sites in the South West, turning animal waste products collected from abattoirs and slaughter houses into tallow and animal feeds. The other firm was Blight & White Ltd., 'metal bashers', specializing in steel fabricating and structural engineering. These two firms challenged the approval of the housing scheme on the grounds that the local authority's motive in allowing housing was really to drive them out by making them vulnerable to nuisance actions by the new occupants (Plate 4.1(b)). The local authority had power to discontinue the applicants' businesses if it chose to but only on paying compensation. The local authority was using an improper method to avoid that expense.

Environmental law

Plate 4.1b Development in Exeter: (b) The industrial surroundings

The judge held that users like the applicants 'have no legitimate expectation that other conflicting uses will never be introduced into the vicinity; the planning permissions which they obtain and enjoy confer on them no right to commit nuisances against their neighbours either now or in the future . . . It is legitimate also for the planning authority to recognise and accept that planning permission for residential development in such an area is likely to give rise to conflicts with those existing users. It is also perfectly proper for the authority to harbour the wider aspiration that the existing users will relocate, whether influenced by carrot or stick (in the form of increased land values if they sell or nuisance actions if they stay)'.

by the courts, particularly during the nineteenth century and the first half of the twentieth century. It enables a landowner to obtain compensation, and often a court order, to stop unjustified interference with the enjoyment of his or her land. Where the land is physically harmed and it is clear who caused the harm, the action will normally succeed, as where plants or fish are killed by severe pollution or a building suffers obvious acid corrosion over a short period (e.g. *Manchester Corpn.* v. *Farnworth* [1930] A.C. 171; *Dulverton R.D.C.* v. *Tracy* (1921) 85 J.P. 217; *Halsey* v. *Esso Petroleum Co. Ltd.* [1961] 1 W.L.R. 683). A nuisance claim may also succeed if there is no physical injury but there is interference with the reasonable enjoyment of land, notably by loud noise or unpleasant smells. What is reasonable depends, in particular, on what is appropriate for the neighbourhood. Today, the character of the neighbourhood is itself determined by planning decisions made by local authorities and by central government in the wider public interest (see box).

The value of private nuisance actions is limited where existing polluting activities are well established and where pollution is generated from several sources, so that it is difficult to show which actually caused harm to any particular piece of land, especially where the harm occurs some distance from a source of pollution. Thus, factories which have used tall chimneys to disperse fumes have avoided nuisance actions by local residents, but may cause worse harm to trees hundreds of miles away, often in a different legal jurisdiction. To some extent these difficulties have been met by the concept of public nuisance.

Public nuisance consists of interference with the rights of the public, including unreasonably noisy or smelly activities which affect a significant number of people. It is a crime. From early days, criminal law was developed to protect private property rights in the interests of social order. Today, a blatant example of pollution damaging someone else's land could be charged as a criminal offence. For example, if a load of toxic waste were wilfully dumped in a wood and killed the trees or were dumped in a stream which fed a fish farm, killing trout in the fish farm, the perpetrator might well be liable to punishment for causing criminal damage. Generally, a criminal prosecution may be brought by any member of the public, although in practice it is usually initiated by the police or some other public body.

Criminal Law has been used from early days to control large-scale pollution by means of the crime of public nuisance where it was unreasonable to expect individual property owners to act. In practice, public control of pollution is generally exercised under Acts of Parliament which identify particular forms of abuse as statutory nuisances and make these specific criminal offences. In cases of statutory nuisance or common law public nuisance, the Attorney General or an elected local authority may step in to obtain an injunction, that is a court order telling the polluter to stop. Private individuals who have suffered special damage, for example, whose gardens have been blighted, may also rely on the law of public nuisance to obtain money damages from the courts.

4.3.2 Private land law as a framework for land management

Private land law makes another, and more positive, contribution to the protection of the environment, through the regimes which it has established over centuries for sharing control of land between numbers of people. Pollution often damages the land where it is produced. Landowners who have a long-term interest in the land are likely to use their property rights to prevent environmental harm and to ensure good management. Thus, parts of a large estate may be let to tenants or sold off. The original owner may include covenants in the lease or sale, which will preserve the land until he or she gets it back, or will protect other land retained nearby. For example, land may be let or sold for limited agricultural or residential use, and industrial activities on it may be prohibited. Sometimes such restrictions become standard practice and will be implied even if they are not spelled out between the parties. Thus the tenant normally is not allowed to 'waste' rented land. Waste could include dumping toxic waste or burning rubbish which emitted fumes and killed all the trees on the land (see box). Traditionally the law of waste struck a balance between short-term profit and long-term conservation of the land.

Effective protection of the environment by means of private land law depends on responsible people having legal interests in large areas of land. To some extent large holdings acquired by special bodies, such as conservation trusts, may be exempt from

> *Re Trevor-Bayte's Settlement* [1912] 2 ch. 339
>
> On 4 June 1912, two years before the First World War, Parker J., a High Court, Chancery Judge, decided a case which illustrates the legal framework for the sound management of a long-established agricultural estate. This case still forms part of English Land Law.
>
> In 1904 family trustees bought an estate of 391 acres in the parishes of Little Hampden and Ellesborough in Buckinghamshire. of which 115 acres was woodland. In 1910 the trustees felled a large number of beech trees for sale. They did this as part of a sound management programme, including replanting and repairing fences. The trust had been made in 1902 as a marriage settlement. It proved life interests for the husband and wife but provided that the property should eventually pass to their children. At the time of the case there was one little girl. The question in the case was whether the proceeds of sale from the timber could be used as income by Mrs Maud Trevor-Bayte or had to be retained as capital for the future.
>
> The judge held that the proceeds were income: 'It is the normal course in the neighbourhood to cut [timber] periodically, and from time to time cut out the larger trees in order to leave room for the growth of the smaller ones, so that in a well-managed beech wood there appear to be trees in three stages of growth, the old ones, the middle aged ones and the young ones. It is in the usual course of management to cut the older trees . . . In order to ascertain what is fairly profit in the circumstances . . . we must take the actual proceeds of sale, and after paying all costs of replanting, repairing and fencing we arrive at what can fairly be called profits of the estate.'
>
> The decision therefore reinforced sound practice of those managing the estate. The person in possession of the land was given the incentive of cutting the trees at the right time because she obtained the profits but had she had to replant and carry out other necessary maintenance as well she could have been prevented from cutting trees prematurely. Legally, this would have constituted 'waste', which a person with only a limited interest in the land is not allowed to make.

such legislation and may effectively continue to use traditional tools of private land law in the long-term interests of the land.

Even when land is managed with an eye to its long-term good, whether by private landowners or by public bodies, there may be problems. Profitable and productive agricultural practice, or even certain conservation policies, may actually compound other forms of environmental harm. A striking example is the conflict caused by the long-approved agricultural practice of adding nitrates to soil planted with arable crops, where the nitrates leach into water systems. Similarly, attempts to diversify the use of agricultural land by tree planting, especially of conifers, may lead to increased acidification of ground water. Again, the cleansing of fields by stubble burning after harvest contributes large quantities of carbon to the problems of atmospheric pollution.

Private nuisance and the whole framework of private land law developed an elaborate system for balancing the competing uses of land. The patchwork of public common law and statutory nuisance curtailed many of the most obvious abuses. Indeed private land law remains the key so far as positive upgrading of the environment is concerned. There, private agreements between landowners and public bodies, supported by incentives to landowners in the form of grants and tax

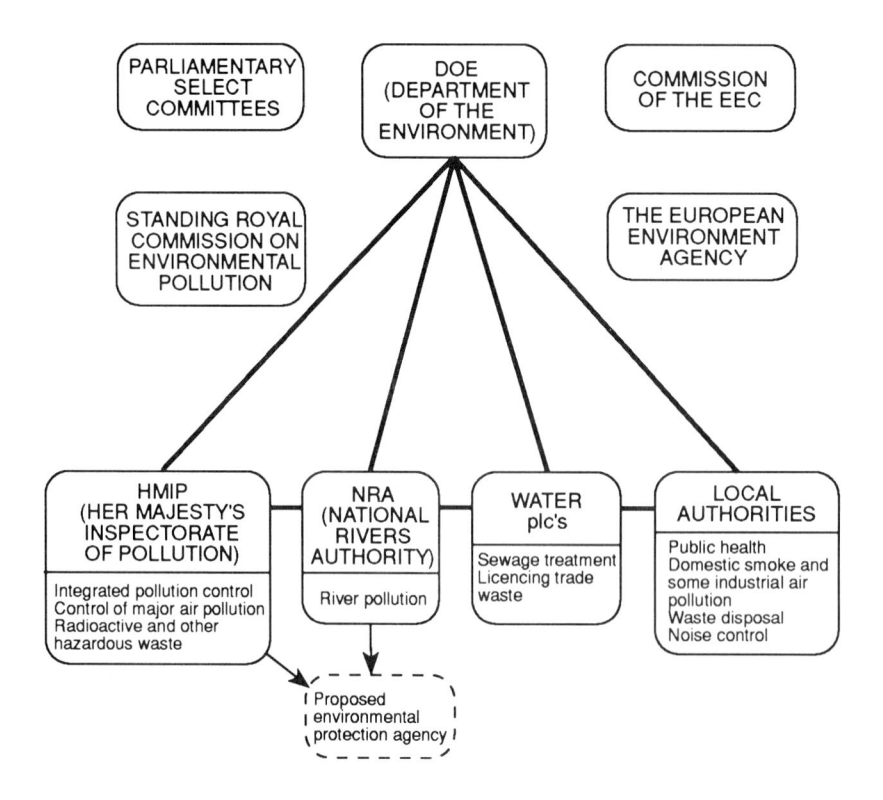

Figure 4.2 The administration of UK environmental statute law

concessions, are invaluable tools. However the second half of the twentieth century has seen rapid increases in polluting activities both old and new, the production of dramatically more dangerous substances and a new awareness on the part of experts, passed on to the public, of the complex effects of pollution on the biosphere, with intractible and sometimes seemingly irreversible effects. The demands of larger and, in much of the world, universally wealthier populations, have imposed huge demands on limited natural resources.

4.3.2 Centralized legislation

New forms of legal control have been sought to provide a comprehensive and integrated framework for managing the environment. In Britain, at least four major influences have accelerated this process. First, the modern system of town and country planning, originally laid down in the Town and Country Planning Act 1947 and now contained in the Town and Country Planning Act 1990, has enabled harmful new uses of land to be prevented altogether or restricted to sites where they will cause least damage.

Second, responsibility for monitoring and regulating pollution has become more centralized. A special inspectorate was set up under the Alkali Works Regulation Act 1906 to monitor and control the most serious forms of industrial pollution of the atmosphere. Separate inspectorates were later developed for radioactive and other hazardous forms of waste. These were integrated in 1987 to form Her Majesty's Inspectorate of Pollution (HMIP). Under the Environmental Protection Act 1990, this inspectorate was given new responsibilities for regulating major industrial polluters under a new system of integrated pollution control. Thus the inspectorate may be able to require the least harmful means of disposing waste in all forms from such sources.

The resources available for controlling pollution in the United Kingdom are still inadequate, and responsibility for administering the system remains awkwardly split between local and central government. In particular, local authorities are responsible for dealing with lesser forms of atmospheric pollution including domestic chimneys and noise. However, further rationalization is promised. In 1989, the British government privatized water resources by selling off the assets of state water authorities to new private companies. However, control of water pollution and water abstraction was retained by the new National Rivers Authority. The regulation of pollution by the National Rivers Authority and by HMIP are now to be combined (Figure 4.2).

Two other influences on the evolution of an integrated system of environmental law have already been mentioned. The emergence of the European Community is now changing the structure of UK law in fundamental ways. Especially with regard to the environment, the EC has imposed new standards on Member States and is forcing them to re-appraise their individual laws on pollution control so as to fill in gaps and harmonize their systems with those of their neighbours. The Council of the Community has passed a Regulation authorizing the setting up of its own European Environmental Agency and the European Environment Information and Observation Network (1210/90). There is delay at Community level in reaching agreement on the location for these institutions. However, there are now new expectations by the public with regard to the environment which provide a stimulus for legal reform and allocation of public funds to make controls work. The public now demands more involvement in decision-making, further research and increased openness of information.

4.4 Law as an instrument for limiting environmental harm

The nature of law itself affects the form that control takes. Good law is clear and simple. Where this is not possible it may be better not to use the law at all. For example, a detailed code might be worked out as to the best means of conserving energy. Optimum times could be set for using electricity, with rules requiring lights and appliances to be switched off when not in use, and only to be used at certain times. Regulations could be laid down for the insulation of buildings. Exceptions could be included so that energy saving would not prevent other aims, for example, if it interfered with safety or security of property.

Realistically, however, the more detailed the rules became, the less satisfactory they would be. It would be harder to ensure that people knew what they were and those who did know would find it more difficult to keep them. If the rules were enforced effectively, the cost would probably outweigh the savings. The law would be resented and people would become less inclined to obey it.

In practice, where detailed regulations are used to protect the environment, they tend to be imposed upon business organizations or professionals carrying out work. The private citizen is most likely to be covered in circumstances where a professional is likely to be called in. For example, building regulations could require all new buildings or extensions to include substantial double glazing and ceiling insulation because such work tends to be done by builders who realistically may be expected to keep up with new rules. Professionals are likely to keep in touch with the regulatory authorities so as to protect their businesses. By contrast, 'do-it-yourself' homeowners are likely to ignore the regulations. On the other hand, they are more likely to comply if public grants are available for energy-saving works and if economic disincentives are used, ensuring that fuel is expensive and so will not be wasted lightly. Rules making it an offence to leave lights and appliances on when not needed would be impractical. However, economic sanctions in the form of high costs could encourage many individuals to regulate themselves, especially if energy-saving equipment, such as devices for automatically turning lights off, were made correspondingly cheap.

4.4.1 Criminal and civil law

Law is probably most commonly thought of as a directly restrictive tool, forbidding or restraining damaging behaviour. This may be because of the popular interest in crime which is fed by the media. Criminal law is concerned with imposing sanctions to punish people who infringe rules laid down to protect the interests of the general public. By contrast, civil law is used to resolve disputes between individuals or between individuals and the state and generally to provide consistent principles for regulating people's affairs. A high proportion of court work is concerned with crime and, in industrially developed countries especially, there is an increasing readiness for activity which harms the environment to be treated as criminal.

However, civil law may also operate as a tool restricting harm. Thus a civil court may order someone to stop a civil wrong which amounts to a nuisance, such as pollution in the form of dust, smoke or fumes from a quarry, factory or waste disposal plant. In common law countries such an order will take the form of an injunction. In Scotland it is called an interdict. If the court order is disobeyed, the matter may be treated very much as if it were a crime and the offender may, as a last resort, be committed to prison for contempt of court.

Restrictive legal rules may be particularly effective when they clearly forbid actions which are serious and readily detectable, for example, a prohibition on depositing noxious substances in natural streams of water. Such a rule is contained in Section 85 of the Water Resources Act 1991 (applies to England and Wales). Punishment can amount to a £20,000 fine and three months' imprisonment from a magistrates' court or an unlimited fine and two years' imprisonment from the Crown Court. By contrast, a rule requiring a landowner positively to keep his grounds 'in good order' would give rise to more problems. It would be difficult to establish what was meant by 'good order' unless detailed maintenance requirements were spelled out.

Public law does sometimes impose obligations to keep property in good order. Thus, a building may be listed as of special architectural or historic interest, under the English Planning (Listed Buidings and Conservation Areas) Act 1991. If the owner allows a listed building to deteriorate, he or she may be ordered to restore it to the state it was in at the time of listing. An owner who fails to comply cannot be punished in a criminal court, but may have the property, in effect, confiscated by the local authority. Compensation is payable but this may be small, especially where the owner intentionally left the building derelict. The owner of a listed building may be obliged to restore it to the state it was in at the time when it was listed even though listing occurred many years before when the building belonged to someone else. In the case of *Robbins* v. *Secretary of State for the Environment* [1989] 1 W.L.R. 201, this meant that the owner of an ancient windmill which had been converted into a dwelling house could be ordered to restore redundant machinery which had become derelict over a period of 30 years when one owner failed to carry out the repairs the building could be compulsorily purchased.

Although general rules for positive maintenance or improvement of the environment are unusual, where a landowner openly accepts such an obligation, for example through taking a grant to plant new woodlands or through buying land subject to a covenant that it will be regularly weeded and tended, the courts will enforce the agreement. Sometimes they will also do so against a person who later acquires the relevant land.

Where law is used as a restrictive tool, it may be applied at various stages in a process and it may be far more effective at certain points than at others. A major difficulty in integrating control is that of finding the most effective levels for regulation so as not to transfer the problem elsewhere. Thus, a law which required petroleum waste in ships to be burnt would reduce marine pollution but could significantly increase atmospheric pollution, unless very expensive and energy-absorbing equipment were required for the burning process. The legal machinery requiring oil slicks to be cleared up after they have occurred will be very different from that aimed at preventing them in the first place. Generally there needs to be appropriate legal machinery to cover both.

A classic analysis of the levels at which controls may be applied and the standards which may be used has been provided by Ogus and Richardson (1979) in research into the legal regime for handling river pollution (see also Chapter 7). They contrast the use of receptor standards, which concentrate on preventing unacceptable levels of harm in particular places, with ambient standards which set overall acceptable levels of contamination in a local or wider environment, rather than concentrating on harm. Standards of both sorts may be used as a basis for restrictions on the actual sources of pollution.

4.4.2 Private proceedings

Traditionally, private law has tended to react to damaging situations by ordering those who have caused harm to pay money damages to those whom they have injured. This has involved setting what amount to very flexible receptor standards. The risk of having to pay damages has then served to deter anti-social activity, particularly because civil damages are generally awarded on objective criteria, and sometimes on the basis of a strict duty to take care. Thus, if a person burns rubber waste on land so that it affects the milk yield of a neighbour's cattle, the person causing the harm will probably have to pay compensation as if a reasonable person in the same position would have anticipated the harm. The polluter will have to pay even if he or she personally did not anticipate harm, perhaps because of an honest belief that appropriate precautions had been taken to avoid any risk.

Strict liability is associated with the leading Victorian case of *Rylands* v. *Fletcher*, which reached the House of Lords in 1868 (L.R. 3 H.L. 330). There, the owner of a reservoir was held to be strictly liable for damage to mineworkings into which water from the reservoir escaped. For this rule to apply, the landowner must bring on to his or her land something which was inherently liable to cause damage if it did escape, despite all the precautions taken to keep it safe. Also, the material in question must have involved an 'unnatural' use of the land. For example, storing drums of toxic waste, which would obviously cause harm if it were to escape, could be regarded as an unnatural use. In practice, however, most industrial activities are now regarded as normal. On that basis they seem to be treated by British courts as natural uses of land so that strict civil liability very rarely provides a means of clearing environmental harm. Moreover, liability under the rule in *Rylands* v. *Fletcher* is strict and not absolute. There is a defence if the pollution or other harm was caused by an unforeseeable natural disaster like an earthquake, or in the more common situation where the escape results from the activities of an unforeseeable trespasser (*Box* v. *Jubb* (1879) 4 Ex Div. 76).

Private law actions and prosecutions, after the event, for causing pollution can help to ensure that harm, once caused, is cleared up. The risk of such proceedings can encourage good practices which avoid pollution in the first place. At least once harm has actually occurred, or when a situation is clearly going to cause harm unless those responsible take steps to deal with it, the courts can grant injunctions ordering the nuisance to be abated or disposed of. However, these methods cannot be as effective in preventing pollution as controls applied at an earlier stage. This is particularly so with the law of nuisance which is not intended to prevent all potential harm so much as to strike a balance between unacceptable levels of harm and the freedom of landowners to carry on potentially polluting activities.

4.4.3 Enforcement: generally

As Ogus and Richardson (1979) have shown, controls may be applied early on in the polluting process by specification standards to regulate materials and plant which could cause or prevent pollution. Alternatively, emission standards may be used to regulate the amount of pollution which is tolerated at source and before it has been dispersed and caused harm. There, the polluter may be left to choose the means of complying with the standard. Generally, under the UK system, standards have traditionally been applied within a discretionary licensing framework. Thus, abstraction of water or disposal of unwanted material from industrial plant into water courses or elsewhere is generally forbidden except under licence. However, licences may be granted on different terms in different places. For example, in heavily polluted rivers, existing plant was long left to operate freely. New sources of pollution on uncontaminated rivers could be restricted more firmly.

One of the consequences of EC involvement in environmental regulation is that more rigorous and universal standards are being laid down to prevent pollution. The European Commission is increasingly setting ambient standards for particular media, for example, pollution levels for water at bathing beaches (76/160/EEC). The United Kingdom has been criticized for failing to impose these satisfactorily in national legislation. Perhaps surprisingly, part of the reason for this may be a greater respect in the United Kingdom than on the continent for the enforcement of law. Traditionally there is a reluctance in the United Kingdom to make regulations forbidding any activity unless the regulations can be effectively enforced. In Europe, law seems to be used more often in a symbolic manner which does not necessarily correspond to reality. Thus, Italy may comply more enthusiastically than the United Kingdom in translating UK directives against pollution from industrial sources into national legislation. However, the Italian laws may be less rigorously and effectively enforced than the rules which have been incorporated in UK law. Where a Member State has actually implemented an EC Directive it is difficult for another state or the EC Commission, let alone a private individual, to challenge that country if it fails to enforce the law, even if it chooses, as a matter of discretion, virtually never to do so.

4.5 Practical aspects of enforcement

Where the law is used as a restrictive tool to control any activity, the relevant substantive rules form part of the general legal system. Their effectiveness will depend on how widely they are observed. The mere fact that the law declares certain aims will encourage many people to observe those aims. Thus the Single European Act Treaty of the EC in 1986 inserted into the Treaty of Rome a new Article 130R, which declares that 'Action by the Community shall have the following objectives: — to preserve, protect and improve the quality of the environment; — to contribute towards protecting human health; to ensure a prudent and rational utilization of natural resources'. This forms part of a new climate which affects other parts of the world. Thus Japan has an international reputation for industrial efficiency, but its environmental credentials have been attacked, for example over its opposition to a ban on destroying the dwindling stocks of whales in international waters. Nevertheless, one of the initial matters of concern expressed by Japanese companies opening up new works in the North East of England has been the likely consequences on the environment, particularly in the light of EC law.

UK legislation also now recognizes the importance of the environment by general policy declarations. Thus, in the Water

Plate 4.2 Mallam Water, Glos., the scene for *Kennaway* v *Thompson* (1981) I Q.B.88

Resources Act 1991, relevant ministers and water authorities are to exercise their powers so 'as to further the conservation and enhancement of natural beauty and the conservation of flora, fauna and geological or physiographical features of special interest'. The Agriculture Act 1986 and various other major statutes concerned with aspects of the environment impose similar general duties. However, general policy statements and even very specific rules forbidding pollution or other environmentally harmful actions such as the destruction of rare species will only be truly effective if the law provides adequate machinery for enforcement.

4.5.1 Remedies and sanctions

Both civil and criminal law have their respective advantages and disadvantages as tools for restricting harm. Major aspects of their comparative effectiveness are the procedural arrangements for obtaining the various remedies they offer and the nature of those remedies. Civil courts are likely to be better equipped than criminal courts to control particular activities by means of specific orders like injunctions. These may be mandatory, as where a person is ordered to restore a building which he or she has damaged, or they may prohibit or restrict future harm.

An injunction may impose very specific restrictions. In the English case of *Kennaway* v. *Thompson* [1981] 1 Q.B. 88, a judge made an injunction limiting speed boat racing on a lake in Gloucestershire so as to protect the amenity of a house on the lake (Plate 4.2). The ban specified in great detail that only one five-day international event and two two-day national events were to be allowed each year, each separated by at least four weeks. Otherwise, club races were allowed on ten days, spread over seven weekends, with further specified gaps between different sorts of event.

The long stop for both systems is imprisonment, either as a punishment in its own right or, for contempt, where a person refuses to obey a civil court. Fines may be a more effective deterrent, especially if they are really substantial. They are the only effective criminal sanction against corporations. However, civil damages may be heavier than a fine and may provide funds directly to remedy the harm which has been caused.

In practice, the administrative systems for regulating the environment provide the most detailed and effective machinery for control, and these will be backed up by criminal prosecutions. However, prosecutions will generally not be brought if a problem can be resolved by administrative means. Thus, British planning law requires those who put up a new building or open a new gravel pit or carry out other forms of development to obtain planning permission first. Anyone who fails to do so may be served with an enforcement notice requiring the land to be returned to its former state. Failure to do this is punishable as a criminal offence (Town and Country Planning Act 1990, Part VII, as amended by Planning and Compensation Act 1991). In practice, however, the situation will frequently be regularized by the offender obtaining planning permission retrospectively. Any planning consent will be subject to conditions which may be very detailed and may involve considerably more control than under a court order. Conditions may include some long-term supervision. A particularly detailed example in English planning law is that provided for mineral abstraction consents, so as to ensure restoration of the land when excavation has finished (Town and Country Planning Act 1990, especially s. 72 and Schedules 5 and 9). The courts will not normally be involved in devising such detailed conditions.

4.5.2 Evidence and standards of responsibility

Once a case reaches court, the rules of evidence become crucial. In civil cases, liability generally depends on flexible standards, such as whether a person alleged to have caused harm has acted 'reasonably'. However, the courts only require the person making an allegation to prove it 'on the balance of probabilities', that is to prove that it is more likely true than not. By contrast, in criminal cases, allegations must be proved to a higher standard such that the court is 'sure' that each element is true. On the other hand, in environmental prosecutions, unlike most criminal cases, there is frequently strict liability and the prosecution does not have to prove that an accused person knew that he or she was doing something harmful.

Thus a firm charged with causing trade effluent to enter a river may be strictly liable if the effluent escapes unexpectedly and despite regular careful inspections designed to prevent such accidents. In the English House of Lords case of *Alphacell Ltd.* v. *Woodward* [1972] A.C. 824, owners of a paper factory were treated as guilty of polluting the river Irwell when material from their processes overflowed into the river after a filter had unexpectedly become blocked by vegetation, despite the fact that a rigorous inspection system had been observed.

Although strict liability may be unfair in some cases, it makes conviction much more easy and causes industrial operators to take greater care to avoid harm in the first place. For a clearly unblameworthy offence a court may impose a nominal sentence. However, the strictness of an offence may also be offset by special forms of defence. Under British environmental legislation, it was commonly open to a person charged with a pollution offence to escape liability by showing that although an emission had occurred, this was despite the use of 'the best practicable means' of avoiding it. For example it is an offence under section 1 of the Clean Air Act 1968 for industrial or trade premises to emit dark smoke. However, there is a defence if it is proved 'that the contravention complained of was inadvertent and that all practicable steps have been taken to prevent or minimise the emission of dark smoke'.

The Environmental Protection Act 1990 has introduced the similar concept of 'best available techniques not entailing excessive cost' (BATNEEC). Thus the Secretary of State for the Environment may prescribe industrial processes which are likely to cause major pollution problems. These may only be carried out on premises which have been authorized for the purpose. It is a condition of any authorization that in carrying on the process 'the person carrying it on must use the best available techniques not entailing excessive cost — (a) for preventing the releases of substances prescribed for any environmental medium into that medium or, where that is not practicable by such means, for reducing the release of such substances to a minimum and for rendering harmless any such substances which are so released; and (b) for rendering harmless any other substances which might cause harm if released into any environmental medium'.

4.6 Law as an instrument for promoting environmental improvement

Pressure on land from increased population, and new uses which can rapidly destroy the land itself, have produced public law regimes for regulating land use. Such regimes may impose new levels of regulation on private landowners, for example, only allowing them to excavate minerals or build on their land or abstract water if they obtain licences from a public authority. Generally public law may provide the planning framework for directing new development to places where it will be least environmentally harmful. Development plans such as those under British town and country planning law may indicate areas where development of particular types will normally be granted. This encourages developers to acquire land there and produce schemes for carrying out development. More drastically, public law may provide public bodies with the power to take positive initiative by acquiring land, if need be by compulsory purchase, for purposes of conservation or so as to direct the form of new development.

Public law regimes may be difficult to relate to the older law, especially as the actual shape of both private and public law is affected by changing economic, social and political factors. Thus, in most parts of the world, exploding populations demand a share in the land which threatens traditional patterns of control by a small number of landowners. However, greater sharing of land to satisfy political demands may make proper management more difficult. For example, there is constant pressure in England for the public to enjoy rights of access to open countryside, but landowners and farmers have resisted this pressure on the grounds that unrestricted access will damage the quality of the countryside. Even open land has to be managed, grass cut and hedges laid. Unrestricted public access could interfere with management or at least deter landowners from bothering.

Particularly in towns, policies designed to increase the availability of housing can reduce the effectiveness of land law in maintaining the environment. Such policies may result in dividing land up by building houses on it and selling off the houses with individual gardens. Apart from the immediate transformation of the landscape, there will no longer be a single landowner who is able to manage the area as a whole and maintain its amenity.

Private law contributes to the protection of the environment by imposing general restrictive standards of responsibilty so as to protect individuals and their property from harm, particularly through the law of nuisance. However, its positive value, in encouraging active conservation and more varied and sensitive forms of development, is being increasingly recognized. In particular, it allows for the creation, under contract, of far-reaching management agreements.

Significantly, public law may be framed so as to encourage positive agreements rather

> *Sweet* v. *Secretary of State for the Environment and the Nature Conservancy Council* (1989) 1 J. Environmental Law 245
>
> In 1985, The Nature Conservancy Council notified Mr Sweet, under section 28 of the Wildlife and Countryside Act 1981, that three fields called Westhay Moor, which he owned on the Somerset Levels, were of Special Scientific Interest. The notice forbade him carrying out any operations specified in a schedule to the notice, without 4 months notice to the Council. The Secretary of State for the Environment supplemented this with an order under section 29 of the Act on the grounds that the Moor was of national importance. Mr Sweet accepted that part of the land was of national importance but he challenged the order in an application to the High Court because it covered additional land which he said did not deserve protection and because the operations forbidden were too wide as they included normal agricultural activities such as cultivation, grazing and burning.
>
> Schiemann J. rejected Mr Sweet's application. He ruled that the site could properly be treated as a single environment which must be protected as a whole. In effect the area with special features could be protected by including a buffer zone which formed part of the same field system. Also the Act could properly be used to restrict ordinary agricultural practice which would not normally need planning consent or any other permission from a public body. Although the effect of the case was to hold up Mr Sweet from draining his land and turning it over to agricultural production he would have been free to do so after giving notice to the Nature Conservancy Council. The purpose of the delay was to enable the Council to reach an agreement with the farmer under which he would manage the land so as to preserve its flora or if necessary to buy the land so that they could make alternative arrangements to manage it.

than simply to prevent enviromnmental harm. An example is the system of Sites of Special Scientific Interest (SSSIs) under the English Wildlife and Countryside Act 1981. Such sites may be designated by the Secretary of State for the Environment. The notice or order will specify operations which are to be restricted on a site so as to protect rare species or the site itself. The scheme provides that where an owner has been notified that land is to be treated as an SSSI he or she must normally give several months' warning before carrying out any of the operations specified in the notice. If the site is of major importance the land may be compulsorily purchased. Normally, however, the aim is to persuade the landowner to enter into a management agreement to look after the land so as to protect it.

In English law, private covenants attached to land have been effective for protecting open country from being built on and for safeguarding the character of an area in other ways, for example, by preventing changes to unsuitable new uses. Such agreements may be made so that they remain attached to pieces of land even after they have changed hands. Generally, however, positive covenants requiring future owners actively to maintain land have not been enforceable against later owners. Economic circumstances may change and such obligations could become oppressive. This objection is not stong where a positive covenant is made with a public body which may be required to act in the public interest and not to treat those bound by covenants oppressively. New powers for landowners to enter into such positive covenants enforceable by local planning authorities may now be made under

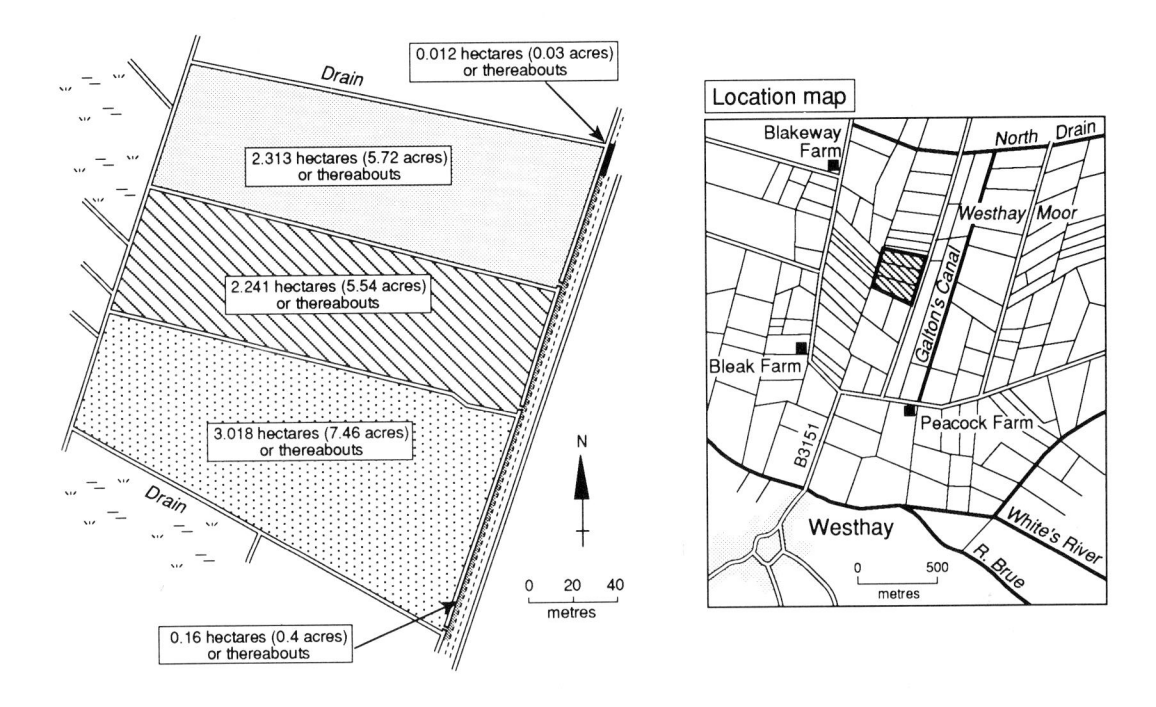

Figure 4.3 The definition of Sites of Special Scientific Interest. The Westhay Moor wetland meadows case carried the notion of damage far beyond the land itself because of the impacts of drainage.

section 12 of the Planning and Compensation Act 1991, providing for the future maintenance of their land by their successor.

4.7 The future of environmental law: public rights and public involvement

In industrially developed countries there is a desperate race against time to introduce integrated national systems of restrictive environmental law with effective enforcement provisions so as to control the overuse of precious natural resources and the degradation caused by pollution. To deal with the immense global threats to the environment, individual states are increasingly improving their own systems of law so as to comply with internationally agreed requirements. Particularly with regard to the oceans, the polar regions and the stratosphere, effective protection may only be obtained by a new and truly international system of law.

At a national level, the breakdown of socialist models of society in Eastern Europe and elsewhere is a reminder of the difficulty of achieving positive improvement of the environment by state direction. However, there seem to be more hopeful prospects for integrating private initiative into a public regulatory framework of environmental law. Public-spirited developers may develop new forms of tenancy or co-operative management regimes. Private associations with environmental concerns may increasingly acquire large tracts of land so as to protect them and make them available for recreational or educational purposes. The English National Trust and English Heritage hold many sites of historic and landscape importance and similar trusts exist in an increasing number of countries. National Trusts of Scotland, Australia and Canada closely follow the English model.

In many countries, important tracts of natural landscape containing rare species are owned by the state as national parks. There are national parks in England and Wales but these remain divided between many private landowners and are protected mainly by especially rigorous planning controls. However, private bodies are taking the initiative to acquire major estates so as to preserve them in the long-term public interest. A notable example has been the acquisition by the Royal Society for the Protection of Birds (RSPB) of a key estate in Scotland's Cairngorms. Bodies such as the National Trust, English Heritage and the RSPB have large memberships of ordinary people who are thus able to influence both the associations themselves and (indirectly) the government.

In England, private individuals or associations to which they belong may influence the effective use of public law by bringing private prosecutions. Such prosecutions by the RSPB of illegal collectors of rare birds' eggs have resulted in heavy fines which may hopefully deter others. The power of private prosecution is not generally available in countries outside the United Kingdom, but most have developed systems of administrative law whereby the ordinary citizen may challenge governmental actions or failure to act, including failure properly to implement criminal law.

A coherent system of administrative law is now being created in the United Kingdom, partly through increasing reference of environmental test cases to the courts. Thus a local authority or a public body such as an electricity company may be challenged if it seeks to carry out development without fully complying with the requirements included in its statutory powers. Increased involvement by individuals and by pressure groups in environmental law is likely to stimulate its development in all the areas discussed in this chapter.

For individual involvement in environmental law to work effectively, more and better information is essential. Such information is increasingly provided by public registers, for example, recording the activities of industries which are likely to cause pollution. An aspect of increasing involvement by the EC in environmental matters has been a Directive by the Council to provide Freedom of Access to Information on the Environment, especially pollution and other threats (90/313 EEC).

Another significant contribution by the EC has been a Directive requiring Environmental Assessment for major new developments which are likely to have significant effects on the environment (85/337). Such assessments were already a requirement in the United States and have frequently been produced on a voluntary basis by responsible developers in many countries.

References

A clear, up-to-date and comprehensive introduction on environmental law is provided by Simon Ball and Stuart Bell (1991), *Environmental Law*, Blackstone, London.

Other introductory works from the perspective of English law, on the general legal background and specifically on environmental regulation are: J.D.C. Harte, (1985), *Landscape Land Use and the Law*, E. and F.N. Spon; David Hughes (1986), *Environmental Law*, Butterworth, London; and Patrick McAuslan (1975), *Land, Law and Planning*, Weidenfeld and Nicolson, London.

Major general reference works containing annotated copies of relevant UK and EC legislation are: Sweet and Maxwell's *Encyclopedia of Environmental Health Law and Practice* (gen. ed. Charles Cross); and Butterworth's *Encyclopedia of Pollution Control* (ed. J.F. Garner).

Journals dealing with environmental law include: *The ENDS Report*, monthly, Environmental Data Services; *Environmental Policy and Practice*, quarterly, Charles Knight; *Journal of Environmental Law*, twice yearly, Oxford University Press; *Journal of Planning and Environment Law*, monthly, Sweet and Maxwell; *Land Management and Environmental Law Report*, alternate months, Wiley.

Valuable texts on particular aspects of environmental law include: John Bates, *Water and Drainage Law*, loose leaf for updating, Sweet and Maxwell, London; Nigel Haigh, *EEC Environmental Policy and Britain* (2nd rev. edn, 1989), Longman, London; G. Richardson, A.I. Ogus and P. Burrows (1982), *Policing Pollution: A Study of Regulation and Enforcement*, Clarendon Press, Oxford; David Vaughan (ed.) (1991), *Current E.C. Developments: E.C. Environment and Planning Law*, Butterworth.

Major articles include: Churchill (1991), 'International Environmental Law and the UK', in Churchill, R., Gibson, J. and Warren, L.M., *Law Policy and the Environment, Journal of Law and Society*; A.I. Ogus and G. Richardson (1979), 'Regulatory approach to environment control', *Urban Law and Policy*, 2.

5

Environmental economics: resources and commerce

Malcolm Newson

Other chapters in this book are written by specialists, as befits the scientific, technical and legal aspects of environmental management. The author's credentials to contribute this chapter are that he is a geographer, a consumer and an environmentalist. To have had an economist contribute would have run the risk of producing a treatise on how conventional economic theory can be adapted or interpreted to include environmental goods even though most of these are not traded and therefore not priced. However, they are valued and in the assignment of value the knowledge-base of an environmental scientist (who is also a geographer, consumer and environmentalist) is by no means irrelevant. Arguably, value is a topic for politics, not economics; the scientist's contribution to value is in the form of assessments of vulnerability, damage, its reversibility and the duration and spread of both desirable and undesirable outcomes of transactions and developments involving the environment. Such is the scope of this chapter but inevitably we must begin with a brief guide to classical economics; economists must forgive the author for inadequacies here, for which he takes full blame despite their published guidance (e.g. Cottrell, 1978; Norton, 1984; Rees, 1985; and Pearce, Markandya and Barbier, 1989 — of whom the latter had the necessary humility to call their guidance a blueprint!).

5.1 Economics: principles

Economics is a science; some of its exponents draw strong analogies with physics (Mirowski, 1989). The origin of such epistemological similarities can be gathered from one of the basic aims of economics — the allocation of scarce resources to achieve the greatest welfare — for humans. Traditionally the 'scarce resources' have been thought of as including capital, labour and land. By the 1930s, however, some economists were adding to this list resources which are scarce in some circumstances — clean air, clean water, etc. — and others began to incorporate the prospect of exhaustion of the more frequently considered resources; these were divided into renewable (e.g. forests) and non-renewable (e.g. oil).

Perhaps the most profound influence on early 'green' economics came in 1970 from the pen of Kenneth Boulding, memorable for his comparison of contemporary resource allocation patterns to the 'cowboy' mentality of limitless, bountiful, empty plains to be settled. He compared the more desirable alternative of carefully balanced resources and wastes to the 'spaceship' economy of small, crowded spaces. We have already seen (Chapter 2) how powerful the 'spaceship Earth' analogy has become in environmentalism. Boulding is less well remembered for his physical analogues — to the laws of

Table 5.1 Neo-classical assumptions

The following are some standard assumptions underlying neo-classical economics, which are made so that economic models are 'workable' (that is, conceptually and analytically tractable).

A. *There exist perfect markets, which implies:*
   Buyers and sellers (agents) have perfect information about the present and the future;
   there is perfect homogeneity and divisibility of goods produced;
   there is totally free entry into the markets by any agent;
   there are an infinite number of agents in both the supply and demands side.

B. *Economic agents maximize utility. Agents are 'rational', which means that they make decisions only by the criteria of maximizing their monetary utility and profits.*

C. *If occasionally there exist some external effects in production and/or consumption, they always can be identified and internalized.*

D. *The markets deal with factors of production which are:*
   Fully employed;
   perfectly mobile from one sector to another;
   react perfectly to marginal changes in the economy.

E. *Markets mechanistically and automatically adjust to changes in economic conditions and thus reversibility is always possible.*

thermodynamics. These allow neither creation nor destruction of matter; in other words, when resources are used to produce the goods and services we need, the same quantity of matter remains in a modified form. This is a powerful signal to us to reform our attitude to waste; the capacity of the planet to assimilate waste therefore becomes a resource in its own right. The laws of thermodynamics also stress that energy use produces dissipation — increasing entropy — and therefore recycling can never be 100 per cent.

5.1.1 The market and rationality

One further contrast brought out by Boulding between 'cowboy' and 'space-ship' economies was that the former emphasizes turnover, whereas the latter maintains its stock in good order. One of the assumptions of pure economic approaches is that there exists a market for goods and services in which trade will produce all the necessary regulation by both positive and negative feedbacks. Table 5.1 lists the other idealized assumptions about the market. We hear a great deal about 'free-market economies' from politicians, both for and against. However, institutional interventions mean that few economies operate with the freedom of the pure models of economists — they are corrupted or distorted in some way and if nation states are the key players in present and future environmental management we should realize this.

Retaining, for the moment, a notion of 'pure market forces', how do these cope with scarcity (either of utilizable matter, 'commons' or capacities to assimilate waste)? Figure 5.1 provides some clues but in the first three labels of the diagram we see the words 'costs' and 'prices'. Thus entry to the beneficial regulatory effects of market forces is apparently debarred to resources which are currently free or have not been costed; we give considerable space below to means of opening this door but initially need to consider whether the 'market optimists' (Rees, 1990) have cause for their optimism for costed and priced goods and services.

Firstly, a prime regulatory process in Figure 5.1 is innovation/substitution. There can be no doubting that mankind's intellectual and manufacturing skills have performed this strategy in the past; however, one of the most credible typologies of environmental attitudes (see section 5.1.5) includes groups of people who are convinced that our options to innovate will become increasingly restricted in the future and like a struggling, hooked fish, we will eventually be guided passively by environmental constraints.

```
                        Scarcity
                           ↓
                  Extraction costs rise
                           ↓
                      Prices rise
          ┌────────────────┼────────────────┐
          ↓                ↓                ↓
   Demand decrease                   Supply increase
          ↑                                 ↑
  ┌───────────────────┐           ┌───────────────────┐
  │ Increase use in   │           │ Increased         │
  │ substitutes.      │           │ viability of      │
  │ Greater economy   │           │ known deposits.   │
  │ in use.           │           │                   │
  │ More recycling.   │           │                   │
  └───────────────────┘           └───────────────────┘
          ↑                ↓                ↑
          │        ┌───────────────┐        │
          │        │  INNOVATIONS  │        │
          │        └───────────────┘        │
          │         ↓             ↓         │
  ┌───────────────────┐   ┌───────────────────┐
  │ Development of    │   │ Search for new    │
  │ new substitutes.  │   │ deposits.         │
  │ Development of    │   │ Development of    │
  │ improved          │   │ methods to        │
  │ conservation      │   │ increase the      │
  │ methods.          │   │ output from       │
  │ Improved          │   │ known sources.    │
  │ recycling         │   │                   │
  │ techniques.       │   │                   │
  └───────────────────┘   └───────────────────┘
```

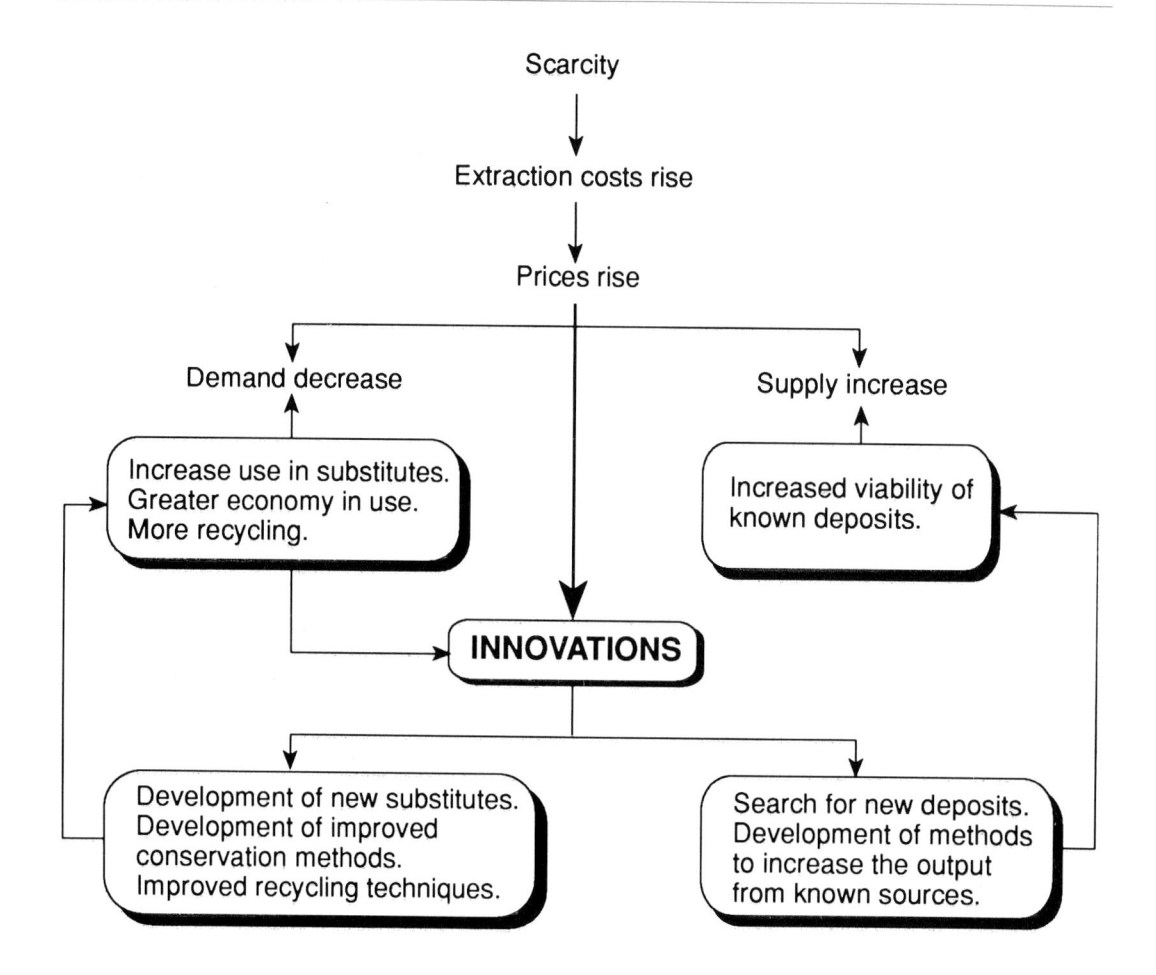

Figure 5.1 Capitalist economies and the modulation of scarcity, price and innovation. (Rees, 1990)

Rees (p. 43) outlines three main classes of challenge to market optimism:

a) The market system is manifestly imperfect,
b) The results of market operations may not conform to the social, economic and political objectives of society,
c) The market cannot cope with — in fact it creates — some forms of natural resource scarcity.

Imperfections in the market are not only created by government interventions but result, for example, from judgements made by key players in terms of market competition. Neither their information about scarcity nor their actions to conserve may be anywhere near perfect; they are, instead, risk-takers. Patterns of investment ahead of exploitation are also unlikely to be rational: a 'good bet' often brings commercial success but has little to do with planet management! There are obvious pressures, both in the political systems of nation states and in large commercial firms, for resource decisions which prioritize 'success' in the short term. One of the major problems of the developing world is that, to be politically acceptable, the market must produce growth which is rapid, a process which depletes rather than conserves resources. Finally, there is nothing in the market system itself which guarantees that a resource is not exhausted. As Rees

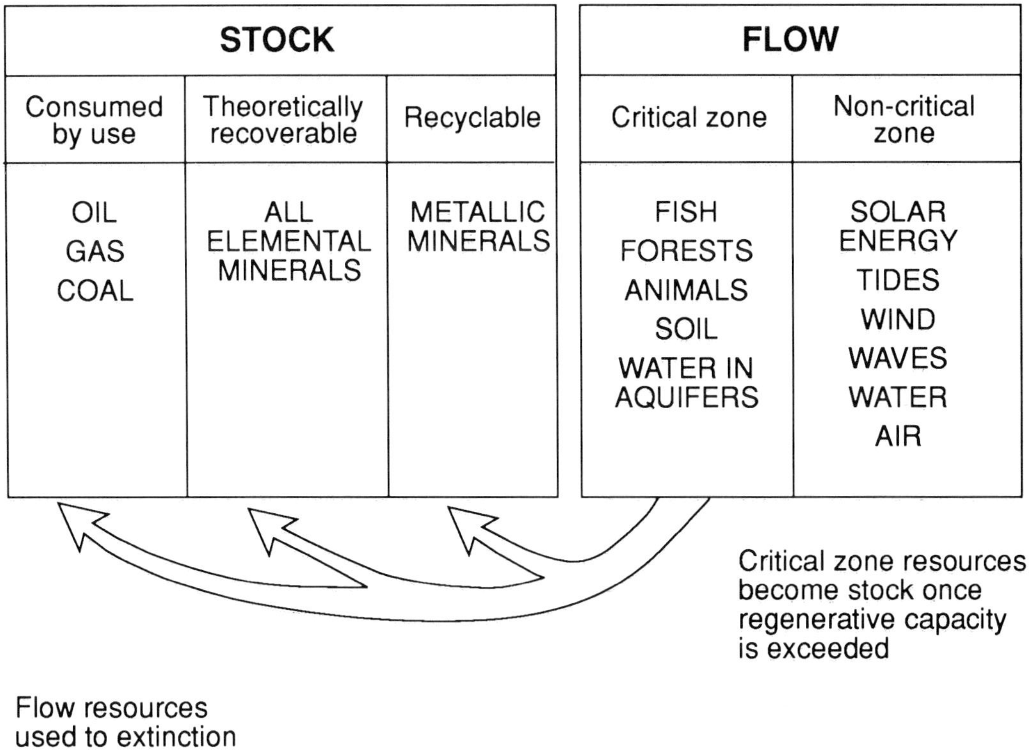

STOCK			FLOW	
Consumed by use	Theoretically recoverable	Recyclable	Critical zone	Non-critical zone
OIL GAS COAL	ALL ELEMENTAL MINERALS	METALLIC MINERALS	FISH FORESTS ANIMALS SOIL WATER IN AQUIFERS	SOLAR ENERGY TIDES WIND WAVES WATER AIR

Critical zone resources become stock once regenerative capacity is exceeded

Flow resources used to extinction

Figure 5.2 Resources classified to guide sustainable exploitation. (Rees, 1990)

claims (p. 47): 'In fact market forces are likely to accelerate the onset of exhaustion since deposits are almost inevitably economically exhausted long before they are physically worked out.'

The market is therefore an imperfect tool for the environment, even for goods and services which are costed and priced. Little wonder, therefore, that the word 'sustainability' has become a signal of desirable improvements, even to the market optimists (see section 5.3).

5.1.2 The 'tragedy of the commons'

Another rallying cry of the 1970s was Hardin's 1968 article in *Science* which addressed, in one simple analogy, the problems of those aspects of the environment which conventionally lie outside market forces — unpriced resources and assimilative capacities. Hardin revealed a picture of the medieval English common pasture, open to general public use for grazing. If one grazier increases his/her herd by one animal the impact on neighbouring users of the common is very small and the grazier has made an increase of income. Magnified to the effects of all graziers taking similar action, the common begins to degrade and erode. As Cottrell (1978) summarizes Hardin, 'all the herdsmen are ruined by the totality of their actions which, individually for each one of them considered alone, is undoubtedly in his own best interest' (p. 8).

In order to be able to explore ways of avoiding such damage to 'free' resources there is a profound job of education; the fact that such resources do not conventionally enter national accounts (see section 5.5) often means that politicians fail to consider them as resources. Figure 5.2 is helpful here in both spreading the definition of resources and categorizing them, with examples, of stock and flow. The message is also clear that if 'cowboys' cannot be encouraged aboard the 'space-ship' (in Boulding's terms) by conserving stock, they should at least pay attention to the vulnerability of the critical flow resources and to the desirability of an increased reliance upon the non-critical flow resources.

5.1.3 Externalities: 'the polluter pays'

The external diseconomy brought about by the action of one of Hardin's herdsmen to the common as a whole was small but the totality was ruin. Generally when stock resources are developed there is an external effect — disruption to soils, vegetation and drainage, pollution and also a measurable social impact. These impacts are not instantaneous, they may be slow to develop and may long outlast the development itself. Rees (p. 261) defines externalities as 'the uncompensated side-effects of any economic or social activity which are not considered by individuals when making private decisions'.

Much of the effort directed at pricing environmental goods and services has gone into getting a calculation of externalities into decision-making protocols such as cost-benefit analysis (see section 5.2). At this point there is again an opportunity to adopt an ethical as well as a practical viewpoint by assessing the scope and definition of externalities of a particular activity as either narrow or wide, short-lived or long-lived. Consider, for example, the effects we now know to have derived from the development of the motor car (Table 5.2); not all of this list can be considered environmental externalities of the development project for the first car but the impacts of technology are clearly made far and wide.

Figure 5.3 illustrates the much more profound point that the elemental resources of air, land and water, together with the living environment, are realistically viewed as 'external' to the behaviour of the economy of capitalist corporations and nation states. Not only are they exploited as substantive resources but their capacity for resilience is also made maximum use of to create carrying capacities for human habitation.

5.1.4 Schools of thought in environmental economics

The environment is managed by individuals, by public or private corporations and by nation states, occasionally co-operating in a global approach. Inevitably, therefore, schools of thought about the role and relations of economics in environmental management classify along similar lines to those of major ethical and political systems. Pepper (1984) has, for example, considered in depth the relationship between marxism and environmentalism. Because, as Pepper says, 'production is the central social process: the defining characteristics of being human' (p. 149), the environment is clearly affected by the relationship of production and how they are politically controlled. Marx separated exchange value from use value for the fruits of production and held that man is part of nature: a promising start for protection of the environment. However, nation states guided by marxist principles have tended to frame materialistic policies in relation to environmental resources and their relatively rapid development from a feudal pre-revolutionary state has caused grave environmental problems, especially in the Soviet Union and Eastern Europe. Nevertheless, critics of capitalist, pluralist systems and the rise of environmentalist thought (see Sandbach, 1980) believe that to ignore class interests makes environmental reform irrelevant.

As this book is being written the best-known example of a marxist economy (the Soviet Union) is disowning the Communist Party in favour of both capitalism and nationalism. Before the profound changes in

Table 5.2 How the automobile has altered our lives (1895 to present)

Values
Geographic mobility
Expansion of personal freedom
Prestige and material status derived from automobile ownership
Over-evaluation of automobile as an extension of the self — an identity machine
Privacy — insulation from both environment and human contact
Consideration of automobile ownership as an essential part of normal living (household goods)
Development of automobile cultists (group identification symbolized by type of automobile owned)

Environment
Noise pollution
Automobile junkyards
Roadside litter

Social
Change in patterns of courtship, socialization and training of children, work habits, use of leisure time, and family patterns
Created broad American middle class, and reduced class differences
Created new class of semi-skilled industrial workers
Substitution of automobile for mass transit
Ready conversion of the heavy industrial capability of automobile factories during Second World War to make weapons
Many impacts on crime
Increased tourism
Changes in education through bussing (consolidated school versus 'one-room country schoolhouse')
Medical care and other emergency services more rapidly available
Traffic congestion
Annual loss of life from automobile accidents about 60,000
Increased incidence of respiratory ailments, heart disease, and cancer
Older, poorer neighbourhood displacement through urban freeway construction

Institutional
Automotive labour union activity set many precedents
Decentralized, multi-divisional structure of the modern industrial corporation evident throughout the auto industry
Modern management techniques
Consumer instalment credit
Unparalleled standard of living
Emergence of US as foremost commercial and military power in world
Expansion of field of insurance
Rise of entrepreneurship
Basis of an oligopolistic model for other sectors of the economy
Land usage for highways — takes away from recreation, housing, etc.
Land erosion from highway construction
Water pollution (oil in streams from road run-off)
Unsightly billboards
Air pollution — lead, asbestos, hydrocarbons, carbon monoxide, nitrous and sulphurous oxides

Demography
Population movement to suburbs
Shifts in geographic sites of principal US manufacturers
Displacement of agricultural workers from rural to urban areas
Movement of business and industry to suburbs
Increased geographic mobility

Economic
Mainstay and prime mover of American economy in the twentieth century
Large number of jobs directly related to automobile industry (one out of every six)
Automobile industry the lifeblood of many other industries
Rise of small businesss such as service stations and tourist accommodation
Suburban real estate boom
Drastic decline of horse, carriage, and wagon businesses
Depletion of fuel reserves
Stimulus to exploration for drilling of new oil fields and development of new refining techniques, resulting in cheaper and more sophisticated methods
Increased expenditure for road expansion and improvement
Increased federal, state, and local revenues through automobile and gasoline sales taxes
Decline of railroads (both passengers and freight)
Federal regulation of interstate highways and commerce as a pattern for other fields
Highway lobby — its powerful influence

Source: Sandbach, 1980

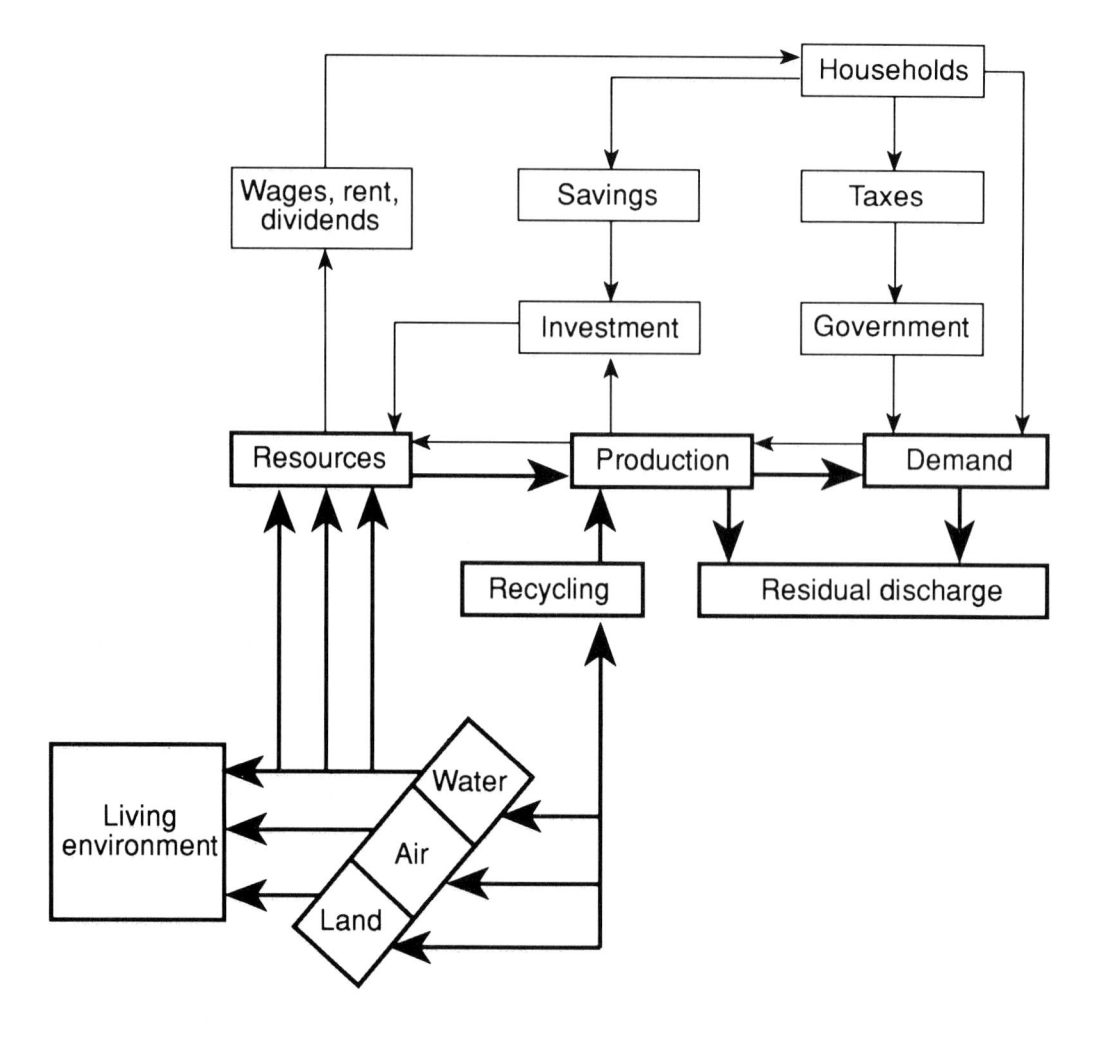

Figure 5.3 Classical economics of the nation state: problems of uncosted resources and wastes. (Norton, 1984)

the Soviet Union the way was laid by a similar shift in Eastern Europe, prefaced by a host of revelations about environmental degradation brought about by outdated technology (much of it supplied by capitalism!) and yoked to the goals of development.

It therefore now seems that political pluralism and democratic decision-making within the context of capitalist corporations and nation states may well be the best structures within which to assess schools of thought in environmental economics. We can therefore consider Figure 5.3 as guidance to both the most common form of the economic system but also to the relationship between that system and the natural system (reduced to water, air, land and living things for ease of depiction).

Modifying O'Riordan's typology, Pearce (1992, see Table 5.3) has listed five categories of attitude stretching away from the 'cornucopian technocentrism' first labelled by O'Riordan as one end of the technocentrist-ecocentrist spectrum. Cornucopian technocentrists believe in the operation of 'free' markets but all the other four categories believe in an increasing degree of decoupling from markets or, at the other extreme, some degree of revolutionary

Table 5.3 Schools of thought within green economics

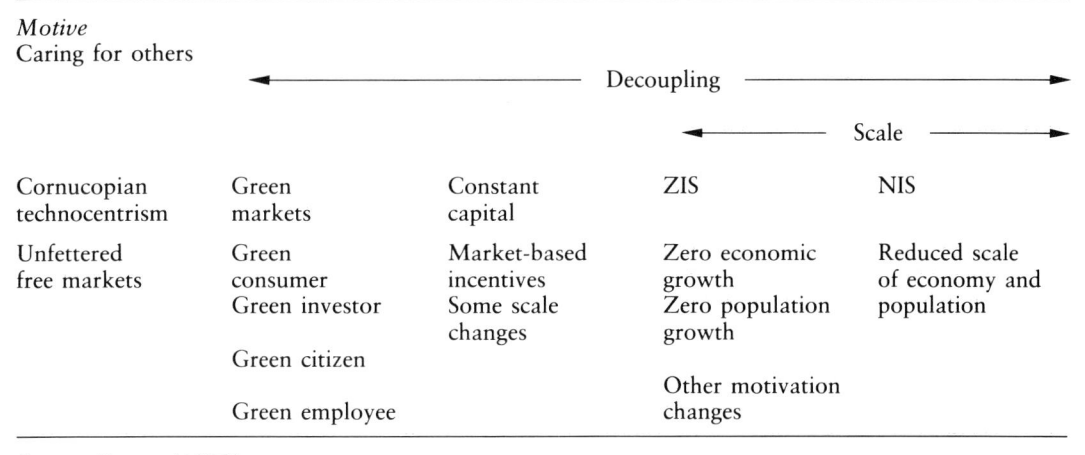

Motive Caring for others				
	←——————— Decoupling ———————→			
			←——— Scale ———→	
Cornucopian technocentrism	Green markets	Constant capital	ZIS	NIS
Unfettered free markets	Green consumer Green investor Green citizen Green employee	Market-based incentives Some scale changes	Zero economic growth Zero population growth Other motivation changes	Reduced scale of economy and population

Source: Pearce (1992)

change of scale in mankind's economic activity. These may be called attitudes to economy whereas O'Riordan's typology consists of attitudes to ecology; the two are clearly related but the basic motives of various individuals classified may well be different.

It is clear that some environmentalists are averse to development as a whole but they have yet to convince nation states of their arguments, many of which inevitably link to a very sensitive area — population control. It may well require a rational approach to the carrying capacity of the earth (see Santos, 1990) before such arguments become more cogent; even with such backing, many environmentalists would be sceptical about the ability to restrain the instincts and abilities of *Homo sapiens* in this way.

Whilst in some sympathy with the views of Sagoff (1988) who rebuffs the attempts by resource economists to invade the sphere of ethics, politics and policy, the remainder of this chapter is devoted to an exploration of mechanisms by which individuals, corporations and nation states may move away from 'free' markets in environmental goods towards 'green' markets and other devices requiring philosophies of politically based economic intervention.

5.2 Attempts to value the environment

Before embarking on this treatment we need to ask whether, at government level, environmental economics is to be used alone to develop management policies. The answer is clearly 'no' since policies are developed by, and through, moral persuasion, law and economics. No one has yet spelled out how the policy 'tool-kit' can be best deployed to solve different problems but it is clear that if environmental economics is to use the market in any form it will be best used to control future actions rather than past actions and larger development projects rather than a plethora of smaller transactions (except for its potential control on consumer patterns).

The first example of economics as an aid to development decision-making comes from flood control engineering schemes in the United States during the 1930s. To obtain 'objective' guidance as to the desirability of providing protection for a particular piece of flood-prone property, the US government assembled all the costs of doing so and compared them with the benefits of doing so. If the ratio of benefits (largely reduced damage from floods) to the costs was greater than unity, the signal was simplistically interpreted as being to proceed with the development. It should be a signal of the vulnerability of cost-benefit analysis that even in the field of engineering for which it was developed it is still criticized and is still being reformed — mainly in terms of defining the true costs and benefits (Pearce and Nash, 1981). Nevertheless, by considering such a simple tool and the marshalling of environmental information to meet its demands we shall acquire insights into the whole problem of costing environmental goods and services (see also p. 47 for a practical example).

5.2.1 Market values of environmental goods and services

Dixon et al. (1988), Pearce, Markandya and Barbier (1989) and Pearce (1991) all provide simple guidance to those in environmental management who need to learn very rapidly the potential of environmental economics to provide regulation of exploitation, pollution and loss of species diversity.

Dixon et al. work exclusively with development projects and rest their advice on project appraisal in the context of the developing world, international finance and, a belated arrival, environmental assessment. The development of good environmental assessment techniques (particularly the 'scoping' element) is especially important as a guide to the use of market values in appraisal. This approach searches out for the points in a web of development impacts where, effectively, money is changing hands in order to cost environmental gains and losses. For example, environmental issues clearly create employment; in other cases externalities of a project reduce employment. Development projects also reduce the opportunities to use, for example, land for another purpose; these are so-called opportunity costs of development (though these are generally applied to conservation of the land intact). Changes in productivity brought about by, for example, pollution externalities can also be costed.

Use of market values is not restricted to individual 'greenfield' projects such as dam construction in Africa; it is increasingly a

Table 5.4a The benefits of pollution control in the USA 1978

	$US billion
Air pollution	
Health	17.0
Soiling and cleaning	3.0
Vegetation	0.3
Materials	0.7
Property values[1]	0.7
Water pollution[2]	
Recreational fishing	1.0
Boating	0.8
Swimming	0.5
Waterfowl hunting	0.1
Non-user benefits	0.6
Commercial fishing	0.4
Diversionary uses	1.4
Total	26.5

[1] Net of property value changes thought to be included in other items.
[2] At one-half the values estimated for 1985.
Source: M. Freeman, *Air and Water Pollution Control: A Benefit Cost Assessment* (Wiley, New York, 1982).

feature of evaluations of all major environmental policies in developed nations. For example, Table 5.4(a) lists the benefits of pollution control in the United States for the year 1978; taking another approach, Table 5.4(b) shows the number of jobs created by nature conservation in the United Kingdom for the year 1986.

5.2.2 Unpriced items: surrogate values and contingent valuation

Clearly project appraisal or policy analysis cannot work if costs and benefits cannot be quantified. Many environmentalists will first have been convinced that 'any value will do' is a dangerous philosophy by the 1971 government appraisal of sites for a third London airport. In this a Norman church was costed at £51,000 because this was its value in fire insurance terms! It may be fortunate in many cases, therefore, that

Table 5.4b Summary of employment figures created by nature conservation, by category

		Full-time FT equivalent Nos.	M.Sc. Special Programmes Nos.
A.	Mainstream nature conservation bodies	1,607 (1) (2)	1,134
B.	General conservation bodies	1,741 (3) (1)	400
C.	Landowners/owning bodies	2,735 (2) (1)	4,414
D.	Capital works, etc.	1,150 (3) (4)	—
E.	Public facilities for nature conservation	3,014 (2) (6)	—
F.	Other visitor services	? (5) (6)	—
G.	Media/publishing	1,400 (6)	—
H.	Production and retail of appropriate goods	1,860 (6)	—
I.	Education and training	150 (6)	? (8)
J.	Research, development, monitoring	600 (7)	? (8)
Totals		14,257	5,948

Source: Dartington Institute (1985) Employment and Nature Conservation. Report to Nature Conservancy Council (ref. CSD Rept. 617).

no information is initially available for use in cost-benefit approaches. In these cases (they are many) two principal alternatives to market values are employed: surrogate values and contingent valuation.

Surrogate values for environmental goods and services, such as clean air, unobstructed views and country rambles, involve the use of an actual market price to indicate a value. They include property values, wage differentials, travel costs and replacement costs. Thus, for example, in order to value a 'home in the country' one might measure the investment made by the proud and satisfied houseowners in the property, the reduction of wages they accepted in order to leave town, or the amount they spend to travel back to town to enjoy urban facilities. If their home were to be threatened by a motorway project one might cost the direct replacement of both the structure and those elements of the 'quality of life' which relate to its desirable rural ambience. Clearly this was not so in the case of the Norman church! A special case of replacement-cost techniques has been developed for projects in which the endangered resource is scarce and highly valued — the use of shadow projects. An alternative piece of development which will provide the same output of goods and services is costed and these costs are added to the costs of the original piece of development. Clearly arguments are rife over the selection of the shadow project because it is then that the search for values attributable to the threatened environment becomes concentrated.

Contingent valuation is a social-survey technique; it is also called hypothetical valuation because it asks 'What if . . . ?' questions of those surveyed. In the basic variant those surveyed are asked their willingness-to-pay or willingness-to-accept-compensation for a change in provision of a good or service. The results are then aggregated to fix the costs or benefits of the change. Problems with the techniques include the fact that respondents may need some form of education for their guesses — such as iteration through a number of scenarios — and may not, even so, be able to offer the hypothetical circumstances for their true value judgements. Variants include 'take-it-or-leave-it' experiments where actual prices are quoted by the interviewer as an offer and 'trade-off' games in which participants are offered 'bundles' of money and environmental resources. An alternative to questioning members of the public is the so-called Delphi technique of asking experts to come up with prices and justifications in the hope that iterations of this will see a convergence by the group to a single, highly considered value.

In a recent study at the University of Newcastle a contingent valuation approach was used to determine the value placed by both local people and visitors upon landscape features of the Yorkshire Dales (Willis and Garrod, 1991). A sample of 600 questionnaires was conducted, revealing a willingness to pay £42 million to preserve the present landscape. This led to the conclusion that the benefits of policies designed to preserve the present landscape, e.g. National Park status, outweigh their cost by a factor of four.

5.2.3 Outside the project appraisal: prices and incentives

In order to effect change in the myriad transactions which make up a modern economy, to educate the population about environmental priorities and to prevent the depletion of the many environmental 'free goods', we can intervene in the market itself. Pearce, Markandya and Barbier (1989) suggest two main means of doing so:

(a) The creation of markets in previously free resources — entrance fees for national parks, dumping fees for offshore waste disposal, etc. These authors associate the strategy with privatization of commons in order to bring about the transactions.
(b) The modification of markets by incorporation of environmental values in market-based incentives.

Pearce and his colleagues have continued to follow up the latter avenue, believing full privatization to be impracticable. The first advantage claimed for market-based incentives is that they are far more efficient than 'command-and-control' systems of environmental standards and enforcement.

The 'proper' price for products is described by Pearce and colleagues as 'one that reflects the wider social costs of production inclusive of any environmental services' (p. 156). Thus externalities of production are added as marginal costs for each unit sold. In order to do this nation states must adopt the 'polluter pays principle', defined thus by OECD (1975):

A. GUIDING PRINCIPLES
Cost Allocation: the Polluter Pays Principle
1. Environmental resources are in general limited and their use in production and consumption activities may lead to their deterioration. When the cost of this deterioration is not adequately taken into account in the price system, the market fails to reflect the scarcity of such resources both at the national and international levels. Public measures are thus necessary to reduce pollution and to reach a better allocation of resources by ensuring that prices of goods depending on the quality and/or quantity of environmental resource reflect more closely their relative scarcity and that economic agents concerned react accordingly.
2. In many circumstances, in order to ensure that the environment is in an acceptable state, the reduction of pollution beyond a certain level will not be practical or even necessary in view of the costs involved.
3. The principle to be used for allocating costs of pollution prevention and control measures to encourage rational use of scarce environmental resources and to avoid distortions in international trade and investment is the so-called 'Polluter Pays Principle'. This principle means that the environment is in an acceptable state. In other words, the cost of these measures should be reflected in the cost of goods and services which cause pollution in production and/or consumption. Such measures should not be accompanied by subsidies that would create significant distortions in international trade and investment.
4. This principle should be an objective of member countries; however, there may be exceptions or special arrangements, particularly for the transitional periods, provided that they do not lead to significant distortions in international trade and investment.

Pollution damage costs which might be incorporated by this approach are shown in Table 5.5.

Moving to mechanisms, Pearce, Markandya and Barbier suggest three:

Table 5.5 Pollution damage costs

	Human health	Fauna	Flora	Natural resources	Materials	Climate and weather
Financial loss	Productivity losses; health care costs including increased research costs to avoid pollution	Lost animal and fish production	Reduced crop production; reduced forest growth; increased aphid predation	Lost production from polluted water or soil; increased treatment costs; reduced flexibility in land use	Reduced life of a material; reduced utility of a material; increased repair costs	Traffic congestion and accident costs due to smogs; reduced agricultural yields and increased lighting costs from decreased sunshine; agricultural losses from shift in world rainfall patterns.
Loss of amenity	Risk aversion: cost of suffering; cost of bereavement; cost of limitations imposed upon an individual, his family and his society	Risk aversion: reduced pleasure from fishing and observing wildlife	Risk aversion: reduced pleasure from landscapes and observation of floral species	Risk aversion: decreased recreation benefits	Risk aversion: endurance of soiled or damaged materials; damage to selected aesthetic monuments and objects	Risk aversion: decreased pleasure from reduced visibility

Source: Based on OECD (1976c)

(a) setting standards;
(b) setting charges or taxes on the polluting product or input to production;
(c) issuing pollution permits in amounts consistent with the standard and allowing those permits to be traded.

Table 5.6 illustrates the steady incorporation of these principles and mechanisms in a number of developed nations. The system is more difficult to apply to international problems such as the enhanced 'greenhouse effect' and stratospheric ozone depletion (see sections 5.3 and 5.5) and few nations have gone as far towards a trade in pollution permits as the United States. The Environmental Protection Agency encourages firms to trade permits within, for example, a level of emissions for a certain urban district; they can also vary their permits between plants (trade with themselves) and 'bank' the improvements over the standards they make to use in future applications for permits. The results in terms of air quality are apparently neutral but the principle is established of modified markets within a command-and-control umbrella.

Table 5.6 Types of pollution charge systems

	Air	Effluent Water	Effluent Waste	Noise	User	Product	Administrative	Tax differentiation
Canada					X	X		X
United States				X	X	X	X	
Australia		X	X		X		X	
Japan	X			X				
Austria		X			X			X
Belgium	X	X			X		X	X
Denmark			X		X	X	X	X
Finland					X	X	X	X
France	X	X		X	X	X		
Germany		X	(X)	X	X	X	X	(X)
Greece	X				X		X	X
Italy		X			X	X		
Netherlands		X	X	X	X	X	X	X
Norway					X	X	X	X
Portugal		X					X	
Spain			X		X		X	
Sweden	X				X	X	X	X
Switzerland	(X)			X	X	(X)		X
Turkey			X					
United Kingdom		X		X	X		X	X

Note: X = applied; (X) = under consideration.
Source: OECD (1989), *Economic Instruments for Environmental Protection*, Paris.

5.3 National and international environmental accounts

When most corporate bodies draw up financial accounts the columns generally record the annual ebb and flow of 'assets' — property bought or disposed of, savings or shares etc. Clearly a properly priced structure for 'commons' such as clean air or water would allow conventional corporate or even national accounts to record better information on the environmental 'deposits and withdrawals' aspects of budgets. However, in order that national budgeting can allow a full approach to environmental assets we need radically to re-think the conventional system: there are glaring gaps in quantification and data-gathering as well as in political thinking on this question. Clearly international trade could well be reformed if, for example, trade in properly accounted environmental assets became a feature of international relations.

5.3.1 Setting up the balance sheet

As stated by Pearce, Markandya and Barbier's *Blueprint*, there are three steps in setting up the environmental balance sheet:

(a) quantify stocks of the chosen resources;
(b) quantify sources, uses and their changes over time;
(c) ensure that stock and flow accounts are consistent.

Possibly the latter task most hinders progress; consider, for example, the different estimates of the value of stocks and flows which might come from conservationists monitoring wetlands and agriculturalists monitoring the expansion of drainage! Nevertheless, for less contentious items it has proved possible, notably in Norway, France and Japan, to develop separate 'satellite' accounts for the environment.

Norway separates natural resources used as raw materials and 'environmental' resources of enjoyment or aesthetics. Clearly the first essential in most national accounting schemes of this sort is a detailed and regular survey of land-use from which, using average inputs, storages and outputs, a good first approximation of the account can be made. There has to be a good monitoring system for some basic resources such as solar radiation and for less physical parameters such as recreation visits. Norway's experience is that the first benefit of such accounting is a more rational land-use planning system.

The addition of monetary values to national environmental accounts is as problematic as it is in cost-benefit analysis of development projects. One can obtain, for example, estimates of the defensive expenditures which are put into household and general protection against pollution and also an approximation of the costs of damage where protection is lacking or incomplete. If these two sums are subtracted from the value of consumption, the 'blueprint' argues, a more realistic value of current welfare is produced. Since 1973 Japan has been attempting to correct its National Income calculations to ones of Net National Welfare by including the problems caused by a wealthy economy to water, air and by waste disposal. Japan's meteoric post-war growth of GDP is considerably humbled by calculating the Net National Welfare in retrospect. The picture may still be too optimistic because there are as yet no standard ways of representing resource depletion over long periods (see also section 5.5). The United Nations tries to co-ordinate the production of Standard National Accounts but it has not as yet grasped this nettle. Instead it has asked governments, for the present, to keep 'satellite accounts' of resources in either physical or monetary terms. Since data compilation and successful updating are the essential prerequisites of any final and successful international accounting system the UN advice is realistic.

5.3.2 Environmental indicators

Clearly the data requirements for national environmental acounting are massive; developed nations should be capable of leadership here as should international organizations with an overview of global data. We therefore briefly review the work of the OECD, the United Nations and the World Resources Institute.

The OECD (Organization for Economic Co-operation and Development) is devoted to 'sound' economic expansion of its largely European members. It has recently published its third *State of the Environment* report (OECD, 1991) in which the criteria listed in Table 5.7 are considered through maps, graphs, tables and essays which link environmental statistics to economic and social progress. The United Nations Environmental Programme (UNEP) has also produced its third Environmental Report (UNEP, 1991) but there, between similar graphs and tables, the emphasis is more on human welfare, including resources, health and natural disasters. The resource theme is very distinctly dominant in the most comprehensive of these publications — the annual review *World Resources* compiled by the Washington-based but independent World Resources Institute (WRI, 1990).

The compilations in Table 5.7 are impressive and represent the power of modern information systems. Nevertheless, analysis and perception is missing from them, other than in prose. What will almost certainly figure in future compilations of this sort is some form of summary calculation for each nation indicating environmental 'health', not only to bring about international comparisons but also to record time trends. In Canada there is progress towards the goal of recording in an amalgamated measure for each of its components 'how the environment did' in a given year. For the moment some insights can be gained from the 'Human Development Index' and its contrast with conventional measures of wealth as shown in Figure 5.4.

Table 5.7 National environmental accounting

Three types of indicator sets are currently under development at OECD in order to contribute to:

(i) measurement of environmental performance with respect to the level and changes in the level of environmental quality, and the related objectives defined by national policies and international agreements. Summary indicators of environmental performance may also be particularly valuable in responding to the public's 'right to know' about basic trends in air and water quality and other aspects of their immediate environment affecting health and well being;
(ii) integration of environmental concerns in sectoral policies. This is done through the development of sectoral indicators showing environmental efficiency and the linkages between economic policies and trends in key sectors (e.g. agriculture, energy, transport) on the one hand, and the environment on the other;
(iii) integration of environmental concerns in economic policies more generally through environmental accounting, particularly at the macro level. Priority is being given to two aspects: the development of satellite accounts to the system of national accounts and work on natural resource accounts (e.g. pilot accounts on forest resources).

5.3.3 World trade and the environment

One of the signal international documents of the last decade was the *Brundtland Report* (WCED, 1987) which highlighted the disparities created by development of different quality occurring at different rates. A basic inequality, that of population and resource demands, distorts world trade and leads (more directly than most will admit) to environmental degradation. The inequality lies in the fact that the developed world houses a quarter of the world's people but these people consume three-quarters of the world's raw materials.

Redclift (1987) stresses that much of the 'natural' environment of developing countries is a reflection of trade relations, transfer of technology and relocation of labour, factors not conventionally viewed as impinging on nature. Redclift cites the clear, adjacent cases of the United States and Mexico, with the environment of the latter reflecting many of the resource and labour demands of the former. Since the concept of 'externalities' in environmental accounting should not respect boundaries, its application might be a start to solving international problems, as might the reformed Standard National Accounts we have described. However, the initial process must be one of education and data gathering in terms of the nature of world trade and the political economy of developments which are promoted to permit nations to participate in it.

Firstly, the demonstration that environmental degradation occurs in the developing world (and that it can be costed) is amply provided by, among others, Barbier (1991). Table 5.8 and Figure 5.5 show the example of watershed degradation in Java. Soil erosion is a particularly useful example of the internationalisation of the environment and illustrates the principles of sustainable use of resources (see section 5.5). Soil is formed by a slow process of weathering of rock and the biological incorporation of the product. Under natural conditions the loss of this resource balances its production, either by gravitational removal of the upper structure or by use of its chemical nutrition by plant growth. Sustainable agriculture considers this natural balance in the longer term but agricultural development is often poorly planned if carried out to produce cash crops for world trade. Blaikie (1985) discusses the political economy of soil erosion, many aspects of which conspire to produce the degraded, useless slopes and silted rivers of eroded terrain.

A further irony (developed by Barrow, 1987, and Newson, 1992) is that soil erosion from ill-planned, rain-fed agriculture in the developing world (especially the tropics) may ruin the grandiose water development/irrigation schemes used by the same nations to further increase food production.

Aspects of the political economy of such degraded 'development' schemes which are most often specified for action are the

GNP PER CAPITA AND THE HDI
(SELECTED COUNTRIES)

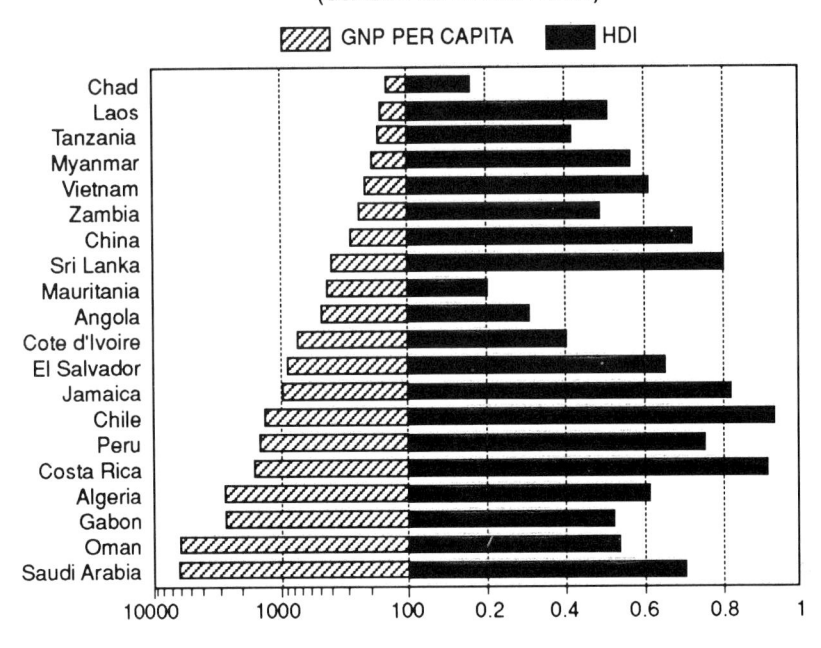

Figure 5.4 Gross National Product (GNP) and Human Development Index (HDI) compared for the developing world. (UNDP, 1991)

Table 5.8 Costs of watershed degradation, Java

The costs of soil erosion on Java ($US million, 1987 prices)	
On-site costs	
Loss of agricultural output in uplands	315
Off-Site Costs	58
Siltation of irrigation	10
Harbour dredging	2
Reservoir sedimentation	46
Total soil erosion	373
Agricultural GDP (%)	2.9
GDP (%)	0.4

Source: W. Magrath and P. Arens, *The Costs of Soil Erosion on Java: A Natural Resource Accounting Approach*, Environment Department, Working Paper No. 18, World Bank, Washington, DC (1989).

disruption of traditional cultures, unsatisfactory land tenure and underestimation of the role of women in rural agrarian society. Traditional societies maintain an inherited risk management strategy with respect to the natural environment. Use values pervade their exploitation of all resources, they respect diversity and their behaviour is adaptive — perfect recipes for the more intensive resource use which is the essential prerequisite of development, but recipes which are often ignored. The foreign 'experts' who often participate in the early phases of development cannot reliably integrate their techniques with these traditions and high on the 'wish lists' of those concerned with a more environmentally-based approach is the so-called 'indigenization of science and technology for development'; a close second is the use of extension workers to place development better into local contexts. International funding agencies support much of the activity defined as 'development projects'

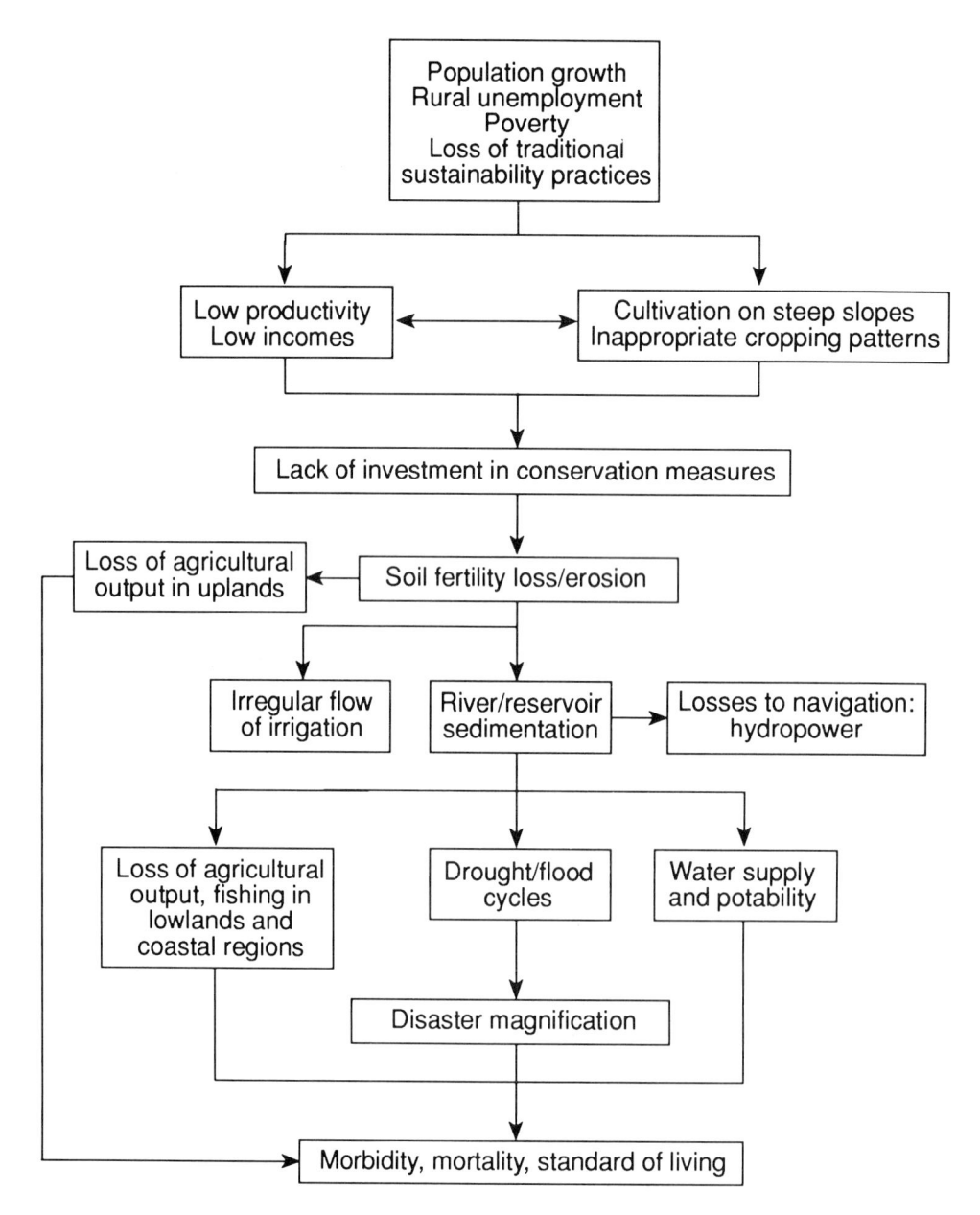

Figure 5.5 Soil erosion in Java — the political economy of development. (Barbier, 1991)

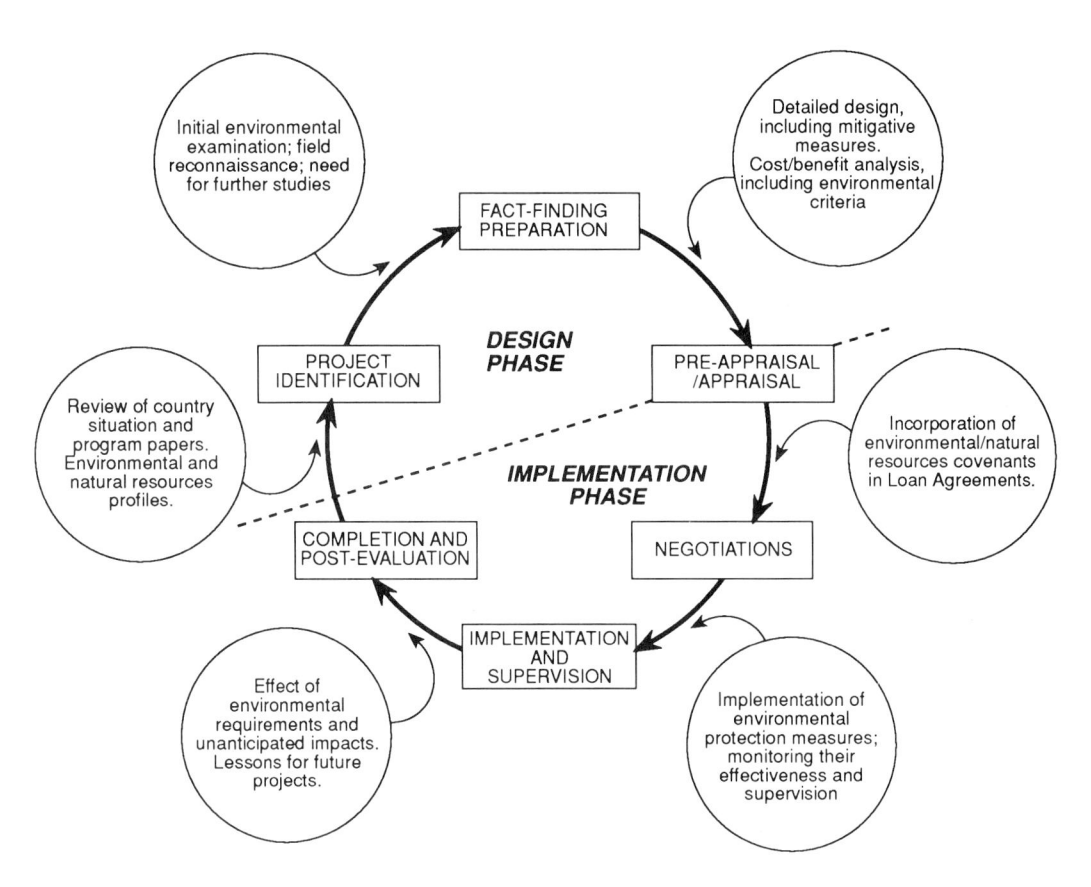

Figure 5.6 Environmental appraisal in the 'project cycle' of development schemes. (Asian Development Bank, 1986, *Environmental Planning and Management*)

in the developing world. As we have seen (section 5.2.2), projects are capable of proper environmental costing and agencies such as the World Bank have now adopted general policies of both environmental impact assessment and realistic cost-benefit analysis in the 'project cycle' (see Figure 5.6).

Problems of the environmental effects of world trade are not the sole responsibility of the funding agencies for development projects; the activities of groups of developed nations, such as the EC countries, also have an impact. The ability of the ECs farmers to specialize in low-risk, high-productivity activities is often condemned for producing food surpluses but it is bought at the expense of realistic prices for developing world products such as meal for cattle.

The internationalization of pollution problems has been emphasized by the enhanced 'greenhouse effect' whose impacts on world climate and sea levels, whilst considered worthy of a precautionary international reaction by nation states, will require controls which are highly problematic in relation to uneven patterns of development.

As Barrett (1991) reveals, the technological assessment of the control problem is multi-dimensional and involves, for example, the economic benefits of forests as carbon absorbers as well as the differential polluting 'value' of coal, oil and gas (ratio 1:0.8:0.6). The most widespread solution considered to date is a tax added to the price of fossil fuels for those who use them. However, there the simplicity of the argument ends for if a nation imposes such a tax

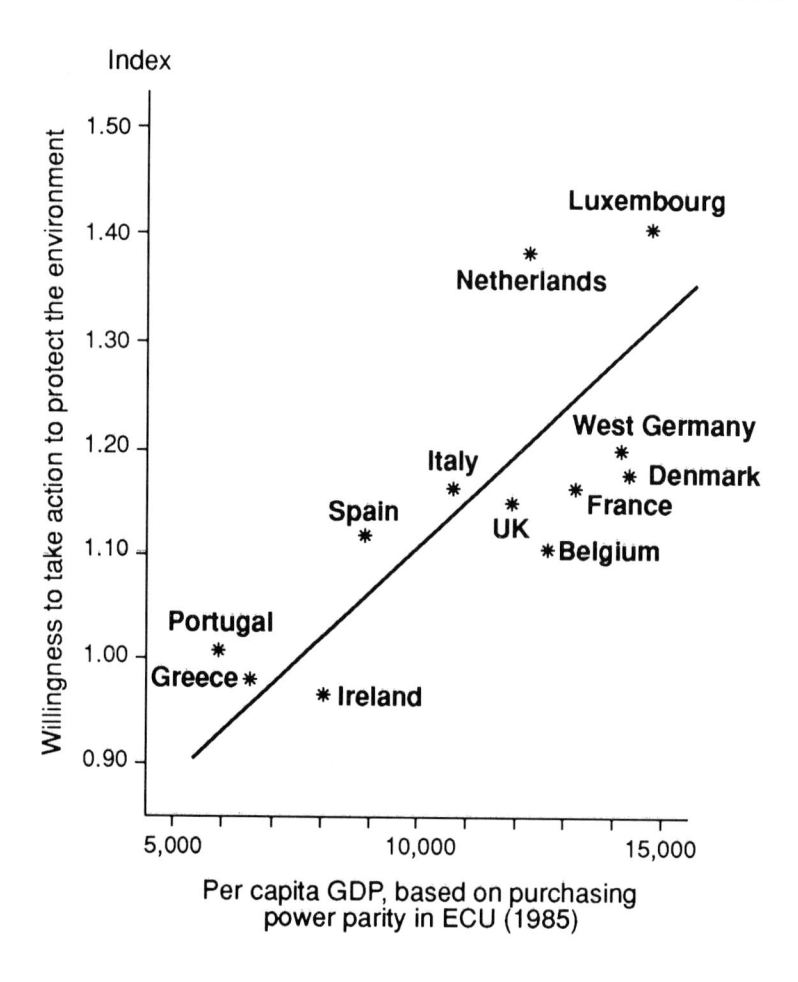

Figure 5.7 The increasing willingness to pay for environmental improvements from national wealth. (Pearce, Markandya and Barbier, 1989)

unilaterally it will quickly put its goods at a trading disadvantage. Equally if all nations introduce an identical tax, the poorer, rapidly industrializing countries will effectively be 'frozen' at their current stage. Markandya (1991) advocates tradeable permits to discharge carbon dioxide, set to reduce and stabilize total emissions.

It is fitting to end this section with a further quotation from Redclift and with a graph (Figure 5.7). Redclift (1987, p. 79) reminds us that:

The environment in the international economy is an internationalized environment and one which often exists to serve economic and political interests far removed from a specific location.

Figure 5.7 illustrates the dilemma that environmental awareness increases with Gross Domestic Product; maybe the 'indigenization' of science and technology and the proper appraisal of both projects and trade can remove the perverse linearity of this plot!

5.4 Commerce and the environment

In developed economies under capitalism many of the transactions having a bearing on the state of the environment occur between manufacturing and retail corporations and

How much more would you be prepared to pay for 'green' products?

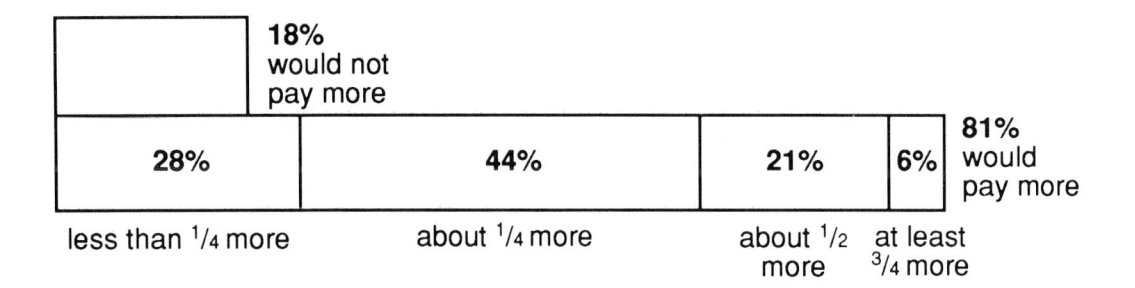

28%	44%	21%	6%
less than ¼ more	about ¼ more	about ½ more	at least ¾ more

18% would not pay more

81% would pay more

What types of 'green' products do you buy and use?

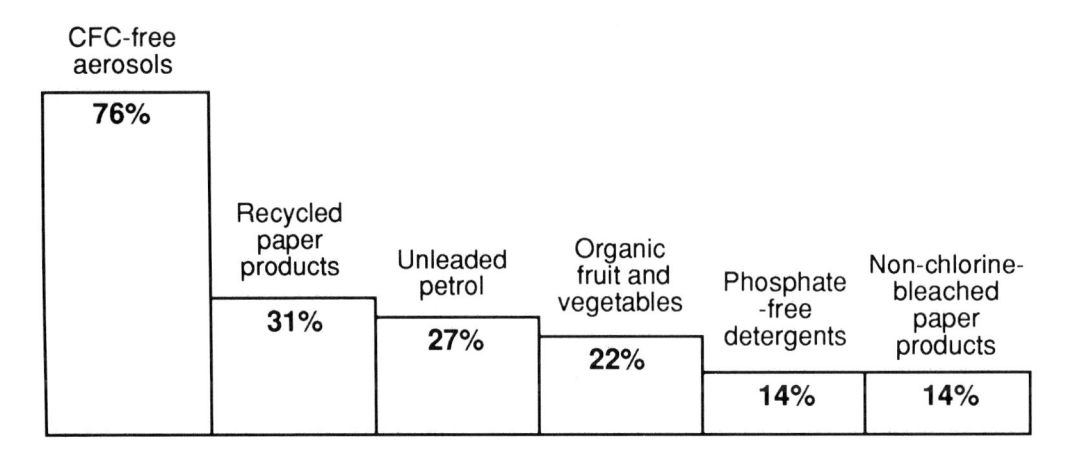

- CFC-free aerosols: 76%
- Recycled paper products: 31%
- Unleaded petrol: 27%
- Organic fruit and vegetables: 22%
- Phosphate-free detergents: 14%
- Non-chlorine-bleached paper products: 14%

Figure 5.8 Consumer surveys reveal how developed world purchasers demonstrate a willingness to pay for environmental qualities. (Consumers Association, 1989)

consumers. Consumerism and environmentalism had contemporary origins in the 1960s when rights to a well-designed product and a healthy living-space went hand-in-hand without a correlation being made between the product and the externalities of its production. This is not the case in the 1990s when *Which* magazine carries pollution guides and a new publication, *The Ethical Consumer*, has a wide circulation amongst environmentalists.

5.4.1 Corporations and their customers

Roddick (1991), writing of her experiences with the 'green consumerism' of her Body Shop enterprise, admits that it took her many years of growing up to realize that she did not object to capitalism but to the corporations at the centre of the system! The Body Shop which, as is illustrated below, has innovated many improvements in corporate behaviour with respect to the environment, has gone even so far as to set up an Anthropological Department. Possibly soaps and lotions for our bedrooms and bathrooms have particular significance to more selfish concerns but the Body Shop is also utilizing the secure knowledge that the average consumer is prepared to pay more for 'environmentally friendly' products, by footing at least some of the bill for the externalities of production or the less profitable production methods necessary to minimize externalities (Figure 5.8). Consumer protests and boycotts have been notably successful in the case of pollution, reaching a peak in the 'issue-attention cycle' in the cases of international pollution impacts such as those of CFCs and pesticides. Other campaigns have shown distaste for the testing of products on animals or the inadvertent slaughter of dolphins by tuna fishermen.

Corporations involved with such issues inevitably respond to consumer sensitivities. Perrier Water removed a million bottles from the retail line after a trace contamination (not pollution) by benzene in 1990. The explanation by Perrier's marketing manager was, 'We are in the business of purity'.

However, corporations are now seeing the commercial advantage of an environmentally friendly image through the all-important medium of advertising, through a healthy and committed work-force and through sponsorship of improvement projects or, more importantly, of research designed to yield 'greener' technologies. In the United Kingdom legislation is also converting the slower-moving manufacturers and retailers with the implementation of integrated pollution control and the duty of care for wastes. Leading organizations such as the Confederation of British Industries produce guidance literature and advice for their members. Many firms now produce environmental mission statements 'setting out their stall' in environmental terms, but two other devices — environmental audits and environmental policies — are a more cogent representation of commitment to a new form of commercialism.

5.4.2 Environmental audits and policies for corporations

The term auditing, borrowed from finance, implies a thoroughness and openness which is essential in a meaningful desire to reform commercial practices. 'Review' is preferred by some because they fear that environmental analysis of a firm's operations cannot be so total as to 'balance the books', as in an audit. However, it is this aspect of the term audit which appeals to the author as a systems-orientated scientist!

Not many examples of either audits or policies are yet publicized; commercial confidentiality is partly responsible as is the high value of this innovative information, though many firms are now prepared to publish, e.g. Procter and Gamble (Hindle, 1990). Forster (1990) summarizes the Body Shop's environmental audit as discovering deficiencies, establishing baselines against which improvement can be judged, establishing good practice, pointing to research needs and promoting awareness. It is interesting in this respect to read the same author's expression of the Body Shop's environmental policy. It covers: campaigns on environmental issues, product development, product testing, purchasing standards, trade with developing countries, waste management, shop premises, packaging, health and safety, transport. By comparison the lists provided by industrial manufacturers are shorter. At Volvo (Boethius, 1989) they investigate the production process, the operation of the car and the eventual disposal of the car; at IBM (Hood, 1989) attention is concentrated on environmental assessment of its manufacturing sites. Clearly the general consumer is unlikely to see lengthy documents expressing the virtue of the good practices adopted by these firms; much more likely is a spread of environmental labelling. The 'blue angel' label in Germany is applied to over 3000 products; the national environmental agency publishes the acceptable standards for this voluntary scheme. Sceptics are concerned that whilst a label may convey acceptable standards for one aspect of production (e.g. use of agrochemicals in the case of food), a full cradle-to-grave analysis is what is really required. The evaluations published by *Ethical Consumer* work on even broader-based standards.

Corporations are now encouraging green investment and so the very basis of the

Table 5.9 1987 estimates of national environmental protection markets

	Total expenditure (ECU bn)	European expenditure (%)	European GDP (%)	Maturity Index
Germany	14.6	36.4	25.8	141
France	7.7	19.2	20.4	94
UK	6.9	17.3	15.5	112
Italy	4.8	11.8	17.5	67
Netherlands	2.0	5.0	5.0	100
Belgium	1.3	3.3	3.3	100
Spain	1.2	3.0	6.7	45
Denmark	0.9	2.2	2.3	96
Greece	0.2	0.5	1.1	45
Ireland	0.2	0.5	0.6	83
Portugal	0.2	0.5	0.8	63
Luxembourg	0.1	0.3	1.0	30
EC12	40.0	100.0	100.0	—
USA	70.2	175.5	105.0	167

Comparable figures for the Far East including Japan are difficult to establish, but indicated levels of expenditure are generally lower than EC12 average or US shown above.
Commercial Union Asset Management Maturity Index: base 100 is when % of European Expenditure equals % of European GDP.
Source: Commercial Union Asset Management (1989).

capitalist development programme is exposed to environmental forces. Matching this pressure with regulative progress, with the value of environmental protection as a business in its own right (see Table 5.9) and with the savings to be made by good practice (3M saved $235 million by its 'pollution prevention pays' campaign), one becomes optimistic that the performance of corporations on environmental issues can be improved. As Anita Roddick (1991) puts it: 'the human race instinctively knows that its spirits will soar when its basic material well-being is provided for in an honourable and humane way.'

5.5 Principles of sustainability

It has been impossible to reach the end of this chapter without using a relatively new word, 'sustainable', to define good practice in relation to man's exploitation of earth's resources. Pearce, Markandya and Barbier (1989) linger on both the broad definition of sustainability, which they liken to biological 'resilience', and the use of the adjective in conjunction with both 'growth' and 'development'. They conclude that sustainable growth is a difficult concept since it is defined by Gross National Product. Therefore GNP should increase over time and not be threatened by feedbacks or social disruption. Because development is, however, defined less rigorously but by notions of the utility, freedom and self-esteem of the individual it can be made sustainable.

5.5.1 Definitions and detractions

Goodland and Ledec (1986) define sustainable development as

a pattern of social and structural economic transformation (i.e. development) which optimizes the economic and other societal benefits available

101

in the present without jeopardising the likely potential for similar benefits in the future.

Barbier (1988, p. 19) emphasizes the processes rather than the aims:

> in general, the wider objective of sustainable economic development is to find the optimal level of interaction among three systems — the biological and resource system, the economic system and the social system, through a dynamic and adaptive process of trade-offs.

It appears that the general enthusiasm for the concept amongst experienced environmentalists has encouraged most of them to seek the elegance of a single definition; the result is a plethora of slightly varying definitions offering critics or sceptics the image of disarray (Pezzey, 1989, identifies over 60 extant definitions of sustainability). More seriously, few authors have yet put down practical strategies for sustainable development; inevitably these will come as sectoral techniques to be operated by established specialisms. There is also a need to vary the procedures for different natural environments. Redclift (1990) takes the example of agricultural development and suggests a classification of environments into those where intensive development is sustainable, those where low-input/high-technology approaches are valid and those where holding the current productivity is the sustainable option. Redclift's principles bespeak a rational approach based upon data but Brindley (1991) emphasizes the approach that will be necessary in the field during development schemes; her ten principles are shown in Table 5.10.

5.5.2 Conventional approaches to the future and intergenerational equity

It would be broad-minded to consider cost-benefit analysis to be a means to sustainability, particularly because of the problems of resourcing it with data, but it does aid decisions on future actions and in one of its components, the financial process of discounting, it incorporates views of the future viability of project investment. Over the economic life of the project (normally shorter than the physical life since replacement will occur) there will be changes in the ratio of costs and benefits and they may be compared with current conditions by comparing with the amount necessary to yield them if invested differently. Clearly the interest rate on the current investment will manifestly affect these calculations. At high interest rates, for example, future benefits will be less attractive than current ones and therefore public long-termism has been encouraged by the choice of (often unrealistic) low rates. There are, however, both private and public costs and benefits over time in most projects and the literature abounds with advice on choice of discount rates.

Even if we agree that the future is given adequate 'respect' in cost-benefit analysis the political acceptability of projects often depends equally on their efficiency and equity, i.e. do they make resources work well and do they spread benefits fairly. It is to this latter concept of equity that those who wish to support sustainable development have made refinements. Intergenerational equity means that, for example, a project must not have high opportunity costs for future generations. A project which, for example, ignores the problems of dismantling, removal, waste, etc. in future years does not display it. The phrase 'for our children's, children's children' becomes a litmus test for development decisions.

5.5.3 Environmental protection: the future role of economics

Writing in 1816 Thomas Malthus warned that overpopulation would see demand outstrip supply in natural resources; despite starvation, disease and war, world population continued to grow. In the 1970s we were warned of 'limits to growth' but again technology has, albeit falteringly, coped. World population now doubles every 19 years and neo-Malthusians are again in print.

Table 5.10 Ten principles of sustainable development

1. Consult with villagers, farmers and all other participants. Reach agreement on both problems and solutions before taking action.
2. Plan small-scale, flexible projects. A plan should be a blueprint, not a prison. It should be able to incorporate new information that emerges during the project.
3. Let the people benefiting from the project make the decisions. The experts' job is to share their knowledge not impose it.
4. Look for solutions that can be duplicated in the hundreds of thousands for the greatest impact on development. But the solutions must still be tailored to fit local needs.
5. Provide education and training, particularly for young people and women, who remain the most effective agents of change because they are bound to the realities of the family's survival.
6. Keep external inputs to a minimum to reduce dependency and increase stability. Subsidies, supplements and inappropriate technology are unsustainable.
7. Build on what people are doing right. New ideas will be adopted only if they do not run contrary to local practice. New technologies must support existing ones, not replace them.
8. Assess impacts of proposed changes. A multi-disciplinary team, ideally including specialists from the same culture, should look at economic, social, cultural and environmental aspects.
9. Consider both inputs and outcomes. The failure of projects focusing on a single outcome, such as agricultural productivity, has proved that more is not always better.
10. Maintain or improve the participants' standard of living. Long-term environmental improvements are unsustainable unless they also address the problems the poor face today.

Source: Brindley (1991).

However, turning the argument of Malthus around and showing distinct optimism for the power of a properly designed and operated system of environmental economics, Pearse (1991, p. 79) concludes that:

we may be putting too heavy a burden on direct regulation to protect our environmental resources, neglecting opportunities to muster the powerful economic forces that have proven so effective in overcoming the limited supplies of natural resources needed for industrial production.

Table 5.11 indicates that Pearse may well be right for other reasons; economic instruments are far cheaper to operate than 'command and control' strategies. Whilst this argument may apply to corporations and nation states it works less well for tackling global problems of such urgency that truly international action cannot await the reform of world trade. It seems likely, therefore, that world treaties and protocols will be needed in advance of internationalized environmental economics to protect critical aspects of global carrying capacities and to bring them under sustainable development programmes.

Known in the popular press as a 'green guru', Professor David Pearce told the British Association's annual conference in 1991 that 'the available economic studies do not bear out the worst fears about the employment, price and income effects of environmental policy'; clearly this is good news but what price sustainable development in a competitive world? He added: 'The reason why the Germans are pressing so hard on environmental issues is that they are laughing up their sleeves at the rest of us. They now make all the pollution equipment, so the harder they push the rest, the more they can sell.'

Table 5.11 Relative costs of command and control and efficient policy instruments

Study	Pollutants covered	Geographic area	CAC benchmark	Ratio of CAC cost to least cost
Atkinson and Lewis	Particulates	St Louis	SIP regulations	6.00[a]
Roach et al.	Sulphur dioxide	Four corners in Utah	SIP regulations Colorado, Arizona and New Mexico	4.25
Hahn and Noll	Sulphates	Los Angeles	California emission standards	1.07
Krupnick	Nitrogen dioxide	Baltimore	Proposed RACT regulations	5.96[b]
Seskin et al.	Nitrogen dioxide	Chicago	Proposed RACT regulations	14.40[b]
McGartland	Particulates	Baltimore	SIP regulations	4.18
Spofford	Sulphur dioxide	Lower Delaware valley	Uniform percentage regulations	1.78
	Particulates	Lower Delaware Valley	Uniform percentage regulations	22.00
Harrison	Airport noise	United States	Mandatory retrofit	1.72[c]
Maloney and Yandle	Hydrocarbons	All domestic DuPont plants	Uniform percentage reduction	4.15[d]
Palmer et al.	CFC emissions from non-aerosol applications	United States	Proposed emission emission standards	1.96

Notes:
CAC = command and control, the traditional regulatory approach
SIP = state implementation plan
RACT = reasonably available control technologies, a set of standards imposed on existing sources in non-attainment areas.
[a] Based on a 40 sg/m³ at worst receptor.
[b] Based on a short-term, one-hour average of 250 sg/m³.
[c] Because it is a benefit-cost study instead of a cost-effectiveness study, the Harrison comparison of the command-and-control approach with the least-cost allocation involves different benefit levels. Specifically, the benefit levels associated with the least-cost allocation are only 82 per cent of those associated with the command-and-control allocation. To produce cost estimates based on more comparable benefits, as a first approximation the least-cost allocation was divided by 0.82 and the resulting number was compared with the command-and-control cost.
[d] Based on 85 per cent reduction of emissions from all sources.
Source: Pearce (1992)

References

Barrow, C. (1987), *Water Resources and Agricultural Development in the Tropics*, Longman, Harlow, 356 pp.

Barbier, E.B. (1988), *New Approaches in Environmental and Resource Economics: Towards an Economics of Sustainable Development*, International Institute for Environment and Development, London, 54 pp.

Barbier, E. (1991), 'Environmental degradation in the Third World', in D. Pearce (ed.), *Blueprint 2: Greening the World Economy*. Earthscan, London, 75–108.

Barrett, S. (1991), 'Global warming: economics of a carbon tax', in D. Pearce (ed.), *Blueprint 2: Greening the World Economy*, Earthscan, London, 31–52.

Blaikie, P. (1985), *The Political Economy of Soil Erosion in Developing Countries*, Longman, Harlow, 188 pp.

Boethius, O. (1989), 'Environmental development from company strategy to action', in *Green Strategies for Business*, Seminar Documentation, IBC Technical Services, London, 20 pp.

Boulding, K.E. (1966), *The Economics of the Coming Spaceship Earth*, Johns Hopkins University Press, Baltimore.

Brindley, B. (1991), 'What is sustainable? Some rules for the development road', *Ceres*, 128, 35–38.

Cottrell, A. (1978), *Environmental Economics. An Introduction for Students of the Resource and Environmental Sciences*, Edward Arnold, London, 66 pp.

Dixon, J.A., Carpenter, R.A., Fallow, L.A., Sherman, P.B. and Supachit, M. (1988), *Economic Analysis of the Environmental Impacts of Development Projects*, Earthscan, London, 134 pp.

Forster, H. (1990), 'Environmental auditing in the Body Shop', in *Environmental Auditing*, Seminar Documentation, IBC Technical Services Ltd, London, 7 pp.

Goodland, R. and Ledec, G. (1986), *Neoclassical Economics and Principles of Sustainable Development*, Office of Environmental and Scientific Affairs, The World Bank, Washington, DC, 60 pp.

Hardin, G. (1968), 'The tragedy of the commons', *Science*, 162, 1243.

Hindle, P. (1990), 'The environmental auditing of a total company', in *Environmental Auditing*, Seminar Documentation, IBC Technical Services, London, 7 pp.

Hood, I. (1989), 'IBM's corporate environmental programme', in *Green Strategies for Business*, Seminar Documentation, London, 8 pp.

Markandya, A. (1991), 'Global warming: the economics of tradable permits', in D. Pearce (ed.), *Blueprint 2: Greening the World Economy*, Earthscan, London, 53–62.

Mirowski, P. (1989), *More Heat than Light. Economics on Social Physics, Physics as Nature's Economics*, Cambridge University Press, Cambridge, 450 pp.

Newson, M.D. (1992), *Land, Water and Development*, Routledge, London.

Norton, G.A. (1984), *Resource Economics*, Arnold, London, 164 pp.

OECD (1975), *The Polluter Pays Principle: Definition, Analysis, Implementation*, Paris.

OECD (1976), *Economic Measurement of Environmental Damage*, Paris.

OECD (1991), *The State of the Environment*, Paris, 297 pp.

Pearce, D. (ed.) (1991), *Blueprint 2: Greening the World Economy*, Earthscan, London, 232 pp.

Pearce, D. (1992), Green Economics, *Enviromental Values*, 1(1), 3–13.

Pearce, D., Markandya, A. and Barbier, E.B. (1989), *Blueprint for a Green Economy*, Earthscan, London, 192 pp.

Pearce, D. and Nash, C.A. (1981), *The Social Appraisal of Projects. A Text in Cost-Benefit Analysis*, MacMillan, London, 225 pp.

Pearse, P.H. (1991), 'Scarcity of natural resources and the implications for sustainable development', *Natural Resources Forum*, 74–9.

Pepper, D. (1984), *The Roots of Modern Environmentalism*, Croom Helm, Beckenham, 246 pp.

Pezzey, J. (1989), 'Definitions of sustainability', Discussion Paper 9, UK Centre for Economic and Environmental Development, London, 48 pp.

Redclift, M. (1987), *Sustainable Development: Exploring the Contradictions*, Routledge, London.

Redclift, M. (1990), 'Developing sustainably: designating agroecological zones', *Land Use Policy*, 7(3), 202–16.

Rees, J. (1985, 1990), *Natural Resources: Allocation, Economics and Policy*, Routledge, London, 499 pp.

Roddick, A. (1991), 'The Body Shop: freeing the corporate spirit', *Geography*, 76(1), 16–20.

Sagoff, M. (1988), *The Economy of the Earth*, Cambridge University Press, Cambridge, 271 pp.

Sandbach, F. (1980), *Environment, Ideology and Policy*, Basil Blackwell, Oxford, 254 pp.

Santos, M.A. (1990), *Managing Planet Earth: Perspectives on Population, Ecology and the Law*, Bergin and Garvey Publishers, New York, 172 pp.

United Nations Environmental Programme (1991), *Environmental Data Report*, Blackwell Reference, Oxford, 408 pp.

WCED (The World Commission on Environment & Development) (1987), *Our Common Future*, Oxford University Press, 400 pp.

Willis, K. and Garrod, G. (1991), 'Landscape values: a contingent valuation approach and case study of the Yorkshire Dales National Park', Countryside Change Working Paper Series, University of Newcastle upon Tyne, UK, 21, 37 pp.

World Resources Institute (1990), *World Resources 1990-91*, Oxford University Press, New York, 363 pp.

Part 2 Practices

6
Patterns of air pollution: critical loads and abatement strategies

Alan Davison and Jeremy Barnes

The generation of energy by burning fossil fuels in power stations and engines, all manner of industrial processes, biodegradation of wastes, and some farming operations lead to the release of thousands of different chemicals into the air. Most have little or no discernible effect on the environment because the concentrations are very low or because they are not toxic to biological systems; some, however, may damage human health or decrease animal production. Some may reduce crop yields or threaten the plants and animals of natural ecosystems. This is not a new situation because there has been air pollution of one kind or another since fires were first used and metals were first smelted. For several centuries there has also been concern about the effects of this pollution but the degree of concern and the pollutants themselves have changed continuously over time. As the cocktail of chemicals in the air continues to change there is an ongoing challenge in recognizing their effects and in devising effective strategies for their control.

6.1 Traditional approaches to air pollution control

Over several centuries, as each new hazard has been recognized there have been attempts to use the legislative process to control it but with varied success. In Elizabethan England, smoke and the stench from fires was considered so offensive that coal burning was banned while Parliament was in session, but that and later legislation did nothing to prevent the massive increase in the use of coal during and after the Industrial Revolution. In contrast, during the last century, the great, unbridled expansion of the chemical industry in Britain led to the introduction of the Alkali Acts of the 1860s, under which registered processes had to use the 'best practicable means' to control emissions. What this meant was that the individual factory could be pressured into employing newer, better technology to control emissions as it was developed. Despite some recent criticisms of the Acts they were the mainstay of control for almost a century and they undoubtably led to an improvement in air quality in the vicinity of many industrial plants. Unfortunately, some industries such as brickworks were not covered by the Acts and they remained important sources of sulphur dioxide, odours and fluoride. However, in the past moves to control pollution have not often been based on altruism or a basic concern for the environment; disasters have played a large part in galvanizing politicians into action, so the epic effects of the London smogs of the early 1950s (Figure 6.1) led to the Clean Air Act of 1956, and with it the use of smokeless fuels and large, centralized power stations. These

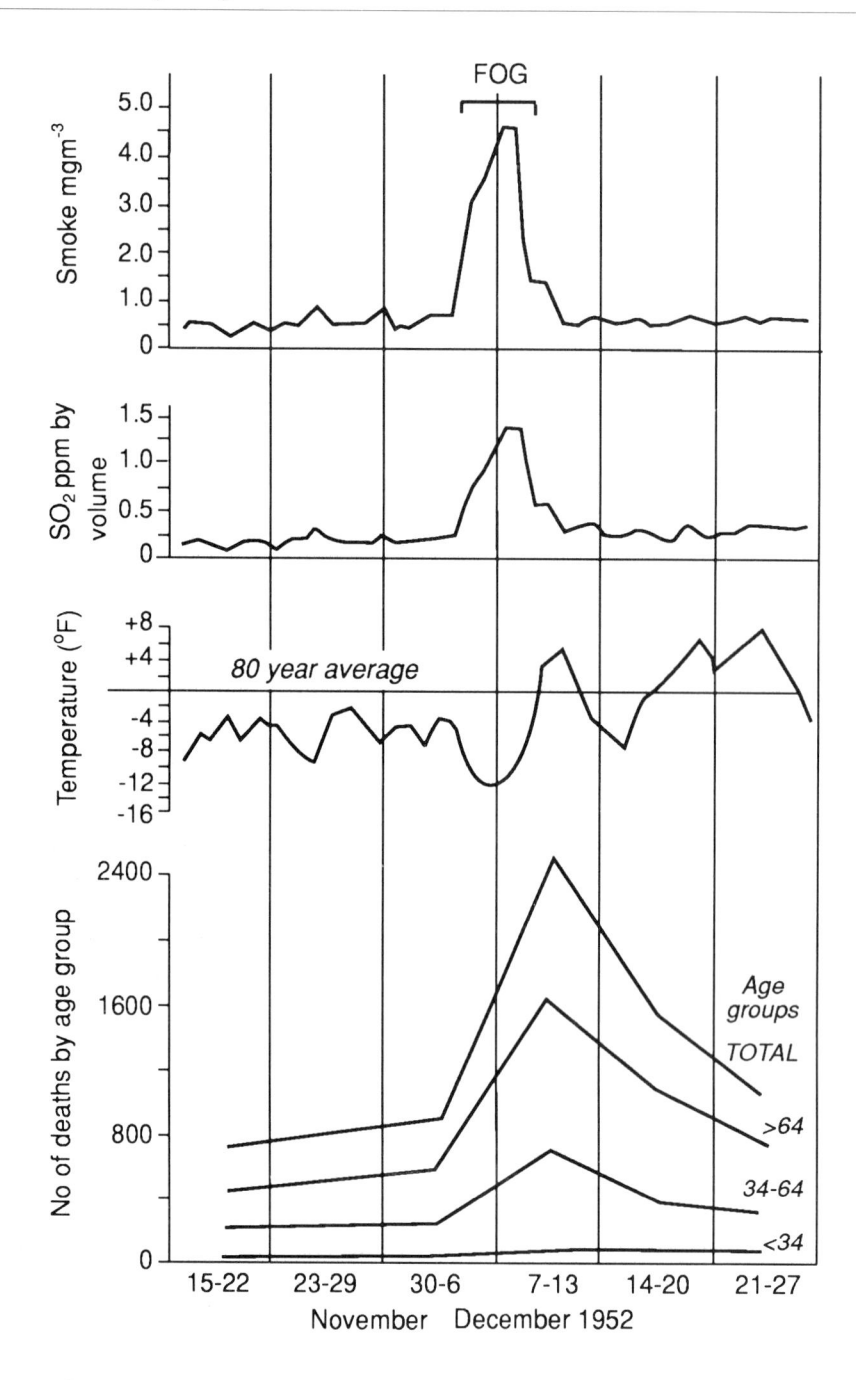

Figure 6.1 The London smog disaster of 1952. The number of deaths rose dramatically during and after an episode in which smoke and sulphur dioxide rose to extremely high levels. Over the period there was an estimated 4000 deaths over and above the number expected for the time of year. Other disasters occurred in 1956 (1000 excess deaths), 1957 (700+ excess deaths) and 1962 (c.700 excess deaths). They led to the introduction of the Clean Air Acts. (Redrawn from Scott, 1953)

emit little or no smoke and they have tall stacks to give maximum dispersal of the gases and therefore minimum concentrations of SO_2 at ground level. The Clean Air Act was tremendously effective in removing acid, sulphurous smog and therefore improving health and the quality of life for millions in British cities but the use of tall stacks to disperse emissions from fossil fuels contributed to the problems of long-distance dispersal and to the acidification of soil and lakes that were to become so prominent an issue less than two decades later. A solution to one problem exacerbated another, illustrating the very basic point that effective control must be based on sound scientific knowledge of all aspects of the problem.

6.2 Point sources: control of fluoride emissions — a success

One of the most instructive examples of successful pollutant control concerns fluoride and, in particular, emissions of fluorides from aluminium smelters. The element fluorine is a component of many of the materials that are used in everyday life, including the surface of non-stick frying pans, aerosols and refrigerants, anaesthetics and pesticides. Fluorides are also used in aluminium- and iron-smelting to enable the process to operate at lower temperatures and therefore to save energy. Coal, clay and some phosphate ores contain appreciable amounts of fluorine; coal-burning, the production of bricks and manufacture of fertilizers may all cause significant release of fluorides into the atmosphere. If the concentrations are high enough fluorides may cause visible injury to plants and a debilitating condition in grazing animals called fluorosis. The effects were quickly identified in the early part of this century because of investigations into the cause of a crippling condition that occurred in certain industries and in parts of rural India where there were very high fluoride levels in the water. During the second World War one of the major sources, aluminium-smelting, expanded rapidly and caused very severe environmental problems.

Smelters of this period damaged forests and limited the farming of cattle over many kilometres. Research between 1940 and 1980 demonstrated the threshold concentrations above which effects on plants and animals develop so it became possible to specify guidelines for air quality around fluorine sources. In North America, where much of the aluminium production capacity was located, and where there was pressure due to expensive law suits, tightening legislation forced the aluminium industry to find engineering solutions to the problem. As a result, fluorine emissions from smelters have been reduced by a factor of almost 200 times in the last 50 years and there should be no significant impact from a modern smelter.

The successful control of fluoride emissions was due to several key features. Perhaps one of the most important is that the individual source of the pollutant can be identified, and even where there are several sources in an area these are rarely very numerous. The chemical and physical nature of the pollutant is also paramount because fluorides are scavenged from the atmosphere relatively rapidly so the effects are essentially local in scale. Even in the worst cases of the large, old smelters, the environmental effects were confined to distances less than something of the order of 20 km from the source. As there is always a recognizable gradient in both the concentration of fluoride and in effects that follow a predictable pattern imposed by wind direction and topography it is possible to trace the relationship between the source and effects.

Another feature that has contributed greatly to successful control is the background research which has determined the threshold concentrations for the most sensitive plants and for grazing animals. This means that it is possible to quantify the environmental effects when they do occur and to specify air quality guidelines to prevent problems happening. Protecting the most sensitive component of an ecosystem will automatically protect all other species. With many pollutants it is very difficult to

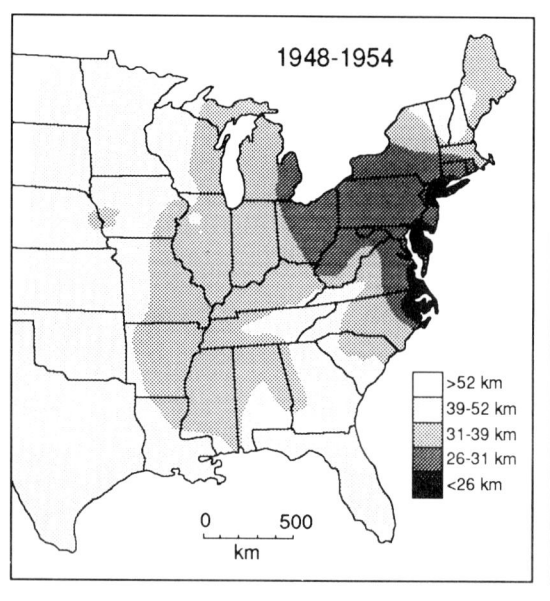

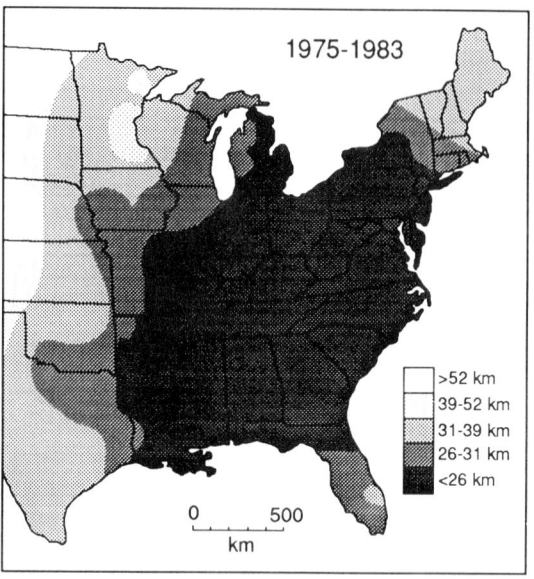

Figure 6.2 Visual range in the Eastern US is reduced by natural dusts and a variety of particles arising from combustion and other man-made sources but the major contributor to impairment of visibility is fine sulphate particles. Visitor surveys indicate that enjoyment of parks is impaired by decreased visual range. (Redrawn from NPS, 1988)

quantify the relationship between the exposure to the pollutant and economic or ecological impact. Fluorides present some problems in this respect but to a lesser extent than other pollutants; effects such as fluorosis in a dairy herd and the condition known as soft suture in peach fruit are readily quantifiable in economic terms.

The final contribution to this success story was the fact that there was an engineering solution to the problem in that it was technically possible to capture and retain over 99 per cent of the emitted fluorides on a day-to-day basis. This was achieved with no apparent deleterious consequences such as an increase in the production of other pollutants.

6.3 Diffuse sources of air pollution

However, in recent years it has become clear that there are pollutants that present more intractable problems. Where a pollutant arises from enormous numbers of sources (see Figure 6.2) such as motor vehicles, the responsibility lies not with one readily identifiable factory but is spread across countless individuals; consequently, control and policing become a massive, intricate undertaking. The problem is compounded when the pollutant is not local but when it travels many hundreds of kilometres, perhaps across national boundaries. Almost everywhere in Europe receives pollutants from other countries, so the legislator is faced not only with difficult legal questions, but also with the problem of determining at each location, the contribution from each country. To make matters worse, there are several pollutants where the effects may take decades to develop so that the link between cause and effect is obscure. There is also a problem where the effects are very difficult to quantify so the environmental or economic cost becomes highly debatable. In such a situation the all important cost-effectiveness of abatement becomes very difficult to assess.

Patterns of air pollution

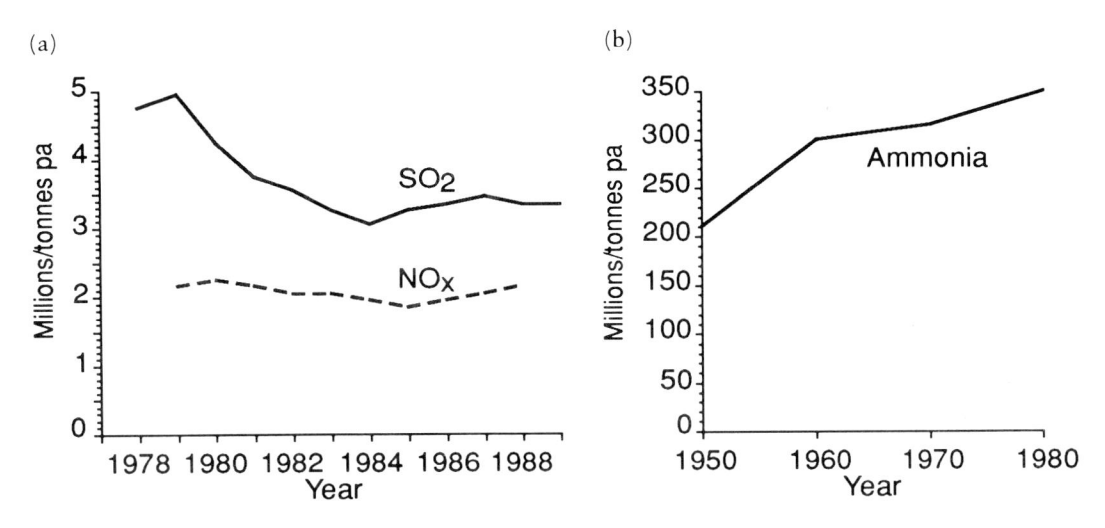

Figure 6.3a and b The nature of air pollution is constantly changing. Sulphur dioxide emissions in Britain have decreased from a peak of over 6 to well under 4 million tonnes per annum but nitrogen oxides are increasing (a) and ammonia emissions continue to rise (b). (Redrawn from UKBERG, 1989; UKPORG, 1990; Kurse *et al.*, 1989)

Two pollutants that fall into this apparently intractable category are the 'acidifying gases' and ozone. They provide a good illustration of the complexity of the problems facing scientists and legislators, and of the way international efforts are being made to find a rational, politically acceptable basis for abatement.

6.3.1 The acidifying gases

The gases that lead to acidification of soil, rivers and lakes are sulphur dioxide (SO_2), nitrogen oxides (NO_x) and ammonia (NH_3). Sulphur dioxide arises mostly from the combustion of the sulphur-containing fossil fuels, coal and oil. The nitrogen oxides are formed during combustion in engines and furnaces, when some of the relatively inert nitrogen gas in the air combines with oxygen. Because they are gases SO_2 and NO_x are taken up by plants, and may have an effect on growth or yield if the concentrations are high enough. In the first half of this century SO_2, probably in combination with NO_x, caused damage to crops and trees in many parts of industrialized Europe. Scots pine, for example, could not be grown successfully in British industrial cities until some time around the 1960s, whilst several species of grass that grow in urban environments evolved resistance to the pollutants. However, concentrations of SO_2 in Britain have been declining over the last decade (Figure 6.3(a)), and the nitrogen oxides are of comparatively low toxicity, so at present these two gases probably cause relatively little damage to vegetation in Britain as a whole. Effects on human health are confined to specific industrial locations and to occasional episodes in major city centres. In other parts of Europe, such as Germany, the situation is similar but in some countries there are still considerable problems with the acidic gases. Regions of eastern Europe where there are power plants that burn large amounts of brown coal pose a large-scale problem of SO_2 pollution while cities such as Athens have a high density of motor vehicles and a topography that traps the pollutants within the confines of the conurbation. This causes episodes of very high NO_x and other vehicular pollutants.

113

Ammonia, on the other hand (Figure 6.3(b)), is released into the air during the decomposition of animal wastes such as manure. Some emanates from the soil and fertilizers. There is some local escape from industrial processes but it is essentially a rural pollutant (Kruse, ApSimon and Bell, 1989). It is taken up by plants and it has caused visible injury near industrial sources and in places such as the Netherlands where there is very intensive livestock farming, but for the most part in its gaseous form concentrations of this pollutant are so low that it is probably of little importance to vegetation and has no health effects on humans. On the other hand, its very characteristic smell has meant that it has been the subject of numerous complaints as a nuisance.

All three gases are carried from their sources by the wind. Turbulence and diffusion dilute the gases and they impact with surfaces such as leaves, masonry, raindrops and the soil so the concentration in the air is constantly decreasing. But they also undergo chemical and physical changes, the gases becoming transformed to small particles suspended in the air; SO_2 is oxidized to sulphates, NO_x to nitrates and ammonia is rapidly changed to ammonium. Eventually this complex mixture of sulphur and nitrogen compounds is deposited by impaction and by being scavenged by rain and mist. Once deposited, they generate acidity so they may cause changes in soil and water chemistry with deleterious effects on plants and animals. In addition, the nitrogen oxides and ammonium provide nitrogen, an essential nutrient, in a form that can be used by plants so they may stimulate the growth of some species. The nitrogen may also be directly deleterious to other plants or alter the competitive balance between species. Increasing the nitrogen content of plants may alter frost hardiness or plant–herbivore relationships so the effects of deposition of nitrogen-containing compounds are potentially subtle and far-reaching.

The relative contribution of sulphur and nitrogen compounds to acid deposition is constantly changing. In the past, sulphur has been the major acidifying source in Britain but it has declined in importance and currently the nitrogen-containing compounds predominate. Furthermore, there is increasing evidence that ammonium compounds are of great importance as acidifying agents in many parts of Britain. This has shifted the focus for the first time towards farming as a source of air pollution comparable in its effects with urban–industrial sources.

Although acid deposition has been researched for almost a century (Cohen and Ruston, 1925), until about 20 years ago it was thought that little of the sulphur and nitrogen was transported over long distances, and that the effects were essentially local. Then reports of acidification of Scandinavian and North American lakes and rivers, with loss of fish populations, led to claims that industrialized countries were damaging rural ecosystems many hundreds of kilometres distant. In Britain these claims were contested for a time by the Central Electricity Generating Board. However, multinational research programmes involving tracking and chemical analysis of air masses, the use of inert marker chemicals to record the trajectories of plumes, and monitoring of deposition on a European-wide scale proved that there is significant long-distance transport of pollutants across international boundaries (Figure 6.4; see also Figure 2.1(b)). The outcome of the Chernobyl disaster brought home to scientists and laymen the scale on which transport can occur and it caused a reassessment of models used to calculate deposition in upland situations. Every country in Europe exchanges acidifying pollutants on a massive scale. This has been graphically demonstrated in Norway where it has been shown that snow layers that are formed when air masses have passed over industrial Europe not only contain acidifying materials but also residues like fly ash and DDT (Elgmork, Hagen and Langland, 1973; Hagen and Langland, 1973).

Costing the effects of this acid deposition in economic, social or conservation terms is complex and potentially controversial. There is no doubt that lakes and rivers have been

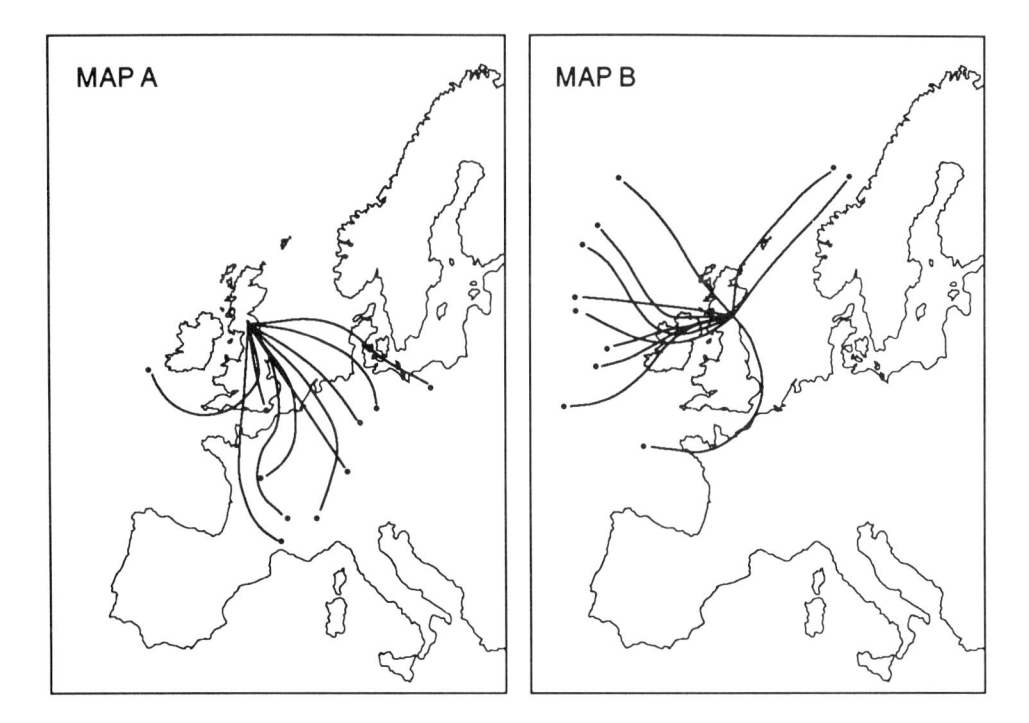

Figure 6.4 Acid-forming air pollutants are transported over very large distances. The acidity (pH) of rain falling at Bush, Scotland, is related to the path of the air mass before it arrives at the site. Acid rain (pH < 4.5) is associated with air that has passed over industrial Europe (Map A). (Redrawn from Fowler and Cape, source to come)

acidified and effectively sterilized (Figures 6.5(a) and (b)) in western Britain, Scandinavia, North America and other countries. A fall in river pH and loss of game fish are relatively easily recorded and costed but in some cases the effects may be more subtle. In Wales, for example, it has been noted that there are fewer dippers (small diving and swimming birds, *Cinclus cinclus*) associated with acidic waters and there is a documented case of the population falling when the acidity increased. Eggshell thickness and weights of eggs are both lower at low pH (Ormerod and Wade, 1990). It is considered that the distribution of dippers relates to food availability and the mineral content of the food. The main prey, insects, molluscs and fish are less abundant in acid waters and furthermore these are also the source of the calcium that is used for eggshells. The clear implication is that acidification of rivers will lead to a decrease in dipper populations. Is it possible to provide a framework for abatement that will take into account the intangible environmental cost of changes in populations of dippers?

Building materials are subject to natural erosion or decay under the influence of chemicals such as carbon dioxide and chlorides but air pollutants accelerate the process (UKBERG, 1989; Butlin, 1991). Granites and some sandstones are relatively resistant to attack by acidic pollutants but the calcareous materials, limestone, marble and calcareous sandstones are very vulnerable. For some specific buildings it is possible to place a monetary value on this in terms of the cost of repair but what is needed as a basis for a rational abatement policy is knowledge of the relationship between decay on the one hand and concentrations of the pollutants and the influence of the weather

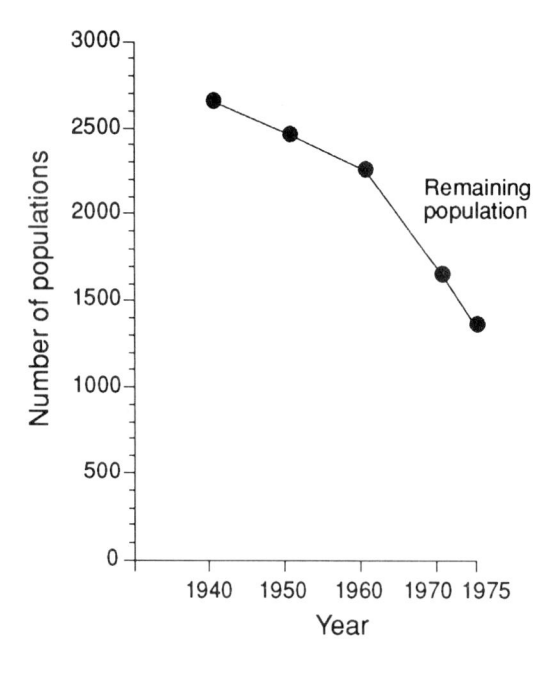

Figure 6.5a The number of species of snail in lakes and rivers is related to acidity (pH) as these data from a survey of 1000 Norwegian lakes show. If acidity increases the number of invertebrate species would be expected to fall. (Redrawn from Environment Committee, 1984)

Figure 6.5b There has been a drastic decline in populations of fish in areas affected by acid deposition, as these data for trout in southern Norway show. (Redrawn from Environment Committee, 1984)

on the other (Butlin, 1991). Such mathematical models are still in the process of development but when they are available it will be possible to judge the effect of abatement against costs. Cathedrals, statues and historical edifices like the Palace of Westminster are affected by acid deposition. The present architect at Cologne Cathedral made a telling point when he commented to the House of Commons Environment Committee (Environment Committee, 1984) that although £1.5 million was being spent each year to replace damaged stone, what he was overseeing was the replacement of a medieval monument. The result is a copy. Few would disagree with the Department of Energy statement on this subject that 'the cultural value . . . cannot be expressed in economic terms' (Environment Committee, 1984).

Over the last decade the layman might have gained the impression from the press that the effects of acid deposition on forests, natural vegetation and crops is well known and understood. However, intensive research over more than 15 years has demonstrated the difficulty faced by all field ecologists: that of assigning simple causes to complex effects. Forest decline is caused not by one factor but has complex multiple causes that include the gaseous pollutants, SO_2, NO_x and O_3, ammonium and modified soil chemistry (Schulze and Freer-Smith, 1991). The weather, particularly dry seasons and hard winters, plays a role, especially in synchronizing damage — and adding thereby to the sense of crisis. It is clear that there are no simple dose–response equations available like those developed for fluoride. Predicting the benefits of reducing acid inputs by a given degree is still being developed through the use of models.

Some of the problems of determining the effects of pollution on natural ecosystems are caused by the fact that there are few long-term records of natural or semi-natural vegetation that were set up in such a way that it is possible to determine if deposited acidity or nitrogen is responsible for any observed changes. Long-term surveillance is not usually seen as exciting science or being immediately relevant yet without that database it is impossible to assess the causes underlying trends.

6.3.2 Ozone

Since media interest developed in stratospheric ozone, and perhaps because of the Victorian myth that ozone is good for human health, many people are confused when they are told that ozone is extremely reactive and potentially toxic to humans and plants. Some tropospheric ozone originates in the stratosphere but it is mostly produced from the action of sunlight on the nitrogen oxides and hydrocarbons that are emitted in fossil fuel exhaust gases (UKPORG, 1987). The importance and toxicity of ozone were first recognized in Los Angeles where the high density of vehicles, the sunshine and the topography lead to high concentrations of a soup of photo-oxidant gases (see Figure 2.5(a)). As the reactions are driven by sunlight and as ozone is also destroyed by chemicals in the atmosphere, concentrations often show diurnal cycles, peaking mid-afternoon and falling as the sun goes down. As it takes time for the ozone to be formed, peak concentrations are found not in the cities but many kilometres from the source of the precursors. Typically, potentially damaging concentrations may be tens to hundreds of kilometres downwind of a city, over rural land or wilderness (see Figure 6.6(a) and (b)). In Britain, the highest summer peaks are found near the south coast, partly due to import of ozone from France, but some of the highest long-term mean concentrations are found on the high hilltops of the Pennines. Because summer sunshine is so variable in Britain the ozone tends to occur in episodes and to vary greatly from year to year.

Once it is inhaled by an animal, taken up by a plant or when it impacts with a surface the reactivity of the gas results in the destruction of the O_3 molecule; it does not persist or accumulate in ecosystems, in contrast with sulphur and nitrogen. Consequently, neither the geographical pattern of dispersal of ozone nor its effects can be estimated by analyzing residues in vegetation, soil or water. Environmental concern is with the concentrations of gas in the air and the duration of episodes rather than the build-up of residues in ecosystems.

The effects of ozone on humans are well known: breathing difficulties, especially in individuals with respiratory complaints, lachrymation and headaches. There is increasing concern about the less obvious, long-term effects of chronic exposure on health. It is potentially toxic to many species of conifer, herbaceous plants and crops at concentrations not far above the natural background level. Concern over the effects of ozone in the United States stimulated a large, integrated study with the objective of providing a credible evaluation of the economic effects of ambient O_3 on US agriculture (Wilhour, 1988). This was done by using experimental data to produce models relating crop yield to O_3 dose and defining the air quality in the agricultural regions. These two studies were then integrated to produce economic predictions (Adams, Glyer and McCarl, 1988) which could be used as a basis for policy decisions (Jordan *et al.*, 1988). Assessing the economic implications is challenging in itself but Adams *et al.* (1988) indicated that a reduction of 1981–3 ambient O_3 concentrations by 25 per cent would result in benefits from increases in yields of eight crops of $1900 million (at 1982 values). This amounted to about 1.9 per cent of annual agricultural revenues.

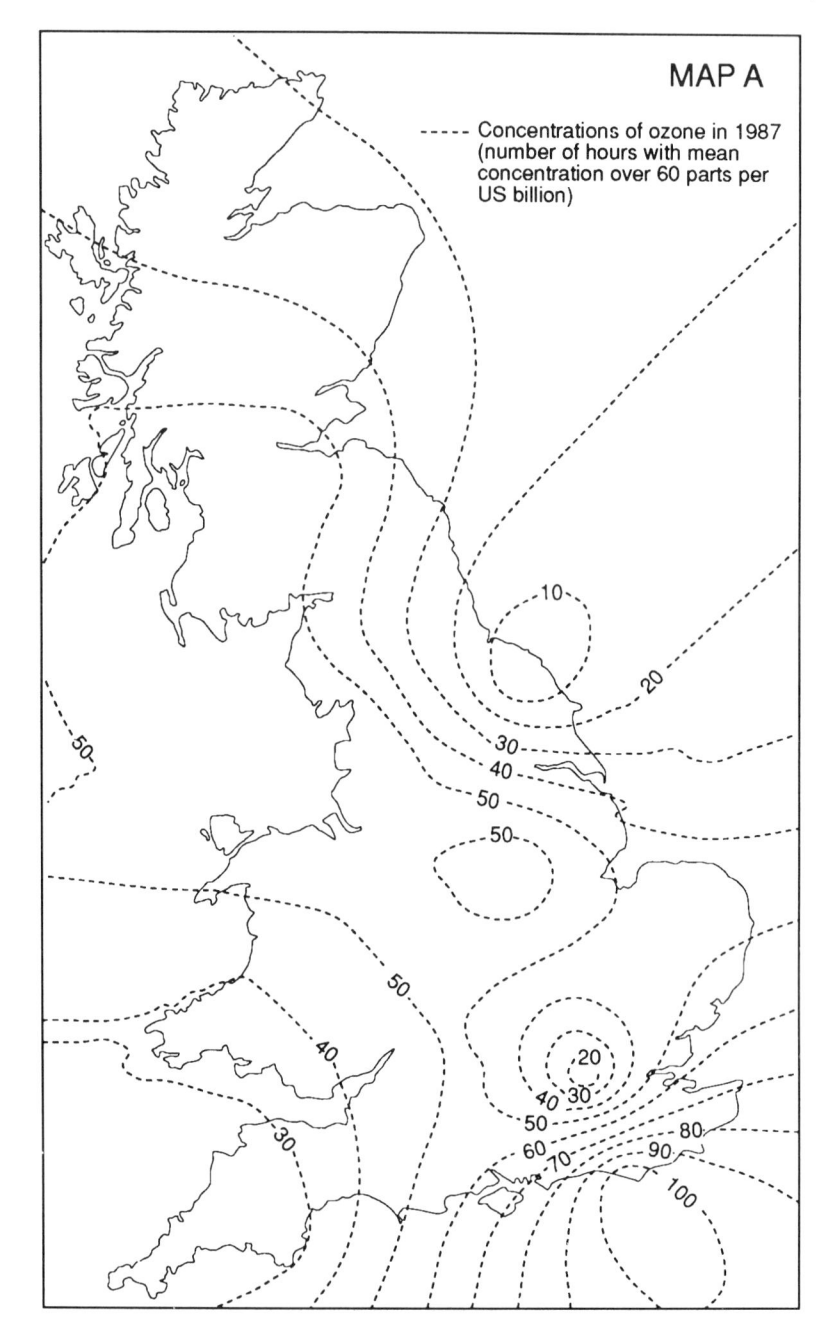

Figure 6.6a and b Concentrations of ozone in Britain expressed as the number of hours with a mean concentration over 60 parts per US billion. Map A is for 1987 a dull, relatively sunless year and Map B is for 1990 when ozone concentrations were much higher. The gradient in concentration out from London is due partly to the time taken for ozone to form and partly to the fact that the precursor gases also destroy ozone. The maps were generated from monitoring sites operated or collated by Warren Spring Laboratory using SURFER software and kriging for the interpolation.

Patterns of air pollution

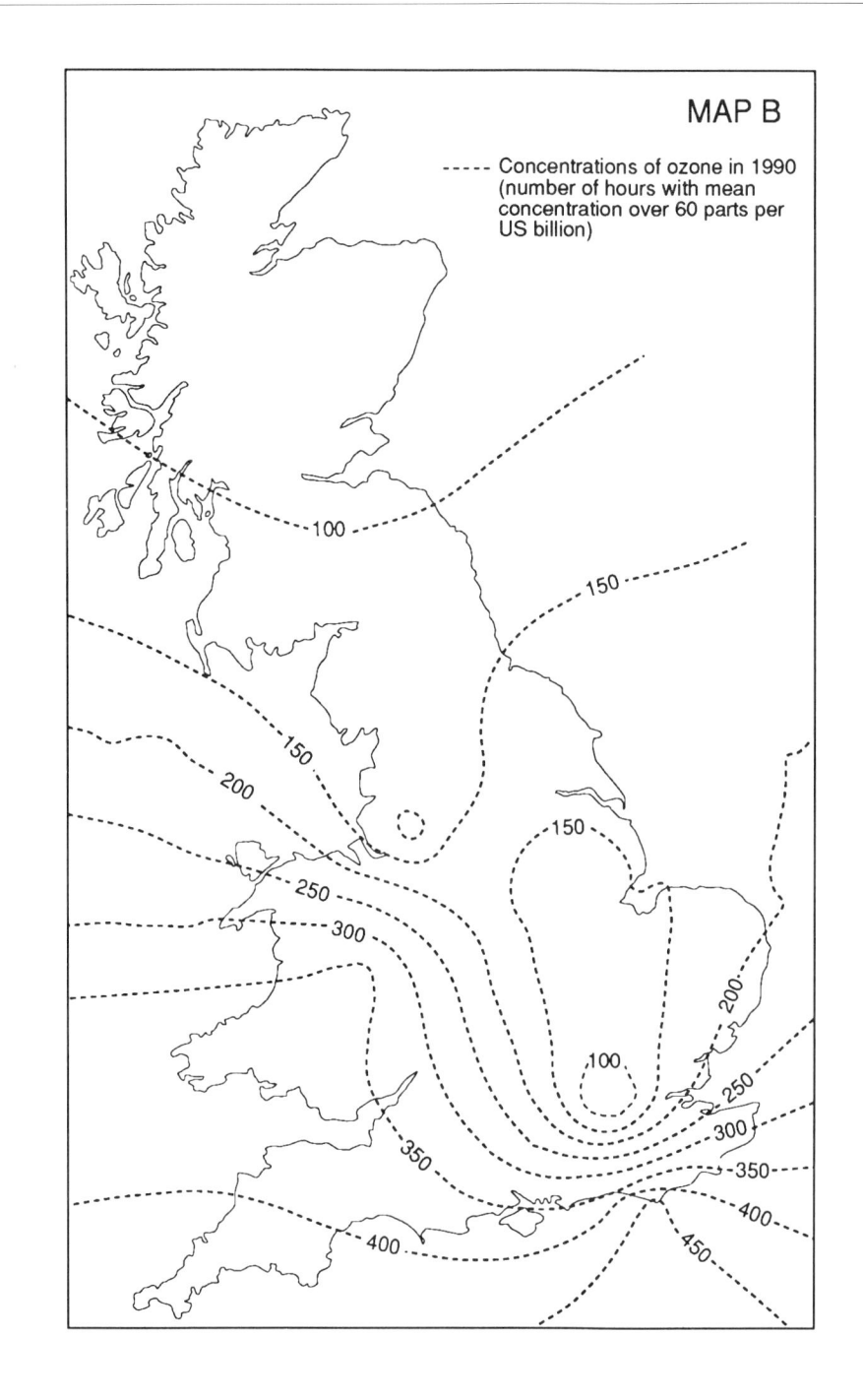

MAP B

----- Concentrations of ozone in 1990 (number of hours with mean concentration over 60 parts per US billion)

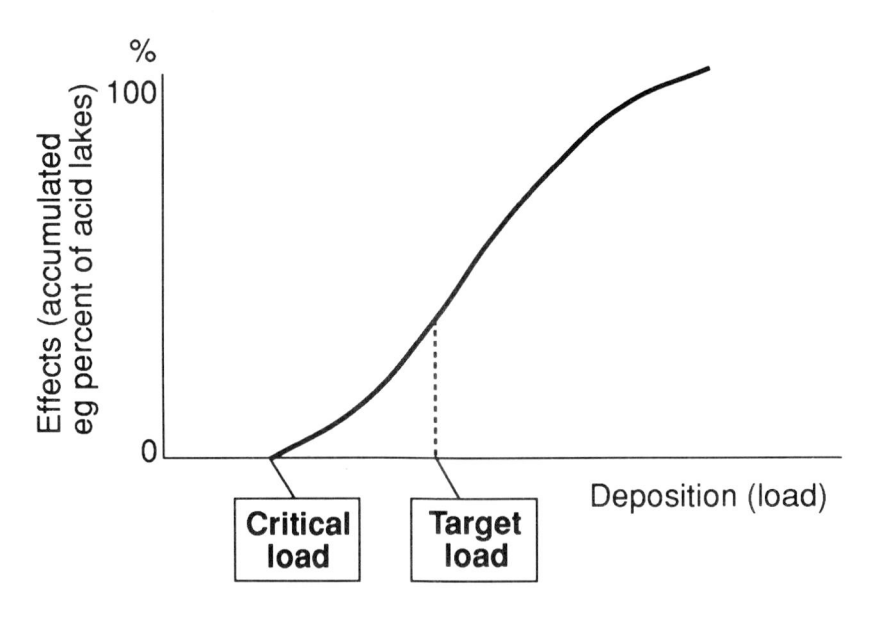

Figure 6.7a Conceptual model of the critical load or level. A receptor, whether it is a living organism, soil or building material is envisaged as being able to withstand a certain degree of deposition or a certain concentration of a pollutant in the air without harmful effect. Above the critical load of deposition or level (concentration in air), there are progressively greater effects. The target value is determined on the basis of the balance between the constraints of technological feasibility, economics and conservation. (Redrawn from Bull, 1991)

However, there has been no similar study for crops in Britain or elsewhere in Europe (Mathy, 1988) and it is not possible to extrapolate from the US study to the European situation (Wilson and Sinfield, 1988). The importance of ozone was recognized much later in Europe than in North America so there has been less research, even on the more important crops. There has still been virtually no research on native plant species of either the United States or Europe, other than a few species of tree. Until recently there have been very few ozone monitors operational so the concentrations and frequency of episodes in Britain and several other countries have been very poorly catalogued. With this inadequate base it is presently impossible to quantify effects on crops or to assess the importance of the pollutant in conservation terms, yet potentially it may be having very serious effects, so attempts at abatement have to start without a firm knowledge of the effects.

6.4 Abatement strategies on an international scale

Clearly, abatement of these two groups of pollutants is a major challenge because of the economic and political implications, so solutions have to have a rational basis. They have also to be cost-effective and be agreed within an international framework.

Currently, under the aegis of the United Nations Economic Commission for Europe (UN-ECE) Convention on Long Range Transboundary Air Pollution (LRTAP), the critical loads and critical levels approach is being considered as a practical basis for future protocols for the control of emissions of sulphur oxides, nitrogen oxides and other pollutants. Under the UN-ECE convention, protocols to reduce emissions are drawn up for a determined period and then they are revised in the light of progress and reassessment of the problems. The present protocol for sulphur dioxide was adopted in 1985 and

CRITICAL LEVELS/LOADS AS A BASIS FOR AIR POLLUTION ABATEMENT

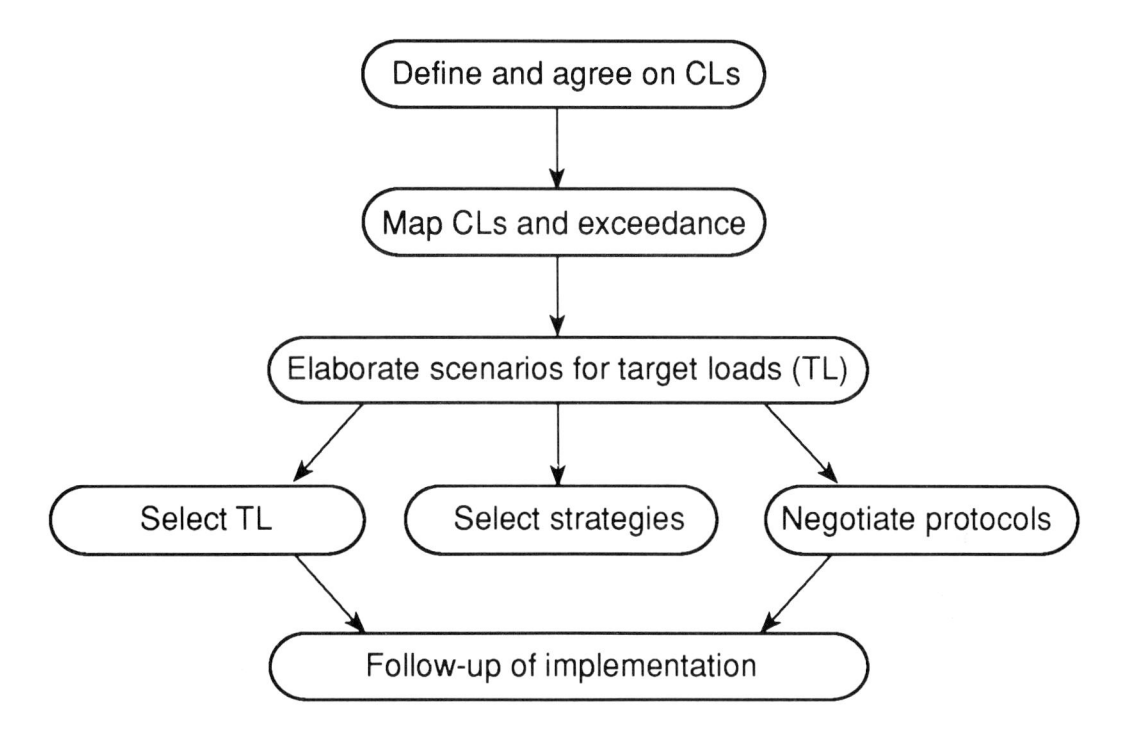

Figure 6.7b Using the critical loads/levels approach to abatement. Steps envisaged by the UN-ECE from the estimation of critical values through the optimization of abatement strategies to implementation.

it is due for revision in 1993. It proposed a reduction of sulphur emissions or their transboundary fluxes by at least 30 per cent from 1980 levels (Bull, 1991). From the start it was recognized that this kind of flat-rate reduction was not satisfactory because it was not known how effective it would be or whether there would be any improvement in the regions where it was most needed. Would a reduction of 30 per cent of Britain's sulphur emissions have any discernible effects on catchments in southern Norway or southwest Scotland? How big would any effects be and how might the benefit be estimated? Which sources would be the most effective to reduce, the cheapest ones or those in the most strategic positions? What were the environmental costs of the methods used to reduce emissions? The flat-rate approach does not have a scientific basis and insufficient was known in 1985 about the effects of the pollutants and of the relationships between rates of emission and rates of deposition/concentrations at the targets.

A little more than a decade ago the term critical load was first used and subsequently it has been developed into a concept that has been adopted by the UN-ECE to form the basis for new protocols. The term is defined in slightly different ways by different authors but a version of that used by Bull (1991) is one of the simplest: 'the critical load is exceeded when the rate of deposition of a pollutant causes harmful effects on a receptor.' The concept resembles that of toxicological limits but at an ecosystem scale – see Figure 2.4. Figure 6.7(a) defines the concept whilst Figure 6.7(b) charts its application, described below.

The critical level followed from the acceptance of the critical loads concept. It covers gaseous pollutants such as ozone; the critical level refers to the concentration or dose (time-weighted concentration) of a gas that has harmful effects on a receptor.

In this simple form the definition of a critical load or level (load or level = value) does not attempt to identify what is meant by 'harmful' and it does not depend on the perceived economic, social or ecological importance of the receptor. Consequently, several variants of the definitions have been introduced for use in calculating values for particular receptors such as soils or fresh water, and there is an ongoing debate and development of the definitions.

A development from the critical load is the target load, which has been defined (Henrikson and Brakke, 1988, in Bull, 1991) as 'the load determined by political agreement'. In his discussion Bull (1991) summed up the role of the target value when he pointed out that it '. . . enables decisions to be made on scientific and social priorities as well as economic constraints'. A target may be set above the critical value when economic or technical considerations are considered of greater importance, and it may be set below to give extra protection to a sensitive receptor when the critical value is uncertain. The target value is not fixed but may be changed in the light of new technology or economic circumstances. Where critical values are already exceeded a flexible target value gives a series of goals to work towards using a timetabled set of control strategies. It is the intention that critical values, maps of the geographical distribution of sensitive receptors and maps of deposition loads or pollutant concentrations will be used to develop target values 'in the light of possible legal, technical, economic and political concerns'. The target values will then be used as the basis for negotiation with regard to internationally accepted targets and emission reduction strategies.

Maps of emission sources, of deposition loads (or concentrations), of sensitive receptors, and of areas where critical values are exceeded will be used in models that will calculate the reduction in emissions required to reduce deposition (or concentrations) below the target values. Complex models will allow the investigaton of the cost-effectiveness of different abatement strategies.

6.4.1 Critical loads: progress and problems

Working groups and task forces within the LRTAP Convention are currently working to produce maps and models that will be used in the revision of emission protocols. There is work in progress refining models of the emission–transport–deposition process so that better maps can be produced of the deposition load on different countries. Numerical estimates of critical loads for different receptors have been calculated using methods that are specified in a draft manual of methodologies (UN-ECE, 1990). The methods vary in the assumptions that they make and the range of input information that they need but most are incorporated into computer-based models of varying degrees of sophistication. Despite differences in method and definition there is a good level of agreement between the estimates that have been produced for critical loads of sulphur and nitrogen in soils (Table 6.1), groundwater and some ecosystems. Initial critical load maps are published and maps are also available of the sensitivity of soils and water to deposited acidity so that areas where critical loads have already been exceeded can be identified (Hettelingh, Downing and de Smet, 1991) (Figures 6.8(a) and (b)). Progress has been so rapid that critical loads have already played an important part in a major planning enquiry (Pembroke power station).

However, some more intractable problems are still being solved, particularly in relation to the nitrogenous pollutants. Nitrogen is a special case because of the fact that it not only acidifies but it is also a plant nutrient that can upset the ecological balance of

Table 6.1 Comparison of critical loads for deposition nitrogen, estimated by different methods

Method	Load kg N ha y-1
1 Nitrogen productivity	15–35
2 Input/output studies	10–15
3 Net uptake by plants	
low–medium productivity	5–15
high productivity	20–45
4 Fertilization experiments	15–25
5 Vegetation changes	10–30
6 Empirical data	>15

Source: Bull (1991)

communities that are nitrogen-limited. Determining the critical load for nitrogen as a nutrient is problematical. The UN-ECE mapping manual refers to two methods, the empirical approach and the mass–balance approach. The empirical approach that was used to produce the first estimates for selected natural ecosystems (Nilsson and Grennfelt, 1988) consisted of identifying case studies where the the best available evidence indicated that the critical load was known to have been exceeded, that is, where the flora or fauna were known to have changed in relation to deposited nitrogen. Some cases were observations of plant communities made over a number of years, others were experiments where nutrients had been added. Rates of deposition were estimated with varying degrees of accuracy and the critical loads were estimated from them.

There are two main limitations of this approach. The first is one mentioned already and that is the perennial ecological problem of assigning changes in flora and fauna to specific causes. Whilst the available evidence pointed to the changes in the case-studies being related to nitrogen deposition, it is unwise to accept such a correlation as establishing a cause–effect relationship without experimental verification. The second limitation is the estimation of the rates of deposition. In none of the case-studies were there data on rates of deposition available for the actual sites over the time period of observation, so the estimates have a large potential error. Bearing these in mind, it is surprising that there is such a good measure of agreement between the methods.

The second method that the mapping manual offers to predict the biological effects of nitrogen is based on a mass–balance. The manual states that it is suitable for managed forests. With this method, the critical load for nitrogen as a nutrient is assumed to be the balance between the yearly mean net uptake of nitrogen in the tree biomass, the long-term net amount immobilized in the soil and the leaching. Any deposition that upsets that balance point is held to be excessive and likely to cause changes in the ecosystem. Despite the fact that the method assumes that there is no biological nitrogen fixation and negligible de-nitrification (assumptions that many biologists would find hard to accept), two recent studies indicate that the method does give useful estimates for some managed forests. Of course, very few forests in Britain are suitable for this approach and forest is a minor ecosystem in terms of the land area of the country. A mass–balance approach has not been suggested for grassland, heath or bog; at present the only approach that can be used with those ecosystems is the empirical one. In view of the limitations of the latter method, there is a strong case for determining critical loads for these other ecosystems by means of experimental addition of nitrogen to plots over a reasonably long time-scale.

Neither of these methods takes into account the fact that plant species react differently to the two different forms of nitrogen that are deposited, nitrate and ammonium. Some species are better able to make use of one form than the other. It is not known whether this means that the biological effects of the same load of nitrogen will differ depending on the balance between nitrate and ammonium. This question has to be answered before critical loads for the biological effects of nitrogen can be safely accepted.

A final question remains on the effects of

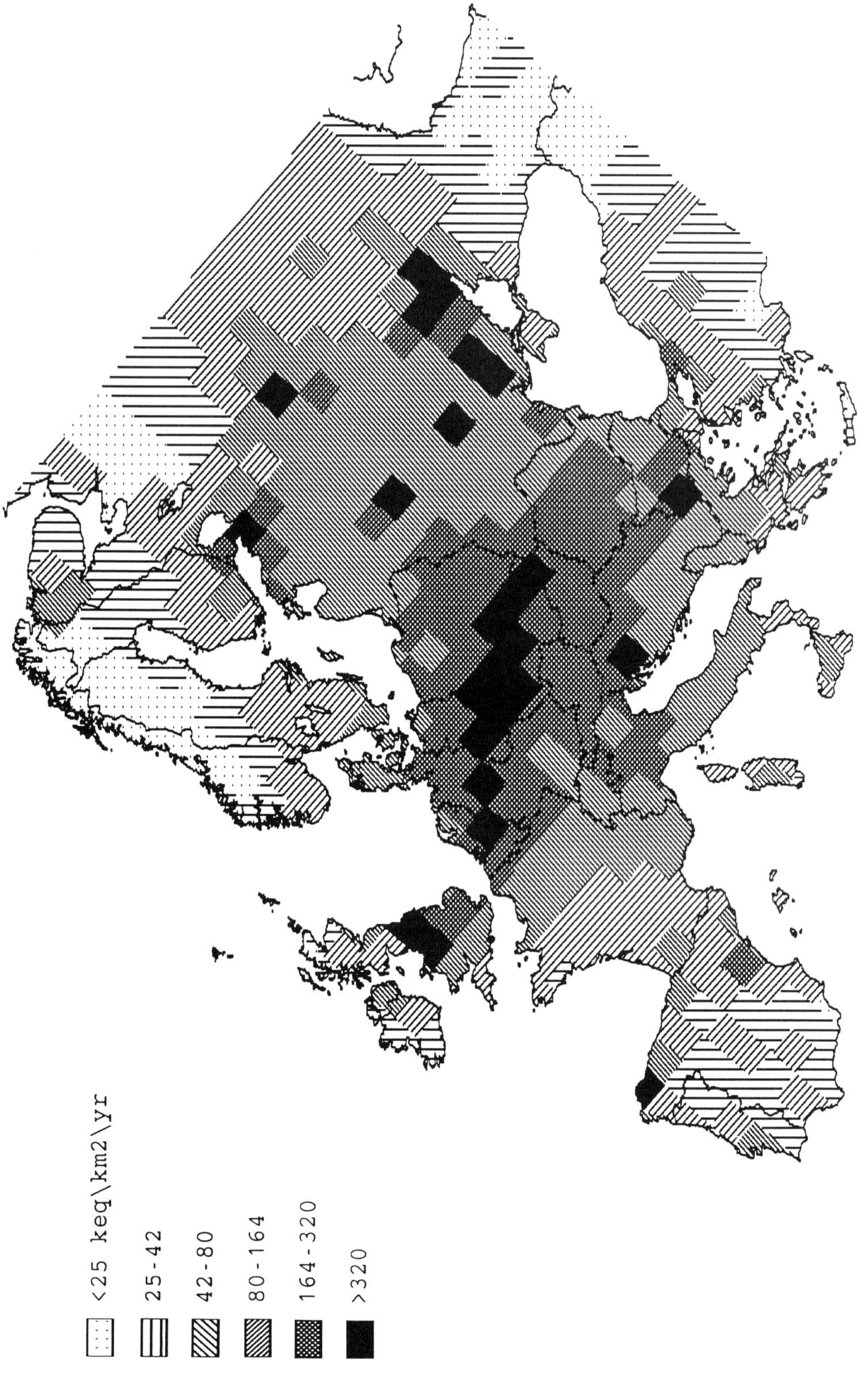

<25 keq\km2\yr
25-42
42-80
80-164
164-320
>320

Patterns of air pollution

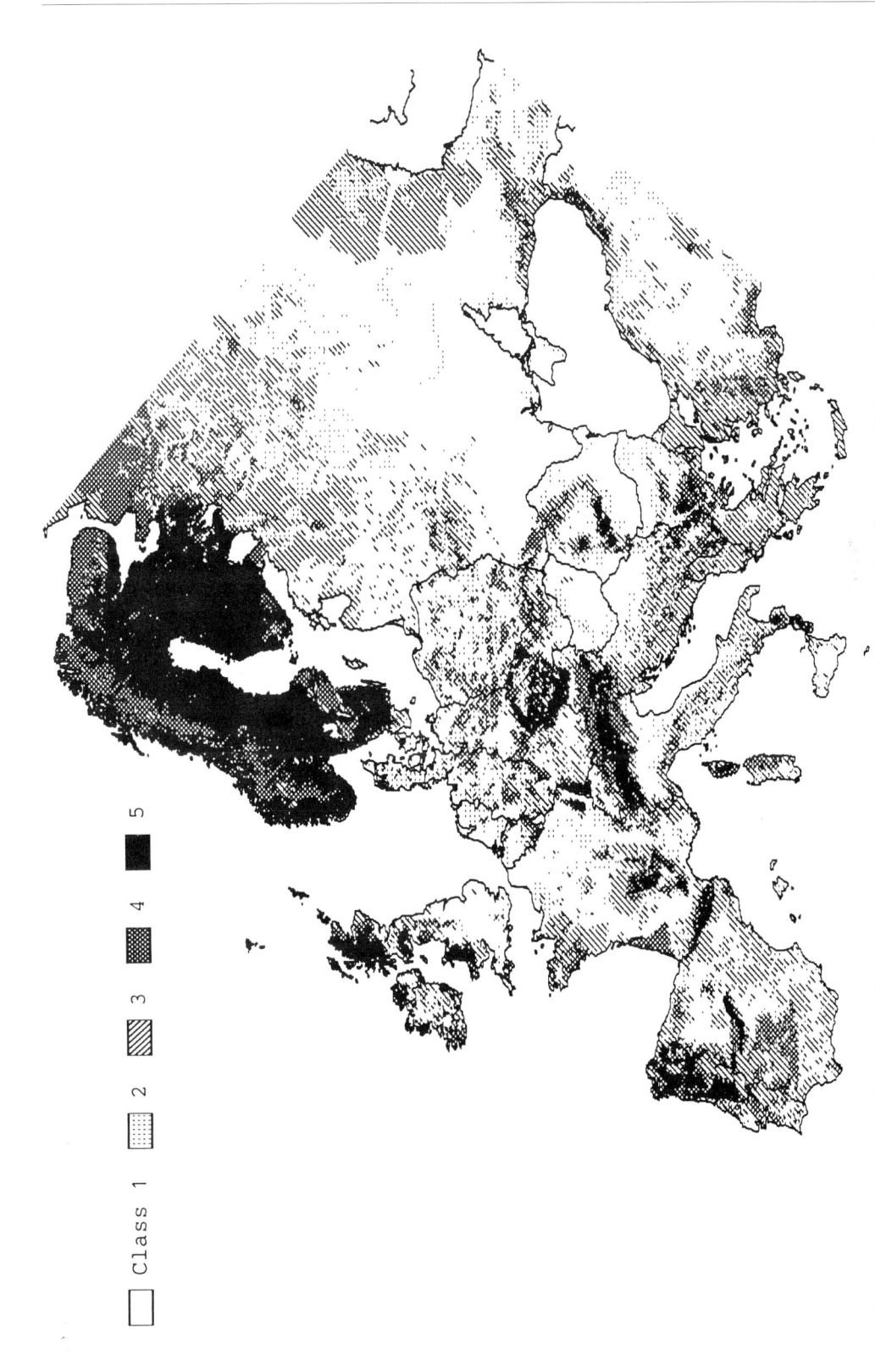

Figure 6.8 (a) Estimated amount of sulphur deposited on Europe, 1985, and (b) the relative sensitivity of each grid square to acid deposition. Sensitivity in this model used by Chadwick & Kuylenstierna, is based on rock and soil type, land use and rainfall. (Redrawn from Chadwick & Kuylenstierna, 1990)

nitrogen as a nutrient. Almost all of the work so far has been with ecosystems in which the soil and flora are nitrogen-limited. That is, nitrogen is in such short supply that it limits ecosystem production or the growth of at least some of the species. But many ecosystems or many of the species in them are not greatly nitrogen-limited. Chalk grassland, for example, a botanically rich ecosystem that covers very large areas of Europe, tends to be limited primarily by phosphorus supply and by drought. What happens when there is a high rate of nitrogen deposition on chalk grassland? Do some species respond or none at all? Does the nitrogen change plant chemistry in such a way as to alter plant–herbivore relations? It is difficult to conceive a method for calculating critical loads for such a system. Direct experimentation would seem to be the only way.

6.4.2 Critical levels; challenging problems ahead

The basic assumption underlying the critical level concept is that it is possible to determine the relationship between the degree of exposure to the pollutant and the response of an organism. The response can be measured in a number of ways such as the appearance of visible lesions, changes in growth or biochemical perturbations, but the degree of exposure is much more problematical. An index is needed that summarizes the combination of concentration and duration of exposure in a way that has a sound biological basis (Hogsett, Tingey and Lee, 1988). This has been the subject of intense debate for many years and there is still no universal agreement on the best approach. The difficulty is caused by the fact that, for example, ozone concentrations vary greatly with time while plant response to the ozone varies with concentration, age, the weather, the degree of water stress and so on. It is usually considered necessary to establish two types of exposure index, one that will provide a measure of the effects of high concentrations

Table 6.2 Critical levels for ozone when present as a single pollutant, estimated from dose–response of sensitive species

Exposure duration (hs)	Concentration (parts per US billion)
0.5	150
1.0	75
2.0	55
4.0	40
6.0	30
Growing period mean (average of 7-hr mean between 09.00–16.00 hr)	25

Note: The levels represent the combinations of concentration and duration of exposure that are considered to be thresholds for effects on sensitive species.
Source: Data from Bull (1991)

over short periods, and one that provides a measure for the converse, chronic long-term exposure. Often, data are used for the part of the day when the ozone is elevated and plants take up the gas at the highest rates. Exposure indices are usually based on the mean maximum concentration over short time periods, the cumulative number of hours above a set threshold or the mean concentration over a defined 'growing season'.

The first estimates of critical levels for ozone were made in 1988 to give a starting point for debate (UN-ECE, 1988) (Table 6.2). These initial proposals were based on a review of published work and on air quality guidelines suggested by previous workers. The method used to quantify the critical levels was to determine combinations of concentration and duration of exposure that were just below the lowest combinations that had previously been reported as thresholds for effects on the most sensitive species. This is the same principle that was adopted for fluoride and other pollutants; protect the most sensitive receptor and all others will automatically be protected. The proposed levels were tentative because the working group recognized several of the shortcomings in the available data. In particular, it was recognized that when other pollutants are present (particularly SO_2) the combined effects of the mixture may be more pronounced than those of ozone alone, so in those circumstances the critical level should be lower. However, one of the the proposed critical levels for ozone, the growing period mean, is already so close to the natural ambient concentration that to reduce it further would in some regions make it lower than this background level, a situation that would clearly be absurd. This difficult position is a consequence of the uncertainties about ozone dose–response relationships and about the most appropriate exposure indices to use when the duration of exposure is very long-term.

The concept of there being a definable 'growing period' is a hangover from research that has been dominated by annual crops and it ignores the fact that in many parts of Europe elevated ozone concentrations occur at all times of the year and that in most ecosystems there are some plants that remain active and sensitive throughout the year. In Mediterranean countries most of the flora is dormant throughout the summer and starts active growth from the onset of the wet season, usually in October. Even in Britain, grasses, clover and mosses remain active and will grow during mild weather from autumn through the winter and into April, as anyone with a lawn will know! The 'growing period' concept needs revision or to be abandoned completely.

The report proposed that more research was needed on the effects on plant communities of low concentrations of ozone over prolonged periods. This was recognition of the fact that almost all of the work on which the critical levels were based was on crops grown in monoculture during the summer months. A few conifers have been investigated but almost all of the work has been relatively short-term and has used young trees that may be physiologically different from mature specimens. Furthermore much of the equipment that is used to expose the plants to the ozone also alters the plant environment (temperature, humidity, light, etc.), sometimes in subtle ways, so interpretation of dose–response relationships even in the case of well-researched crops is difficult. There are relatively few studies where plants have been exposed to ozone for longer than a single summer, but in nature plants are exposed for longer periods and they are subjected to the stresses of competition, winter frosts, drought, pests and disease. All of these may modify the effects of the pollutant so it is a very large step to use the results of the existing published experiments for the definition of critical levels for the protection of natural ecosystems.

The shortcomings of the proposed critical levels are illustrated by comparison with ozone concentrations recorded in the United Kingdom in the last few years (see Figure 6.6) because at least one of the proposed combinations of concentration and exposure time has been exceeded everywhere in Britain, including the remote hills and northern Scotland. This does not mean that our wilderness areas and National Parks are necessarily at risk from ozone; it is more an indication of how much we still need to learn about ozone before we can define reliable critical levels. The way forward is undoubtedly through more, carefully targeted research and an iterative process of discussion and revision.

6.5 Conclusions

The critical loads/levels concept has developed into an all-embracing, integrated approach to the problem of abatement of transboundary pollutants. It has a sounder scientific basis and should lead to more cost-effective control than the previous, flat-rate approach that led to the '30 per cent club'. In regions where the critical value is already exceeded, the definition of target values will allow a phased, stepwise progression towards critical values as technical or economic restrictions permit. International efforts under the aegis of the LRTAP

Convention of the UN-ECE have led to considerable progress in the process that will lead from the definition of critical values, through mapping, modelling, optimization of control strategy and technical improvements in emission control to the most cost-effective schemes for abatement. Challenging problems remain, particularly in relation to the biological effects of deposited nitrogen on terrestrial systems and in defining critical levels of some gaseous pollutants such as ozone.

Acknowledgements

The author wishes to acknowledge valued comments from Dr K. Bull. The work on critical loads and levels in the United Kingdom, including that of the authors, is financed by the Department of the Environment.

References

Adams, R.M., Glyer, D.J. and McCarl, B.A. (1988), 'The NCLAN economic assessment: approach, findings and implications', in W.W. Heck, O.C. Taylor and D.T. Tingey (eds), *Assessment of Crop Loss from Pollutants*, Elsevier Science Publishers Ltd., Essex, 552 pp.

Bull, K.R. (1991), 'The critical loads/levels approach to gaseous pollutant emission control', *Environmental Pollution*, 69, 105–23.

Butlin, R.N. (1991), 'Effects of air pollutants on buildings and materials', *Proceedings of Royal Society of Edinburgh*, 97B, 255–72.

Chadwick, M.J. and Kuylenstierna, J.C.L. (1990), *The Relative Sensitivity of Ecosystems in Europe to Acid Deposition*, Stockholm Environment Institute at York University, York, 65 pp.

Cohen, J.B. and Ruston, A.G. (1925), *Smoke: A Study of Town Air*, Edward Arnold, London.

Elgmork, K., Hagen, A. and Langland, A. (1973), 'Polluted snow in Southern Norway during the winters 1968–71', *Environmental Pollution*, 4, 41–52.

Environment Committee (1984), *Fourth Report from the House of Commons Environment Committee, Acid Rain, Volume I*, HMSO, London, ISBN 0-10-008664-0, 86 pp.

Hagen, A. and Langland, A. (1973), 'Polluted snow in Southern Norway and the effects of the meltwater on freshwater and aquatic organisms', *Environmental Pollution*, 5, 45–57.

Hettelingh, J.-P., Downing, R.J. and de Smet, P.A.M. (1991), *Mapping Critical Loads for Europe*, CCE Technical Report No. 1. Co-ordination Centre for Effects, National Institute of Public Health and Environmental Protection, Bilthoven, Netherlands.

Hogsett, W.E., Tingey, D.T. and Lee, E.H. (1988), 'Ozone exposure indices: concepts for development and evaluation of their use', in W.W. Heck, O.C. Taylor and D.T. Tingey (eds), *Assessment of Crop Loss from Pollutants*, Elsevier Science Publishers Ltd., Essex, 552 pp.

Jordan, B.C., Basala, A.C., Johnson, P.M., Jones, M.H. and Madariaga, B. (1988), 'Policy implications from crop loss research: the US perspective', in W.W. Heck, O.C. Taylor and D.T. Tingey (eds), *Assessment of Crop Loss from Pollutants*, Elsevier Science Publishers Ltd., Essex, 552 pp.

Kruse, M., ApSimon, H.M. and Bell, J.N.B. (1989), 'Validity and uncertainty in the calculation of an emissions inventory for ammonia arising from agriculture in Great Britain', *Environmental Pollution*, 56, 237–57.

Mathy, P. (1988), 'The European open-top chamber programme: objectives and implementation', in W.W. Heck, O.C. Taylor and D.T. Tingey (eds), *Assessment of Crop Loss from Pollutants*, Elsevier Science Publishers Ltd., Essex, 552 pp.

Nilsson, J. and Grennfelt, P. (1988), *Critical Loads for Sulphur and Nitrogen*, UN-ECE/Nordic Council Workshop Report, Skokloster, Sweden, March 1988, Nordic Council of Ministers, Copenhagen.

NPS (1988), *Air Quality in the National Parks*, US Department of the Interior, National Park Service, Air Quality Division Report, Denver, Colorado (no ISBN, pages not numbered).

Ormerod, S.J. and Wade, K.R. (1990), 'The role of acidity in the ecology of Welsh lakes and streams', in R.W. Edwards, A.S. Gee and J.H. Stoner (eds), *Acid Waters in Wales*, Kluwer Academic Publishers, Dordrecht, 337 pp.

Schulze, E.-D., and Freer-Smith, P. (1991), 'An evaluation of forest decline based on field observations focussed on Norway spruce, *Picea abies*', *Proceedings of Royal Society of Edinburgh*, 97B, 155–68.

Scott, J.A. (1953), 'Fog and deaths in London, December 1952', *Public Health Records*, 68, 474–79.

UKBERG (1989), *The Effects of Acid Deposition on Buildings and Building Materials in the United Kingdom*, Report of the Building Effects Review Group, London, HMSO, 106 pp.

UKPORG (1987), *Ozone in the United Kingdom*, Interim Report of the United Kingdom Photochemical Oxidants Review Group, prepared at the request of the Department of the Environment, Publication Sales Unit, South Ruislip, Middlesex, ISBN 0-7056-1145-X, 112 pp.

UKPORG (1990), *Oxides of Nitrogen in the United Kingdom*, Second Report of the United Kingdom Photochemical Oxidants Review Group, prepared at the request of the Department of the Environment, Publication Sales Unit, South Ruislip, Middlesex, ISBN 0-7058-1616-8, 104 pp.

UKRGAR (1987), *Acid Deposition in the United Kingdom 1981–1985*, Second Report of the United Kingdom Review Group on Acid Rain, prepared at the request of the Department of the Environment, published by Warren Spring Laboratory, Stevenage, Department of the Environment, Publication Sales Unit, South Ruislip, Middlesex, ISBN 0-85624-457-0, 104 pp.

UN-ECE (1988), *Critical Levels Workshop Report*, Bad Harzburg, FRG, March 1988, UN-ECE, 146 pp.

UN-ECE (1990), *Draft Manual on Methodologies and Criteria for Mapping Critical Levels/Loads*, prepared by the task force on mapping, UN-ECE, 99 pp.

Wilhour, R.G. (1988), 'Introduction' in W.W. Heck, O.C. Taylor and D.T. Tingey (eds), *Assessment of Crop Loss from Pollutants*, Elsevier, Essex, 552 pp.

Wilson, R.B. and Sinfield, A.C. (1988), 'Policy implications from crop loss assessment research: a United Kingdom perspective', in W.W. Heck, O.C. Taylor and D.T. Tingey (eds), *Assessment of Crop Loss from Pollutants*, Elsevier Science Publishers Ltd., Essex, 552 pp.

7

Patterns of freshwater pollution

Malcolm Newson

7.1 Sources, pathways and targets in the freshwater environment

Human development requires that we use the assimilating power and capacity of land, air and water for waste disposal; rivers cope with this load through a number of natural processes shown in Figure 7.1. The geographical framework for the analysis of pollution problems and remedies (Chapter 2) becomes, for the freshwater environment, easier to appreciate where surface drainage waters form the pathway and more difficult where groundwaters predominate. Surface waters flow in artificial pipes and sewers, often from known pollution sources, or in open channels of various degrees of naturalness. Flows are contained and visible; they can be gauged and monitored. Waters entering the soil and percolating to groundwater bodies are invisible, their routeways are harder to predict and they often carry pollutants introduced across broad areas of topography. This separation of flowpaths also allows us a very simple division of pollution problems into point and diffuse (or nonpoint) sources.

7.1.1 Point sources: sewerage, domestic and industrial

The most obvious point source, localized and identifiable in the freshwater environment is the outfall of a sewerage system to a surface stream; it often discharges foul-smelling or discoloured water and clearly pollutes the stream or other receiving water (Plates 7.1(a) and (b)). Sewerage (the engineered system of collecting pipes beneath our homes, factories and streets) collects sewage (the waste from domestic properties, industry and, in older systems — see Figure 7.2, street drainage or stormwater) for processing at a treatment works. The aim of treatment is to remove pollutants to various degrees of perfection (see Table 7.1 and Tebbutt, 1983). Whilst purification processes exist which will purify sewage sufficiently for humans to drink the discharge from the STW (sewage treatment works), such perfection is rarely necessary. Instead the physical, chemical and biological nature of the final effluent is set by standards fixed by toxicological studies or by the judged capacity of the receiving waters to carry out natural processes of purification (see Figure 7.1). Indeed, where the receiving waters have traditionally been viewed as having an infinite capacity to purify, 'raw' sewage has been released, for example, into estuaries and the sea. This view has become less acceptable in recent years as environmental capacities of such systems have been exceeded.

The laws and standards for sewage discharges were developed towards the end

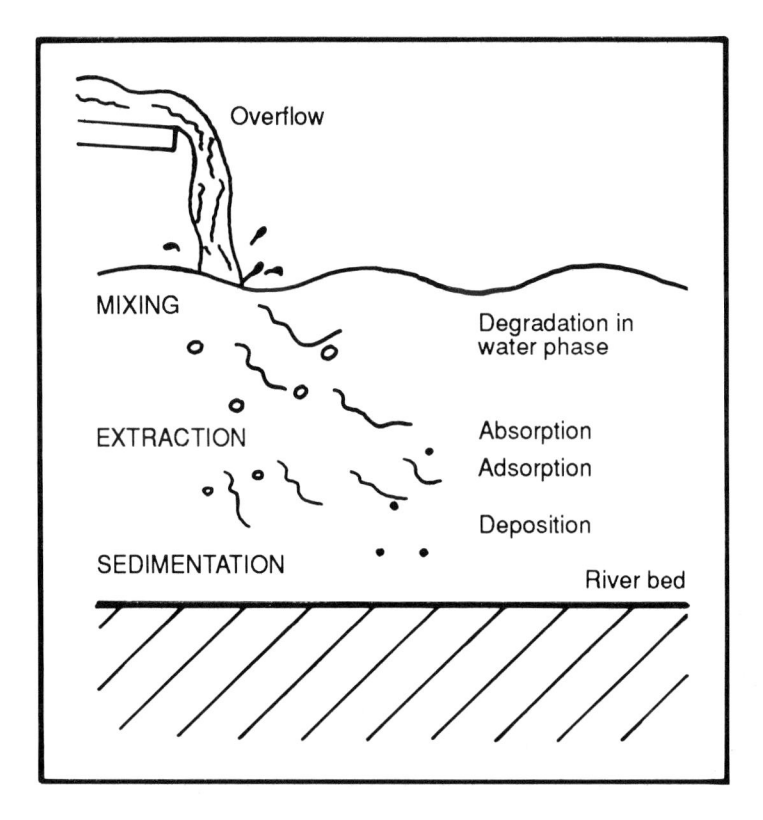

Figure 7.1 Some of the natural physical processes leading to the 'purification' of discharges into flowing waters. (Modified from Crabtree, 1988)

of last century in the United Kingdom when it was realized that the solution of one problem — urban drainage and removal of the products of sanitary improvements — became a problem in its own right; this is what we might now call an externality (see Chapter 5) of the health improvements in Victorian cities. Until 1815 use of the sewers for human excrement was forbidden — they were for the disposal of surface water and kitchen waste; after 1847 it was obligatory to connect lavatories to the sewer network. The addition of factory wastes to sewer systems became more common at the end of the nineteenth century and the 'cocktail' of pathogens, poisons and other contaminants entering sewers has grown in complexity ever since. For sewage treatment to be effective requires careful scrutiny of the waste entering the system and of the conditions to be met in the river, lake or sea into which the final effluent is discharged. It is often difficult to anticipate or detect new substances entering sewers; on the one hand, industrialists fear that secrets of manufacturing processes will be revealed by chemical analysis of their wastes and, on the other, the collection of stormwater by so-called combined sewerage systems may bring an uncontrollable and unpredictable load of road filth, spillages, etc. One category of load entering such systems is dog faeces, amounting in the United Kingdom, it is claimed, to 17 $g.m^{-2}.yr^{-1}$.

Sewer systems which receive stormwater are not designed to carry runoff in flood conditions; they 'leak' at pre-designed sites into the river network. Similarly, STWs are not expected to cope with high peak flows — they provide only minimum treatment for flows in excess of six times the dry weather flow of the system. In fact these storm sewer

Plate 7.1a A point source of pollution, Causey Burn, County Durham. The source of the foam is the sewage treatment works for the town of Stanley

overflows are a nice example of traditional judgements about the capacity of an environment to cope with a pollution load (Crabtree, 1988). The surface stream in flood is swollen in volume, flows faster and is more turbulent; as a result it is able to perform an enhanced role of natural purification. However, it is now known that such discharges cause biological stress in some streams, even where chemical standards are maintained (see Section 7.2.3).

Sewers and sewer discharges are not the only point sources of pollution of the freshwater environment, though they represent a controllable element which is politically understandable to communities in the developed world. Until recently rural point-source pollution was considered rare; monitoring systems for river quality were concentrated in urban areas receiving major STW discharges. However, the intensification of the livestock farming sector has led to a rapid escalation of farm pollution incidents in recent years. The waste from the production of silage as cattle feed and the mixture of dung and urine from livestock housing ('slurry') are both many times more damaging than raw human sewage, particularly through their deoxygenating effects on water (Figure 7.3(a) and Conway and Pretty, 1991). Point sources of pollution therefore occur at leaking silage clamps and from overflowing slurry stores (Plates 7.2(a) and (b)); these, together with dirty washing water from farm yards/buildings, dominate the tally of agricultural pollution incidents

Plate 7.1b Downstream from the outfall of the Stanley works you may play in the country park but not in the stream

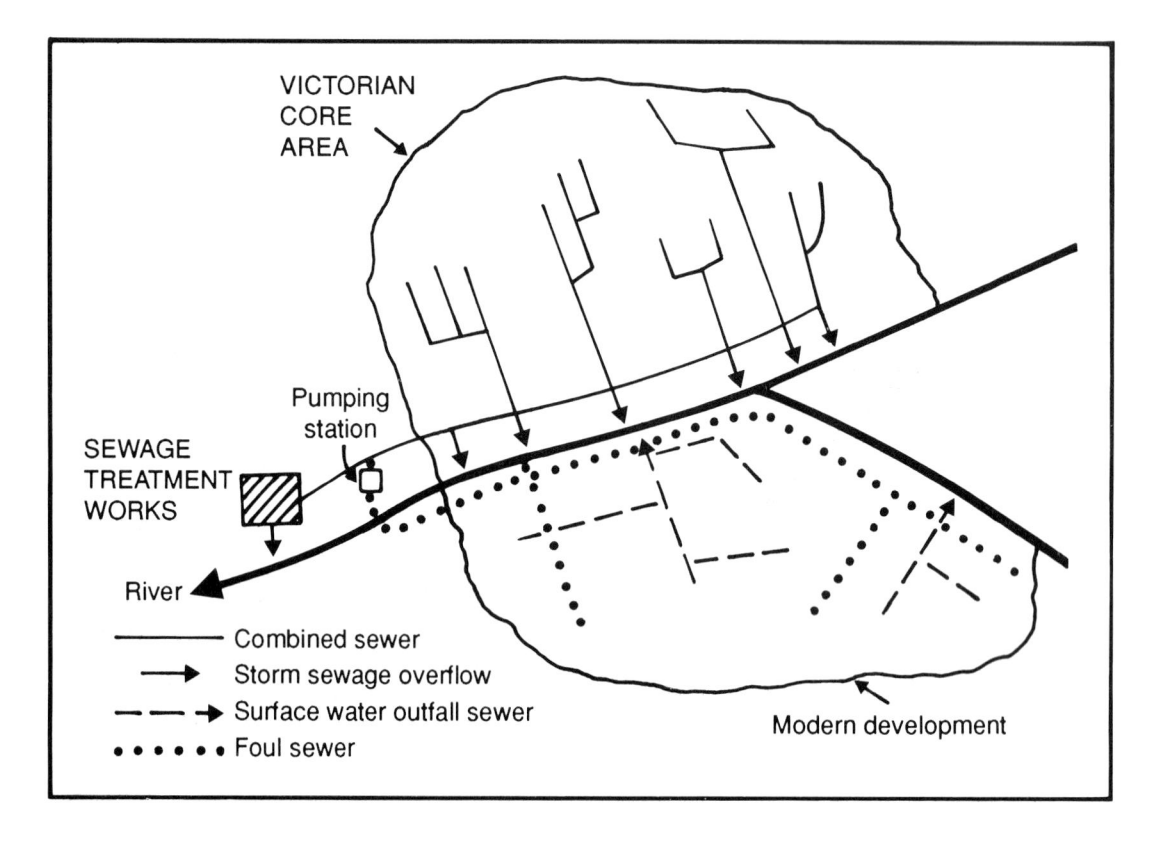

Figure 7.2 The planform of sewerage systems in urban Britain. Only recently have attempts been made to separate foul sewers from surface drainage. (Crabtree, 1988)

Table 7.1 The aims of sewage treatment in chemical terms

Determined	Content in 'raw' sewage (mg/l)	Content in river under EC regulations (mg/l)
Biochemical Oxygen Demand (BOD)	400	7
Suspended Solids (SS)	500	(10 under UK regs)
Ammoniacal nitrogen	25	2
Nitrate nitrogen	0	(11.3* — sources are mainly diffuse)
Phosphate	10	0.7
Bacterial coliforms/100 ml	10^7	10^4

* Nitrogen, c.f. 50 mg/L nitrate.

Patterns of freshwater pollution

Plate 7.2 Point sources of agricultural pollution if spillage occurs (a) a silage 'clamp', (b) a slurry store

(a)

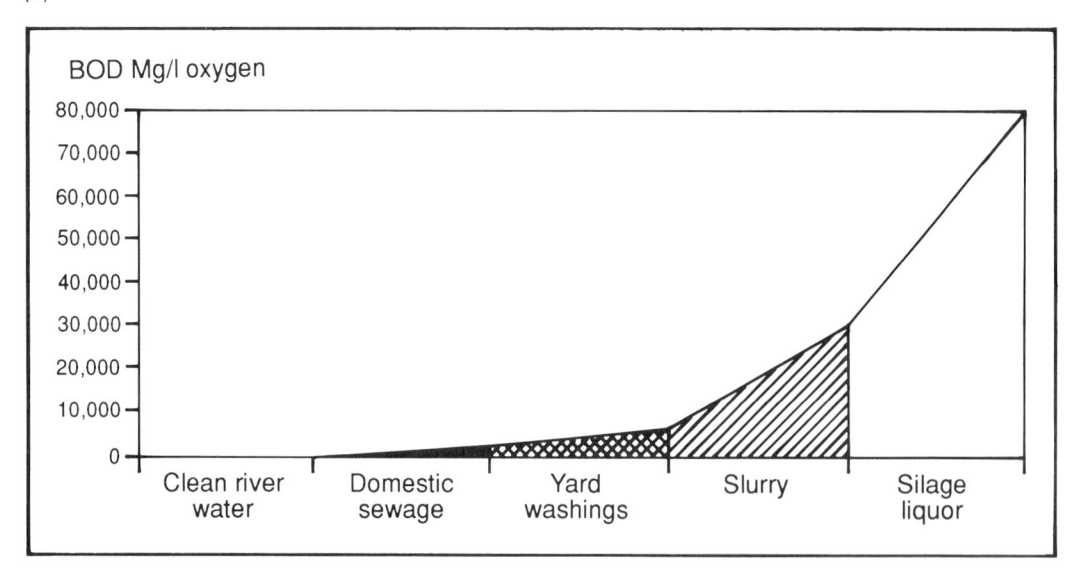

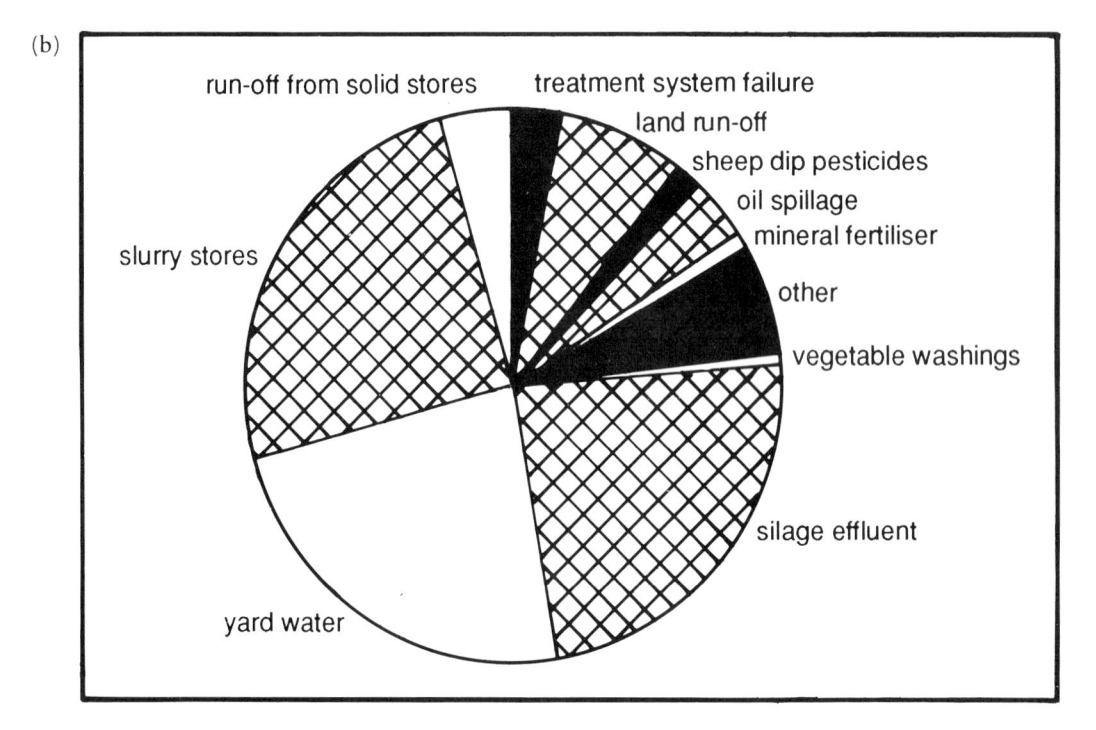

(b)

Figure 7.3a The polluting power of some agricultural wastes expressed as Biochemical Oxygen Demand. (Water Authorities Assn., 1988)
  b Sources of agricultural pollution incidents in England and Wales. (Water Authorities Assn., 1988)

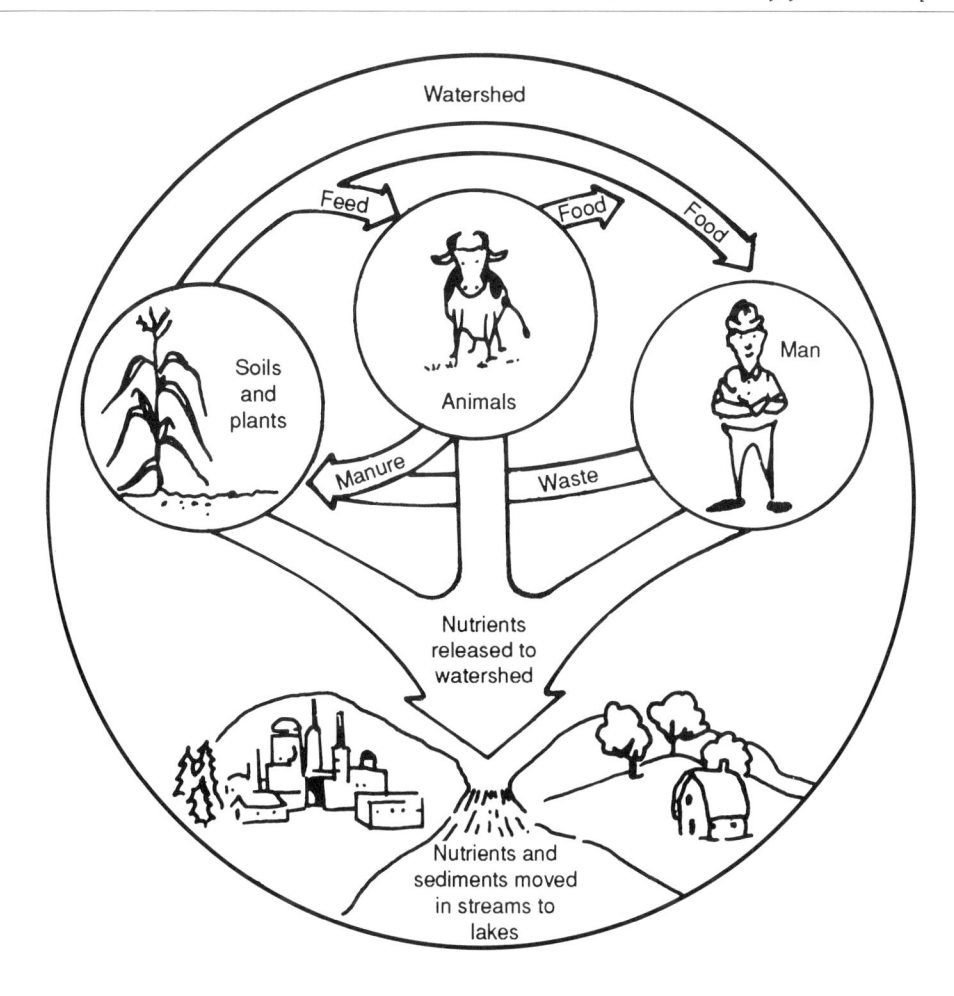

Figure 7.4 The major components of the watershed nutrient cycle leading to cultural eutrophication. (Porter, 1975)

recorded by the rivers authorities in England and Wales (Figure 7.3(b)). The time may well come when small treatment works may be necessary on farms to purify livestock wastes (and perhaps generate useful by-products such as methane and compost). However, during the 1980s and 1990s the emphasis in Europe has been to reduce agricultural productivity — legislation to curb farm point-source pollution has therefore been brought in within a context of de-intensification as a means of discarding the problem at source. There have also been improvements in farm buildings to retain pollutants, codes of good practice to fix distances between stores of pollutants and streams, and big improvements in stream monitoring for rural pollution, including the use of biological indicators (Newson, 1992a).

7.1.2 Diffuse sources: agricultural and atmospheric sources

Only in recent decades, thanks to improved monitoring systems, a better understanding of hydrology and environmental concern, have diffuse sources of pollution risen in prominence in environmental management. Diffuse sources which have been highlighted include agriculture, for its use of fertilizers and pesticides, and forestry where the canopy is seen to act as a site for the

Table 7.2 Trophic states of river systems

Reach	Oligotrophic	Mesotrophic	Eutrophic
Headwaters	Flood-prone coarse sediment Mosses/algae Fringing herbs and grasses	(transitional)	Lowland rivers on soft rocks Ditches or dykes in areas of intensive agriculture Herbs/grasses clog channels
Middle reaches	(transitional)	Deposition and softer rocks produce better substrate for channel plants	Submerged species with fringing reeds and emergent herbs
Lower	Deposition may allow channel plants		Nutrient pollution may eradicate some species — turbidity and algal growth

deposition of air pollutants, especially acids (see also Chapters 3 and 6).

To obtain the productivity desirable in modern intensive agriculture it is necessary to enrich all aspects of the growing medium and to prevent the improved output being diminished by pests. The result, in terms of freshwater pollution by nutrients, is known as cultural eutrophication (Figure 7.4). Freshwaters naturally occur in three main classes of nutrient status (see Table 7.2). Thus eutrophication can be a natural process as biological productivity increases (e.g. as a land surface 'ages' after glaciation). Cultural eutrophication is more rapid and more damaging. The loss of nitrate from intensively farmed landscapes into both groundwater and surface streams has also become perceived as an important and dangerous diffuse source of pollution to human drinking water supplies. The issue is a difficult one because toxicological arguments rage about nitrate as a pollutant; limits for concentration are therefore problematic. There is a possible connection between excess mineral fertilizer and human cancer; however, empirical evidence for the link is at best circumstantial. What is known is that high-nitrate drinking water produces 'blue baby syndrome' (Methaemoglobinaemia) in young children but there have been relatively few cases (House of Lords, 1989). The reasons for the European Communities taking a highly precautionary approach of limiting concentrations to 50 mg.l^{-1} (cf. World Health Organization limit of double this) appears to be related to the politics of controlling agricultural production (Conrad, 1990).

The control of pests is an important part of most profitable uses of land but the application of pesticide chemicals (directly designed to be toxic to biota) is difficult to control (both technically and legally). Whilst applications to crops can be made under weather conditions which avoid drift or runoff and point-source pollution can be prevented by avoiding stream margins, there is bound to be a general increase in soil residues and in the biological system which leads to non-point pollution, particularly of surface waters. Once again, in Europe a highly precautionary concentration limit of 0.1 μg.l^{-1} has been fixed for individual pesticides in water.

Chapters 3 and 6 deal with the causes of surface water acidification, a 'more diffuse' source of pollution than in the case of cultural eutrophication because of the extra element of dispersal provided in the additional atmospheric pathway. However, the relatively predictable trajectory of both air masses and their aerosol and particulate loads makes regional assessments possible.

7.1.3 Accidental sources: spillage

Despite the pressures of development and the drive for profit in the capitalist system only a tiny minority of pollution of freshwaters is deliberate. Occasionally the media record examples, such as the use of toxic chemicals by poachers wanting a large illegal kill of fish. Much more common are cases of simple human error (as in the recent occurrence of aluminium pollution of the water supply at Camelford, Cornwall, which resulted in both human illness and fish kills). Nevertheless, there is strict liability (Chapter 4) for such 'accidents'.

A broad category of accidental spillages involves the transport of toxic materials by rail, road or waterway (see also Chapter 10); crashes occur and whilst they may not be over or near water the volume and timing of the spill may mean that pollution occurs by rapid downhill drainage. River managers therefore have prepared emergency procedures such as pumps and booms; in cases where river regulation flows are available from reservoirs these are utilized to dilute the 'slug' of pollutants — the artificial equivalent of a flood flow.

In recent years spillage of pollutants has occurred under unusual weather conditions, for example, from farms which are storing silage or slurry under otherwise appropriate conditions. Controls include bunds and traps for these eventualities and the same principles are being applied now to the storage of industrial and agricultural chemicals — both sound basic construction and an emergency facility (and preferably a location well away from sensitive watercourses and/or proofed against groundwater percolation).

7.1.4 Pollution of groundwater: an area of neglect

We have already discussed the fact that diffuse sources of pollution are a relatively recent arrival on the scene of water management; part of the problem has been our relative lack of understanding and monitoring of the groundwater system. This is an invidious situation especially given our dependency on groundwater. Groundwater has the special virtues of being slower to respond to drought than surface waters and, where inputs of pollutants are relatively low, of being able to filter and purify water to a state fit for human consumption with the minimum of expenditure.

The first legal cases of groundwater pollution in the United Kingdom came during the nineteenth-century metal-mining era; on the Mendip Hills the deterioration in quality of paper at a paperworks fed by a cave spring was traced to a lead mine waste heap many miles away where the stream entered a sinkhole. Most groundwater flows are, however, much more complicated than in cavernous limestones but have the benefit of longer residence times and greater filtration than can occur in caves. Passage through the soil offers the first opportunity for purification; secondly the saturated pores in the rock below the water table dilute, disperse and often chemically purify the slow flow towards springs, wells, boreholes and streams. However, certain pollutants have been recorded in dangerous concentrations deep into groundwater systems (and see also Chapter 8, Figures 8.2 and 8.5). Organic solvents, nitrates and radioactive isotopes have all been identified during the more detailed approach now being taken in the developed world to groundwater pollution.

7.2 Chemical and biological monitoring and control of water pollution

Every system of pollution control requires a parallel system of monitoring, preferably a system which allows detection of point and diffuse sources, both continuous and episodic conditions, and which detects damage to both human and other biotic targets. Such a system must also be robust, repeatable, serve legal requirements and build in the ability of the polluted system to recover.

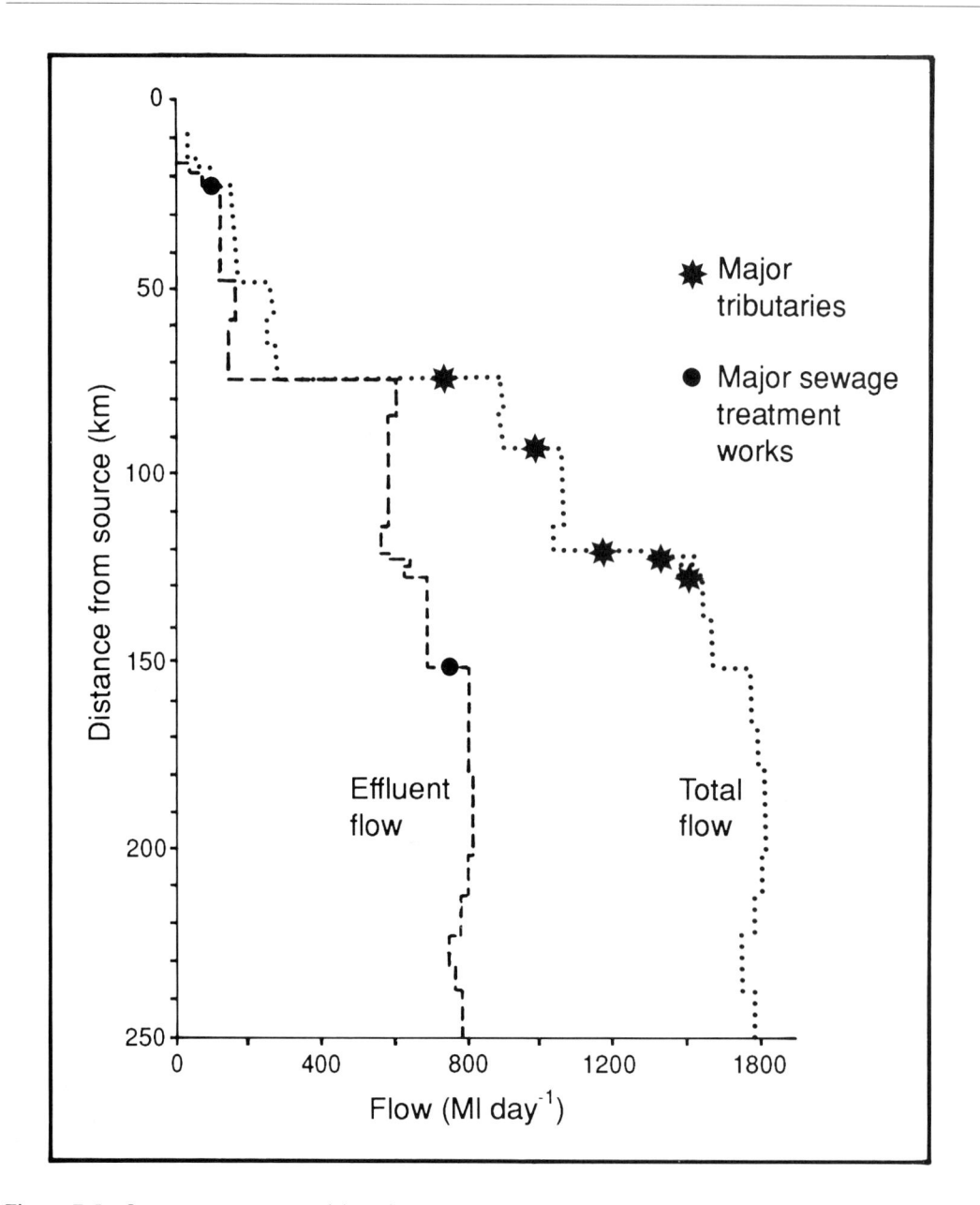

Figure 7.5 Sewage treatment and low flows in the River Trent, England. (Farrimond, 1980)

7.2.1 Environmental capacities: dilution, dispersion, exchange in water

We have referred briefly to the capacity of natural freshwater systems to bring about apparent purification (often merely storage) of polluted discharges. Clearly, to the Victorian health reformers with no vision of our aesthetic, recreational and conservational view of clean rivers, the river (and thence the sea) provided an ideal, convenient and apparently boundless purification capacity for human waste, drainage from streets, and wastes from industrial processes. At first only biological and human health indicators were available to show the folly of this miscalculation; the important freshwater fisheries (notably for salmon) were damaged and received legal protection against pollution (as early as 1861 in the United Kingdom) whilst successive epidemics of water-borne disease forced a rapid improvement in purification technology for human supplies (e.g. chlorination).

That we have been too optimistic about the capacity of the freshwater environment to cope with pollution is shown up by the fact that at low, dry-weather flows many of our major rivers are receiving much of their water from sewage treatment works (Figure 7.5). Together with industrial effluent outfalls there are some 7000 point pollution sources on the rivers of England and Wales; a further thousand outfalls discharge heated water from power stations and water from mines, pumped to keep them open.

At first, when standards were set for discharges from treatment works they were set in the chemistry laboratory and it has taken nearly a century for direct biological monitoring to return to prominence. The basis of original, chemical standards for sewage discharges (developed in England and Wales by the Royal Commission on Sewage Disposal in 1912, but not enforced until 1936) was full utilization of the diluting properties of the river to remove the threat of pollution. A standard was set for Biochemical Oxygen Demand (BOD — the demand exerted by organic matter in the sewage), Suspended Solids and Ammonia, though only the two former categories became of widespread significance, with levels set at 20 mg/l and 30 mg/l respectively. This was calculated using an almost meaningless average dilution factor between the outfall and the lowland river. In practice, sewage treatment works discharging into estuaries and the open sea have had very few controls until the very recent past and were allowed to discharge 'raw', untreated sewage because of the alleged superiority of these environments to disperse and dilute the pollution. This might have been true chemically but the solid load of faeces, sanitary towels and condoms in modern sewers has made the situation untenable in a world increasingly devoted to recreation in or near water, whether on rivers or at the coast. Treatment standards are set to rise (and the costs of treatment) extremely rapidly thanks to European legislation. However, the overflow of sewer systems in storm conditions appears unlikely to be cured — the pipe is not made yet which will contain all flows — and river systems are more efficient carrying systems when in flood.

Implicit in the choice of effluent standards for sewage treatment works was a dilution factor between the polluting liquid and the 'clean' river from upstream; in fact this may not be met at low flows and clearly deteriorates over time as a river basin becomes populated and industrially developed. Secondly, assumptions were also made about the flow characteristics and biological significance of the channel into which the discharge occurs: dispersion may not occur, as the 'streaking' of rivers downstream of some works clearly demonstrates. Chemical exchanges in the receiving waters also vary according to the flow, the stage of basin development and even the time of day (water temperature affects oxygen demand and exchange). As Figure 7.6 shows, a considerable deterioration in river water quality occurs below any polluting outfall. It is around this simple scenario that the 'argument' between

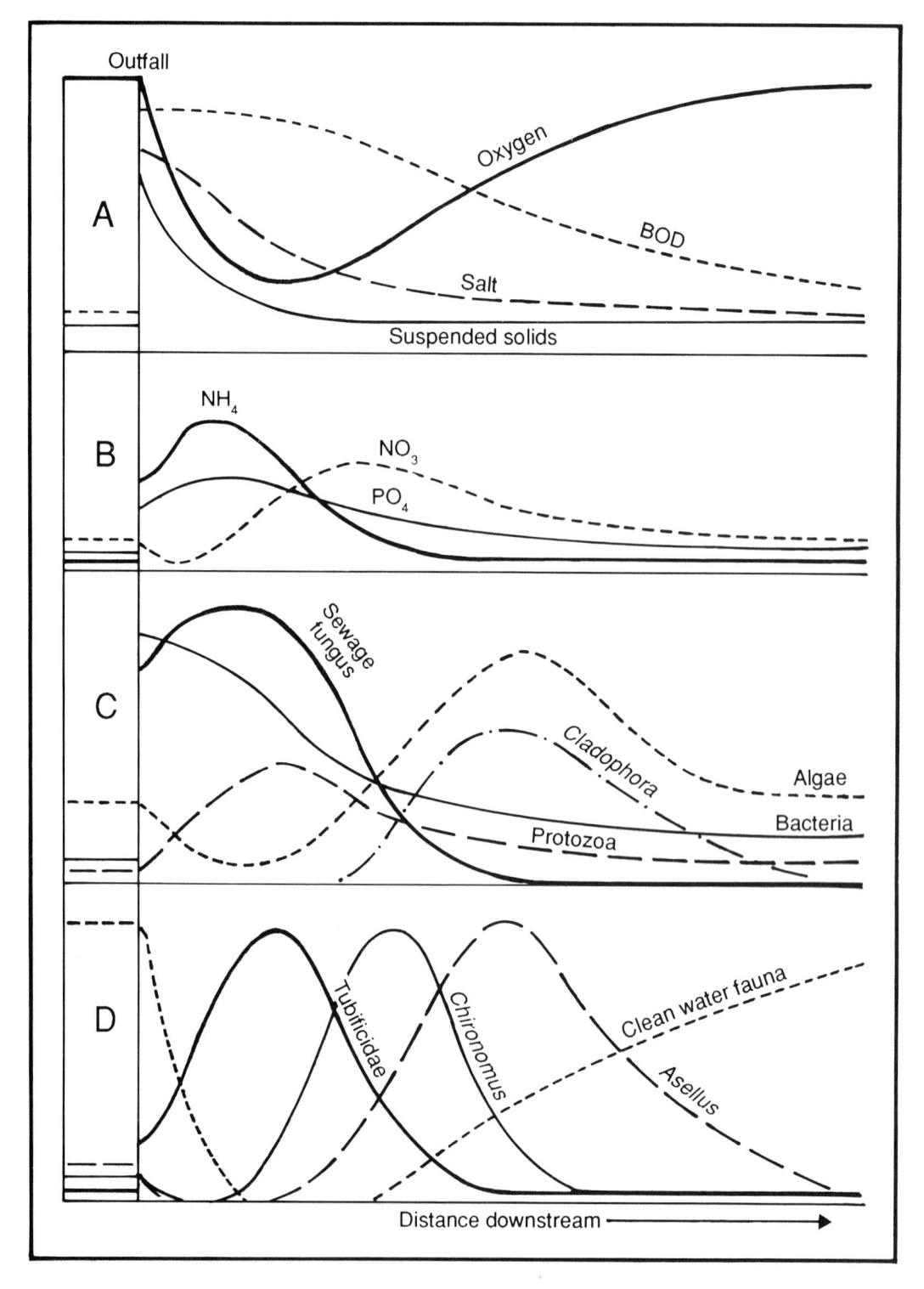

Figure 7.6 Four components of the decline and recovery of river water quality downstream of an outfall from a sewage treatment works. (Mason, 1981)

chemical and biological monitoring and the political debate between emission controls and environmental objectives are set (see also Chapter 2).

7.2.2 Chemical monitoring within the policy context of control

The chemical constituents and disease-producing organisms of human sewage are relatively well understood; the threefold standard for sewage discharges after treatment have, therefore, been little attacked (although the frequency of monitoring and the point at which a legal infringement has occurred are much in contention — see Rees, 1989; Birch, 1989). It has also been suggested that direct assessment of ammonium should be carried out on discharges — a persistent and growing problem and compounded with nitrate pollution. The disease-producing capacity of water is normally indicated by human gut bacteria such as *Esherischa coli* (Mason, 1991); at the coast there have been recent calls for assessments of viral contamination of bathing waters.

The policy context for chemical sampling has progressed beyond a matter of technical trust on operators of sewage treatment works to the point where successive pollution laws have set up the principle in the United Kingdom of licences or consents to discharge. In these the regulatory body for rivers sets the conditions to be met by each applicant to discharge and then monitors both the discharge waters and the receiving waters to check on compliance. Two more features of recent legislation in the United Kingdom are of interest here; one is that consent conditions and monitoring results are made public and the second is that a system of 'charges for discharges' is levied to repay the costs of administration and monitoring — the 'polluter pays principle' in action.

It alarms many environmentalists that pure practicality and realistic costing demand a relatively restricted range of parameters to be monitored as part of consents and relatively infrequent chemical sampling (as little as once per month). Further alarm concerns the number of occasions on which pollution law officially 'turns a blind eye' to episodes of pollution by demanding that chemical tests 'pass' on 95 per cent (not 100 per cent) of all occasions sampled. Further 'unofficial' practices creep into regulation such as works which permit pollution when they know they are not being sampled or inspectors who sample with a view to intimidating a polluter into future compliance (a cheaper procedure than prosecution — see Hawkins, 1984).

Two other major concerns over chemical sampling procedures are expressed by environmentalists. The first is that industry's rate of innovation, and use of chemicals means that regulatory bodies may be ignorant of the pollution potential of much of the waste entering sewers; industry may be unwilling to disclose new chemicals for fear of losing commercial leadership in a particular process. The second, in the United Kingdom, is that a system of flexible consents, in which the discharge is matched to environmental objectives for the river, is far too relaxed to prevent long-term river and coastal pollution and contravenes the European Community's principle of strict control on emissions. These are two fundamentally different principles of environmental management; it can be argued that the UK approach is more realistic, even more geographical (Newson, 1991), but it is highly dependent on a thorough knowledge of natural purification processes (in the short and long term) and of all the components of the discharge. Chemical sampling alone does not and cannot assure us on these grounds at a reasonable cost.

7.2.3 Biological monitoring: integrative in space and time

Other than a long-standing concern for the viability of fisheries (the operation largely of self-interest) we have shown little official concern for stream biota in relation to pollution.

Table 7.3 Chemical and biological monitoring of river pollution: comparative advantages

Criterion	Chemical sampling	Biological sampling
Coverage	Only chosen determinands are registered	All potentially damaging chemicals registered by their effects but not specifically
	'Frozen' to sampling point — can't trace easily to source	'Trail' of damage upstream may lead to 'culprit'
Time-scales	Limited to the 'snapshots' of the chosen sampling intervals except where continuous sampling	Integrates over time — lasting damage after chemical quality returns to normal
Cost	High for instruments, laboratory time and travel	Lower for techniques but expertise costly
Legal acceptability	Precise; ceremony of sampling (like being breath-tested!); replicated samples allow independent tests	Statistical/subjective; may depend on other conditions

So long as our own water supply for domestic use was chemically pure and biologically inert ('free of bugs'), we have failed to identify stream and lake biota as targets of pollution. We now realize, however, that freshwater organisms have both an intrinsic value, to be conserved, and two applied values — as part of the natural system of water purification (microbiological) and as indicators of pollution (macro-organisms). In the latter role biota have many advantages by comparison with chemical analyses (Table 7.3).

Of the advantages claimed for biological monitoring perhaps the most important is that it potentially records the history of recent pollution episodes in any given habitat: the Achilles' heel of chemical sampling is that, unless it can be made continuous (progress is slow in designing such equipment) across a very wide range of determinands, it cannot achieve this integration. In parts of the river network where monitoring for compliance with consents is irrelevant — for example, rural areas — biological monitoring provides possibly the only indication of pollution from, for example, a faulty slurry store. Identification of the culprit farm and acceptability of biological evidence (which has been hard to gain in the courts) are essential for a prosecution under the pollution laws in such cases.

Hellawell (1986) provides a highly detailed review of biological monitoring techniques; in a similar way to chemical monitoring there are debates as to which organisms and which methods of expressing their abundance or diversity make up the best system. Recently prominence has been attained by a multivariate classification of freshwater invertebrates which is related to certain habitat features of the stream channel where they are found. The RIVPACS system (Wright, Armitage and Furse, 1989) can then be used in predictive mode; in a new reach the characteristics of the stream are used to predict what invertebrates 'should be' present — discrepancies can then be considered to indicate pollution, either present, past or continuing. Much more progress and refinement of such procedures is necessary before the extensive reach and relative cheapness of biological monitoring makes the technique as powerful, especially in legal terms, as chemical analysis. Some industrialists and farmers may well nurse anxieties that biologists are inherently more 'green' than chemists and therefore bring a bias to regulatory activity!

7.3 Towards geographical patterns of management: river basin scale

Figure 7.7 distinguishes between the issues faced by point-source freshwater pollution of one river reach and those raised by the context of that reach in a river basin. The basin is an interlinked ecosystem which bespeaks integrated or holistic management of its constituent elements (Newson, 1992(b)). The challenge to managers is to make the assumptions of the Royal Commission of 1912 about dilution factors into a modern, mathematically modelled system of real-time controls on effluent discharges. The treated sewage outfalls from farms, factories and homes would be planned in relation to the water quality objectives for the various reaches of the river network. Nevertheless, as the figure shows, over the longer period there is a threat from the unchecked growth of pollution from non-point or diffuse sources which will eventually force further expensive tightening of controls on point discharges; controlling non-point pollution is an equal challenge to river managers to that of modelling point sources in the network.

7.3.1 Limits to growth?

We have already discussed the fact that the environmental anxieties of the early 1970s were characterized by pessimism about the exhaustion of basic natural resources if growth and exploitation continued unabated. In some senses the last two decades have seen a falling-away of such absolute claims as individual pollution and resource problems and their scientific treatment have come and gone in the issue-attention cycle. In the near future we shall need to re-evaluate the power of computer modelling armed with the much better data now available. Will this exercise reveal limits to growth and will the imposition of such limits prove politically acceptable? The topic of freshwater pollution proves interesting from this viewpoint, especially in a river basin context and when we further stretch the frame to include the coastal zone and shelf seas.

In some ways European and North American water pollution controls have already limited growth in one sector — agriculture. There is an interesting correlation between such pollution control devices as Nitrate Sensitive Areas (where crop choice and fertilizer use are restricted), vulnerable zones, buffer strips and the fact that agricultural production in the nations which have adopted them was already in surplus — growth itself was part of the problem and restricting it is therefore a politically acceptable solution. Punishing fines on farmers for accidental pollution of rivers might not have proved so acceptable during, for example, the period of war-time and post-war food shortages!

Industry has, it may be claimed, been better treated, with harsher regulatory controls on river pollution being phased in, made non-retrospective and equipped with 'derogations' permitting timed compliance after a period of investment in controls. Whilst nations like the United Kingdom are equipped with public policies of planned development they have been generally unwilling to see planners, rather than scientists and technologists, operating pollution control mechanisms. The geographer would argue, however, that the limits to growth of population, agriculture and industry are spatially variable and the upshot of this variability (of environmental capacities and cultural attitudes) provides an ideal platform for a greater involvement of rational planning.

7.3.2 River basin management institutions and pollution control

In a way not available to management of air or land degradation through pollution, the river basin with its name and clear boundaries, permits a unitary framework within which pollution is one component of environmental management. This strong identity offers basin managers an advantage in the way they work since they also have, if political will permits, control over the

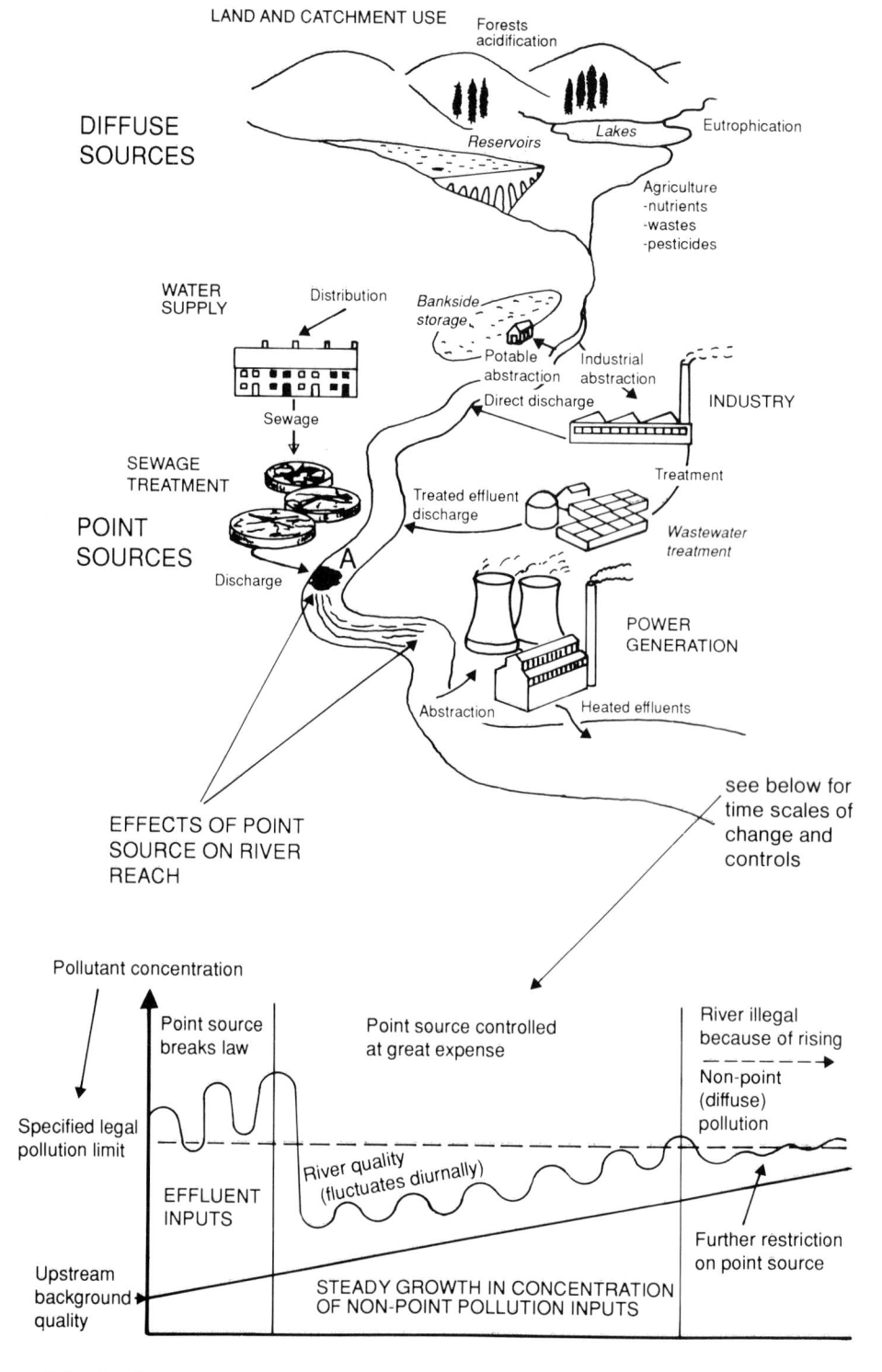

Figure 7.7 Catchment management problems caused by freshwater pollution from both diffuse and point sources. In the graph below the decreased purity of upstream flows is seen to complicate the regulation of downstream water quality

Table 7.4 River and canal water quality classification applied to quinquennial surveys in England and Wales

Description	Class	Current potential use
Good quality	1A	Water of high quality, suitable for potable supply abstractions; game or other high-class fisheries; high amenity value
	1B	Water of less high quality than 1A but usable for substantially the same purposes
Fair quality	2	Waters suitable for potable supply after advanced treatment; supporting reasonably good coarse fisheries; moderate amenity value.
Poor quality	3	Waters which are polluted to an extent that fish are absent or only sporadically present; may be used for low-grade industrial water supply; potential for further uses if cleaned.
Bad quality	4	Waters which are grossly polluted and are likely to cause nuisance

resources, extremes, conservation and public enjoyment of the river system as a whole. As Figure 7.7 suggests, this balancing of the river basin as a total resource must, theoretically, extend to land, since 98 per cent of the precipitation which finds its way into rivers first falls on land.

The recent rise in the issue-attention cycle of diffuse or non-point water pollution and the parallel growth of concern for conservation and recreation within river systems offers the opportunity for river managers to extend their interest in pure engineering issues such as dam-building or pure chemical/biological technology (such as sewage treatment works) to that area of broader planning of both land and water within the basin. In the United Kingdom, the Water Act of 1989 established an organization capable of this broader activity — the National Rivers Authority. Part of its brief is 'catchment planning', first to bring about integration of its own sectoral operations — water resources, flood defence, pollution control, navigation, recreation, conservation, fisheries — and later, with less political certainty, an extension into local government planning processes. The second step is already partly implemented in the Nitrate Sensitive Areas regulations, in the 'good practice' required by farmers and foresters, and in the inputs to development control in flood-prone areas. The National Rivers Authority's conservation function also potentially allows it to promote the restoration of such natural purification features as ponds, lakes and wetlands.

Another advantage of strong river basin management institutions, purely in terms of pollution control, is that monitoring schemes additional to those for compliance with the law can be instituted to record the national status of river purity, discharges from rivers to neighbouring seas and both of these features through time. Five-yearly surveys of this type have been carried out in England and Wales since 1958 (and a similar system now operates in Scotland); the classification scheme used to assign each river reach to a quality (which is then mapped nationally) is shown in Table 7.4, whilst Figure 7.8 shows the achievements of pollution control strategies in bringing improvements to the worst cases of river quality degradation. However, such monitoring results give recent cause for alarm in that some of our better-quality rivers appear to be 'slipping back', partly as a result of new and uncontrolled rural pollution sources. In 1990 the pollution survey included biological indicators of river water quality and in future the biological status of a reach will be able to 'override' purely chemical indications. The next big challenge to river managers is to incorporate the estuaries and coasts of Britain in the much stricter limits for pollutant concentrations set by the European Community.

Already a 'Red list' of chemical toxins has been controlled by emission standards, raw sewage discharges are to be drastically curtailed, and curbs on cultural eutrophication will be implemented by 1995. These

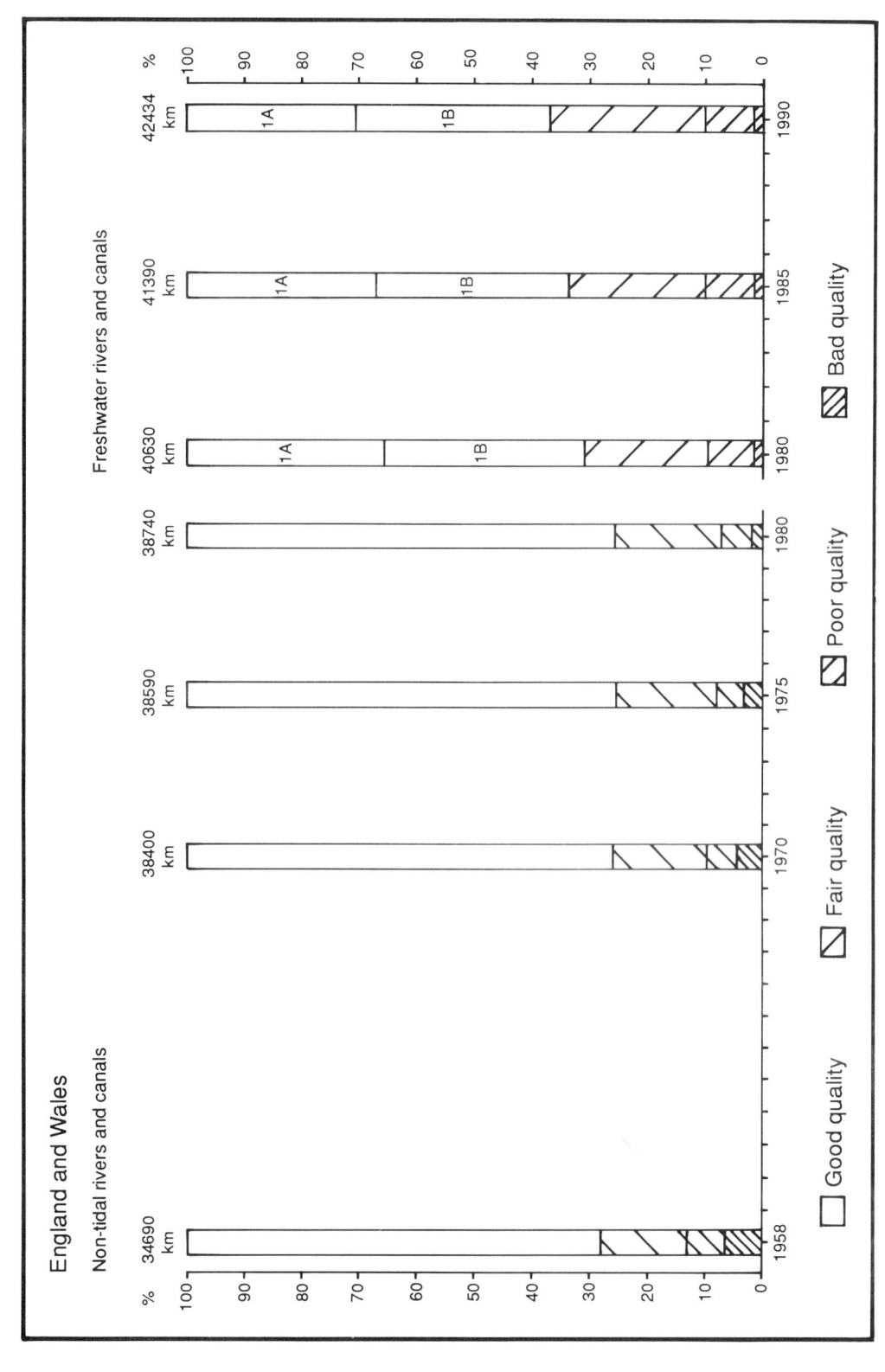

Figure 7.8 River water quality in England and Wales 1958–1990. The criteria were modified in 1980 and so two outcomes are shown

changes not only involve an extension to the field of marine pollution — they also involve land and air pollution. The extra amounts of sewage sludge produced by improved treatments will need to be disposed of on farmland or by incineration — a clear case for integrated pollution control (see Chapter 14).

References

Birch, T. (1989), *Poison in the System*, Greenpeace, London, 107 pp.

Conrad, J. (1990), *Nitrate Pollution and Politics*, Gower, Aldershot, 82 pp.

Conway, G.R. and Pretty, J.N. (1991), *Unwelcome Harvest: Agriculture and Pollution*, Earthscan, London, 645 pp.

Crabtree, R. (1988), 'Urban river pollution in the UK: the WRC River Basin Management Programme', in J.M. Hooke (ed.), *Geomorphology in Environmental Planning*, John Wiley and Sons, Chichester, 169–85.

Hawkins, K. (1984), *Environment and enforcement: regulation and the social definition of pollution*, Oxford University Press, 168 pp.

Hellawall, J.M. (1986), *Biological Indicators of Freshwater Pollution and Environmental Management*, Elsevier, Barking, 546 pp.

House of Lords (1989), *Nitrate in Water*, Report Select Committee on the European Communities, HMSO, London, 51 pp.

Mason, C.F. (1981), *Biology of Freshwater Pollution*, Longman, London, 250 pp.

Newson, M.D. (1991), 'Space, time and pollution control: geographical principles in UK public policies', *Area*, 23(1), 5–10.

Newson, M.D. (1992a), 'Land and water. Convergence, divergence and progress in UK policy', *Land Use Policy*, 9(2), 111–21.

Newson, M.D. (1992b), *Land, Water and Development*, Routledge, London.

Rees, J. (1989), *Water Privatisation*, Dept Geography, London School of Economics, 289 pp.

Tebbutt, T.H.Y. (1983), *Principles of Water Quality Control*, 3rd edn, Pergamon, Oxford, 235 pp.

Wright, J.F., Armitage, P.G. and Furse, M.T. (1989), 'Prediction of invertebrate communities using stream measurements', *Regulated Rivers: Research & Management*, 4, 147–55.

8

Patterns of land, water and air pollution by waste

Terry Douglas

8.1 The waste 'stream': sources, types, volumes

Waste has been simply defined as 'something for which we have no further use and which we wish to get rid of' (Barron, 1989). Unglamorous though it is, waste has been the focus of considerable environmental attention during the last quarter of the twentieth century as communities the world over have begun to recognize the hazards that its mismanagement can entail. Critical in our better management of waste is an understanding of the 'flow' of the waste 'stream' from its source to its disposal and beyond. It is a measure of the infancy of the study of waste geography and a token of the problems in waste management that for many areas of the world little data exist on the volume and type of waste generated or on disposal routes. It is also difficult to compare data from different countries as definitions of waste categories vary, there being little consensus, for instance, of what constitutes hazardous waste (British Medical Association, 1991). Such data that do exist are often merely estimates. A breakdown of waste arisings for England and Wales by weight is shown in Table 8.1, although, as with so much data on waste it needs to be treated with caution.

Table 8.1 Waste arisings in England and Wales

Waste type	Quantity (million tonnes per annum)
Liquid industrial effluent	2000
Agricultural	250
Mining and quarrying	130
Industrial	50
(Hazardous 3.9)	
(Special 1.5)	
Domestic and trade	28
Sewage sludge	24
Power station ash	14
Blast furnace slag	6
Building	3
Total	2505

Source: House of Commons Environment Committee, Second Report, Toxic Waste (1988/9)

8.1.1 Industrial sources and constituents

Industrial waste as a category is extremely broad and can encompass materials ranging from the inert to the most hazardous. Waste volumes and types are often highly dependent on the nature of the industrial enterprise, the type of manufacturing process and the extent to which low- and non-waste technology has been adopted.

A survey of 320 industrial waste producers in the Tyneside area of northeast England

Table 8.2 Waste per employee for selected industries in Tyne and Wear

Manufacturing activity	Waste per employee (tonnes)
Metal manufacture	20.35
Chemicals	16.60
Paper, printing and publishing	3.44
Vehicles	2.53
Mechanical engineering	2.12
Metal goods	2.06
Timber, furniture	1.80
Textiles	1.35
Bricks, pottery, glass, cement	1.18
Shipbuilding	1.12
Electrical engineering	0.54
Clothing and footwear	0.49

Note: Data from survey of firms undertaken in 1983.

Source: Tyne and Wear Waste Disposal Plan: *Draft Consultative Document* (1985)

demonstrated a wide range of waste arisings when measured as waste per employee (Table 8.2). The distinction between industrial waste as a general category and hazardous waste is an important one. It is hardly surprising that the chemical industry produces a large volume of waste given the nature of the manufacturing processes. However hazardous waste is defined, its geography often closely matches the distribution of the chemical industry. Thus in the United Kingdom the counties of Cheshire, Cleveland and South Yorkshire figure prominently as producers of the sub-set of hazardous waste named 'Special' waste which is regarded as a particular threat to public health (Gatrell and Lovett, 1986). Neither is it any coincidence that one of the most celebrated causes of mishap with regards to the disposal of toxic wastes took place at Love Canal near Buffalo, a major chemical production centre in the United States (see below, section 8.4.2). This case raised public awareness to the issue of siting industrial waste facilities as well as triggering legislative measures to regulate such wastes. A recent US poll showed that over two-thirds of respondents cited actively used hazardous waste sites as a top environmental concern (Environmental Protection Agency, 1990a).

Table 8.1 shows that 'special waste' represents a relatively small proportion of industrial waste. There are no internationally agreed principles which determine the classification of an industrial waste as 'hazardous', 'poisonous' or 'special', but in general the following criteria, which can apply to solids, liquids or sludges, have been adopted:

(a) inflammability — potential to catch fire;
(b) toxicity — harmful or fatal when ingested or absorbed;
(c) persistence — potential for bioaccumulation in the environment;
(d) corrosivity — capacity to corrode metal containers.
(e) reactivity — unstable under normal conditions.

With over 80,000 chemicals estimated to be in international commerce and with about 1000 additions each year (Simpson, 1990), the task of assessing their impact on health and environment is immensely difficult.

The United States Environmental Protection Agency (EPA) has drawn up lists of hazardous wastes under three categories: source specific wastes from industries such as petroleum refining and wood preserving; generic wastes from common industrial processes such as solvents and degreasing operations; and commercial chemical products which includes specific products such as some pesticides.

In the United Kingdom, the Department of the Environment has drawn up a similar list of 23 designated substances, the 'Red List', which require particularly strict controls when discharged into the environment. The list is dominated by pesticides, polychlorinated biphenyls (PCBs), mercury, cadmium and their compounds (Department of the Environment, 1990).

Table 8.3 Composition of household waste in the UK

Component	Potentially recoverable	% wet weight
Paper	✓	31
Vegetable/putrescible		10
Ferrous metals	✓	7
Glass	✓	9
Non-ferrous metals	✓	<1
Textiles	✓	2
Rubber/leather/wood		4
Plastics	✓	3
Moisture		25
Fines/dust/ash		8

Source: Barron (1989)

8.1.2 Municipal solid waste sources and constituents

Municipal Solid Waste (MSW) or domestic waste is that portion of the waste stream which arises from households and is produced as part of day-to-day living. Its collection and disposal are usually the responsibility of local government. It is an extremely heterogeneous mixture of constituents which tends to vary according to season, the social characteristics of the neighbourhood and which has changed in response to evolving lifestyles. Data from the Netherlands indicate a strong correlation between indices of consumption (total consumption and food and beverage consumption) and MSW production (Nagelhout et al., 1990). In the western world a 'garbage crisis' is commonplace and can be seen as the result of relatively affluent 'throwaway' societies. A typical composition of MSW is shown in Table 8.3.

Data on MSW are poor but often better than those for industrial waste. The United States, Canada and Australia feature at the top of the list of per capita domestic waste generation with European countries at a slightly lower level and virtually no data available for Third World countries, many of which have no MSW collection outside major urban centres (Table 8.4).

The management of MSW is beset by two problems. Firstly, the vast volumes that are produced by western societies have required considerable space, usually in the form of landfill sites or tips (Plates 8.1 and 8.2). Despite the initiation of reduction and recycling measures, MSW volumes are still increasing in most countries (Figure 8.1), although

Table 8.4 Selected MSW statistics for various countries

Country	Annual per capita production (kg)	% disposed by landfill	% disposed by incineration
Australia	681	98	2
Austria	216	57	19
Canada	642	94	6
Denmark	420	64	32
France	289	33	32
Germany (W)	447	83	9
Italy	246	38	20
Japan	342	28	67
Netherlands	502	66	19
Sweden	300	52	38
Switzerland	336	13	49
UK	332	80	6
USA	744	na	na

na = not available

Source: United Nations Environment Programme: Environmental Data Report (1990)

Patterns of land, water and air pollution by waste

Plate 8.1 Domestic waste tip for an island community of less than 200 people, Raasay, Inner Hebrides, Scotland. (Photo: T.D. Douglas)

Japan is a notable exception. Secondly, although MSW is generally regarded as being a lesser risk to the environment than the 'hazardous waste' element of the industrial waste stream, it does contain significant components which present a more serious risk to the environment. These can include the remains of paint and solvents, cadmium or mercury batteries and insecticides. Furthermore, the propensity of the organic fraction of household wastes to produce leachates which have high pollution potential is well documented (Gray *et al.*, 1974).

The challenge of managing MSW has been to find disposal methods to cope with the rising volumes. As most of the waste has traditionally been dealt with by landfill, this has meant the selection and acquisition of sites near to urban sources of MSW. In many places this has placed a lot of stress on land around the peripheries of cities and the bad perception often associated with MSW has led to countless outbreaks of the 'NIMBY' syndrome — Not In My Back Yard! (McKechnie *et al.*, 1983; Furuseth and Johnson, 1988; Douglas, 1988).

8.1.3 Construction and demolition wastes

At the least harmful end of the waste spectrum comes those wastes which are relatively inert and from which there is little risk of serious environmental pollution. In the United Kingdom, these wastes generally fall outside the category of 'Controlled Wastes' as they are biologically and chemically stable. These wastes include subsoil, rock and brick from building sites and are often disposed to landfill as well as being used in

Plate 8.2 Municipal Solid Waste being tipped at the Keele Valley Landfill on the outskirts of Toronto, Canada. This is the largest landfill in Canada with the potential to produce 1785 million m^3 of landfill gas in the next 30 years. (Photo: T.D. Douglas)

reclamation schemes, landscaping and other civil engineering works.

8.2 Disposal technology: the alternatives — landfill, incineration, treatment and recycling

8.2.1 Holes in the ground: environmental capacities and the sanitary landfill

Landfill has been a favoured method of waste disposal for a number of reasons, often because it is the cheapest available method and also as the result of the availability of holes in the ground (Table 8.4). The 'void space' available through extractive operations in a country such as the United Kingdom with a large quarrying industry is likely to be larger than the requirement for new tipping space. Baillie (1984) has estimated that the voids created in the United Kingdom by mining and quarrying every year could accommodate five times the United Kingdom's MSW production. The problem arises, however, in that the void space is not always in close proximity to the sources of waste materials. Neither is the nature of many of the voids created appropriate for the landfilling of many wastes, which, by the nature of their pollution potential, must be contained and isolated from groundwater. Many metropolitan areas have insufficient landfill sites to meet their needs and void space scarcity is a major problem for virtually all cities. Perhaps most graphic is the New York (USA) landfill on Staten Island which, unable to expand laterally, is being

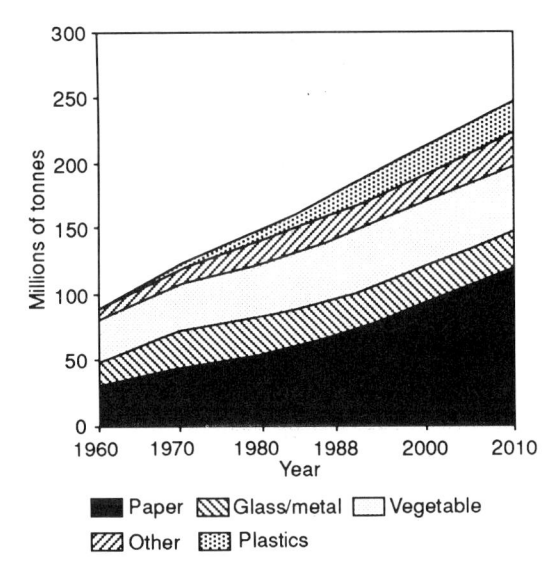

Figure 8.1 The 'Garbage Crisis' in the United States: generation of materials in MSW between 1960 and 1988, with projections to 2010. (Environmental Protection Agency, 1990b)

piled up skywards (landraising), so that by the year 2000 it will form the highest point on the coastline between Maine and Florida. Long-distance haulage of MSW is often the result of landfill shortage with dramatically increased disposal costs. New Jersey hauls more than half its wastes out of the state to West Virginia, Pennsylvania and Ohio over distances of up to 500 km. London similarly has insufficient landfill sites to meet its needs; the Greater London area produces just over three million tonnes of MSW annually. Over 90 per cent of this is landfilled, but only 9 per cent is deposited directly at a landfill by collection vehicles, the vast bulk is taken to a landfill site via a transfer station and completes its journey via road, rail and Thames barge.

The reclamation of quarry sites is potentially attractive as mineral extraction has led to the production of more derelict land than any other form of industrial activity. Properly managed, derelict voids can be reclaimed by a process of sanitary landfilling, ultimately bringing the land back into productive use and providing much needed waste disposal sites.

Sanitary landfill as a technique has replaced the open tipping which characterized landfill disposal before the 1950s. Landfill sites are now commonly 'engineered' and operated so that wastes are placed in layers 1–2 m deep and compacted by metal-wheeled vehicles. An uncontaminated cover material, usually soil, is spread over the wastes daily and blowing litter and pests controlled. A local supply of cheap covering material can prove to be a limitation for sanitary landfilling: large quantities are required with UK guidelines stipulating uncontaminated material to a depth of 0.15 m to cover all newly deposited waste at the end of each working day (Department of the Environment, 1986).

8.2.2 'A burning issue': domestic waste incineration — dream fuel or pollution nightmare?

Incineration as a treatment for MSW is not new. Its principal advantage, the reduction of bulk, has long been recognized and increasingly represents a desirable policy goal as landfill space becomes scarcer and more costly. The main disadvantages of burning waste include the cost of establishing and running the plant and the pollutant emissions to air.

In the United Kingdom at the beginning of this century, there were many more incineration plants than there now are. 338 incinerators were reported in 1914, 295 of which had boilers for waste heat recovery (House of Lords, 1988/9). With the development of easy access to landfills, these numbers declined and at present some 35 are in operation in the United Kingdom, mainly in urban centres. Together they burn no more than 6 per cent of the MSW arising in the United Kingdom and only a handful recover waste heat. Most of these incinerators are nearing the end of their design lives and will either have to undergo expensive fitting of emission control systems

Table 8.5 MSW disposal by incineration for European countries

Country	% MSW incinerated	No. of MSW incinerators	No. which can recover energy
Belgium	25	28	13
Denmark	32	40	36
France	32	280	64
Italy	20	94	16
Netherlands	19	11	6
Spain	58	8	3
UK	6	35	6

Source: House of Lords Select Committee on the European Communities (1988/9)

to meet new air pollution standards or they will have to close, diverting the raw waste which they treat to landfill. Yet the search for renewable sources of energy has revived UK interest in waste to energy schemes with several being planned where waste catchments are large, notably in London. Incineration is a more popular technology elsewhere in the Western World (Table 8.5). For instance, the high population densities and limited landfill opportunities in Japan have made incineration the normal method of MSW treatment. Currently over 40 states in the United States are operating, building or considering 'waste-to-energy' plants (Hershkowitz, 1987) and the pattern in western Europe shows that many countries incinerate a growing proportion of their waste and recover energy too. The attractions of turning rubbish into energy are clear. Armson (1985) has calculated that a typical city of half a million people would produce 150,000 tonnes of MSW each year. If this is incinerated and two-thirds of the heat so generated is recovered, the resulting heat energy would be sufficient to provide 10 per cent of the city's heat requirements. Yet the construction of a new incinerator is expensive, at over £50 million, and beyond the capabilities of many UK local authorities, particularly as landfill costs, though rising, are still relatively low.

The choice facing the waste manager considering whether to adopt incineration as a disposal option is complex and often controversial. The 'dioxin debate' is certainly threatening the more widespread use of incineration incorporating energy recovery. Locally the choice is often between the increasing unpopularity of landfill and the risks associated with toxic emissions from incineration.

A recent House of Lords Committee Report summarizes the position as far as the United Kingdom is concerned: 'The Committee judge from the evidence that they have received that combined incineration and heat recovery schemes are desirable. (They) enhance the attractiveness of incineration on grounds of economy and energy conservation.' The report concluded that incentives to develop such schemes should be considered to bring the United Kingdom into line with other European countries (Table 8.5).

8.2.3 Other forms of treatment

One of the principles in dealing with waste and MSW in particular is the reduction of the volume and increase in the density of the waste. Clearly this is particularly important where landfill void space is scarce. Several techniques are available but they add to costs as they involve waste transfer and special plant:

Compaction. This option is quite widely used where waste has to be transported over long distances (e.g. out of urban areas).

Powerful hydraulic rams are used to compress the waste into closed containers suitable for road or rail haulage.

Shredding and baling. Mechanical shredding of MSW is a normal prerequisiste for baling, a good method of densification. The shredded waste is compacted into cubic bales of about 1 m; these are then bound with wire mesh and can be handled easily during transfer and emplacement at the landfill. The method is environmentally more acceptable as it reduces the litter and dust nuisance and has been used at landfills in close proximity to housing. Densities of 1.0 to 1.3 tonnes per cubic meter can be achieved which represent improvements of at least 25 per cent over untreated waste.

Pulverization and recovery. Pulverization is a high energy technique which uses mechanical means to break up the waste in the presence of a lubricant, usually water. It can be used to separate the organic and inorganic fraction and is thus a necessary but expensive treatment in the production of refuse-derived fuel (RDF) and for separating out the organic component for composting.

Desirable though many of these techniques of treatment are, their cost has meant that they are not as widespread as they might be. In the United Kingdom, for instance, 79 per cent of all MSW was disposed of to land without any treatment in 1988–9; of the remainder, 13 per cent was treated by compaction or baling, 6 per cent by incineration, and less than 2 per cent was subject to other methods including resource recovery (Barron, 1989).

8.2.4 The 'Four R's': reduction, reuse, recycling and recovery

The 'Four Rs' represent a convenient shorthand to illustrate the alternative approach to waste management, namely to deal with the causes rather than the effects of waste. They were coined in North America where the 'throwaway' aspects of the consumer society had produced a garbage crisis and they have become a rallying call to those trying to find green solutions. They represent a hierarchy with the first R, reduction, being the most desirable policy and the fourth, recovery, the least (Johnson, 1990):

Reduction. Keeping materials out of the waste stream usually requires a change in consumers' and producers' habits and philosophy: substituting durable goods for disposable ones and reducing waste in production systems.

Reuse. Reuse involves extending the life of materials and perhaps is best illustrated by refillable beverage containers and rechargable batteries.

Recycling. Although recycling does help to conserve resources and reduce waste streams, there are economic and environmental costs associated with the collection and processing of recycled materials. Ambitious targets have been set for the recycling of MSW, but the process has been slow in having any appreciable effect although extensive trials have taken place (Institution of Civil Engineers, 1991).

Recovery. Whereas successful recycling requires separation of waste at or near its source, many schemes now exist which recover both materials and energy from post-consumer waste. This may involve relatively simple processes such as magnetic separation to recover ferrous metals or complex integrated reclamation plants such as that at Byker, Newcastle (Plate 8.3) or several in the United States.

8.3 Emissions from waste disposal: leachate from landfill, air pollution from landfill and incineration

It is a mistake to view all landfills as a final resting place for wastes — the out-of-sight, out-of-mind mentality. Escapes of materials are inevitable: materials carried by water as leachate and gaseous and particulate emissions to the air. The concept has emerged of the landfill as a 'reactor' (Baccini, 1989), in which wastes are treated and the products of

Plate 8.3 The Byker Reclamation Plant, Newcastle. Opened in 1979 as a pilot development to promote waste recovery. The plant handles 50,000 tonnes of domestic refuse per year and recovers metals and other materials. It produces over 12,000 tonnes of refuse-derived fuel pellets per year. (Photo: T.D. Douglas)

the treatment or reaction can escape to the environment or be managed by collection and venting systems.

8.3.1 The hydrogeology of 'holes in the ground'

The protection of groundwater has been identified by environmental agencies as a priority task. Groundwater resources are seriously threatened by improper waste disposal, namely poor selection of landfill sites in areas of groundwater vulnerability and the legacy of old landfills dating from the era before adequate regulation (see also Chapter 7).

Precipitation falling on an actively used landfill site will invariably infiltrate the waste materials and percolate through the waste, leaching organic and inorganic constituents as it passes. This product is known as leachate and will vary in its polluting capacity according to the amount and type of waste, the amount of water infiltrating (the effective precipitation) as well as the nature of the biochemical and physical breakdown processes within the body of the wastes. Leachate volumes can be large: at Pitsea in Essex, one of the largest UK landfills, where industrial waste, much of it liquid, is co-disposed with domestic waste, there is an estimated 1.5–2.0 million cubic metres of leachate in a waste mass of about 10 million tonnes (Knox, 1985). Whereas the potential to pollute from hazardous waste sites is

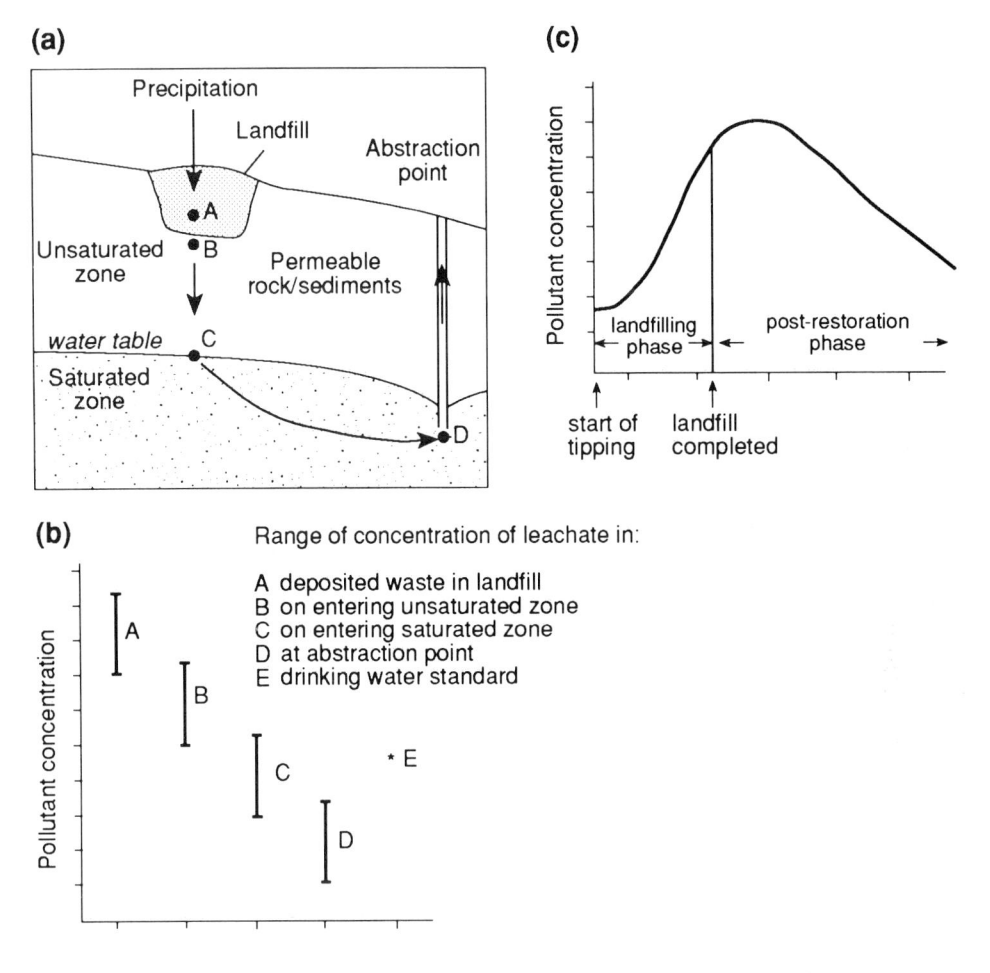

Figure 8.2 Schematic diagrams to show hydrogeological relationships below a landfill site in permeable rocks. (a) Cross-section showing pollution pathways; (b) illustration of the way in which a pollutant concentration can be attenuated through the groundwater system; (c) schematic graph of pollutant concentration through time. (Source: Department of the Environment, 1986)

obvious, leachate from domestic waste landfills can be high in suspended solids, organic and inorganic content.

Historically, there have been two different approaches to the protection of groundwater below and surrounding landfill sites. In reality these can best be represented as being at two ends of a spectrum with a variety of intermediate combinations. At one extreme is the most widely accepted philosophy of containment which seeks to isolate the waste from groundwater by establishing an impermeable barrier. In this way, leachate will be contained on-site. Should the natural geology not afford adequate protection because permeability rates are too high or leachate could escape through fissures in the underlying rocks or sediments, it is possible to line the landfill with imported natural materials (usually clays) or by using artificial liners.

The other end of the spectrum is represented by the *dilute and attenuate* philosophy, increasingly under attack. The principles underlying this approach recognize that the leachate will become attenuated as it migrates through the unsaturated zone and will be further diluted by mixing with

Plate 8.4 Leachate collection lagoon, North Carolina, USA. This artificially lined structure is used to collect leachate from a landfill in which incinerator residue is tipped. (Photo: T.D. Douglas)

groundwater. The system is illustrated in Figure 8.2, where the pollution pathway is shown in (a); a schematic example of how pollution concentration can be attenuated in (b); and the graph in (c) shows the decay of pollution concentration at a point in the system with time. Clearly such a strategy affords protection only when the hydrogeological information is sufficiently good to allow the accurate prediction of the rates of movement of leachate within the unsaturated and saturated zones and when the biochemical and physical processes leading to the attenuation of the leachate can be adequately modelled. It will be evident that these conditions are not often met and adherence to the precautionary principle would suggest that the essence of the technique, the dispersal of pollutants into a large pollution plume controlled by the vagaries of a largely unseen geology, can be a risky enterprise. The siting of many old landfills without due regard to these factors has indeed proved disastrous.

However, the other end member of these opposing philosophies, *concentrate and contain*, is not without disadvantages. As the substrate is impermeable, the landfill, if inadequately covered, can become saturated with infiltrating rain and the leachate can spill overland. Leachate can be controlled with the installation of a collection system which comprises a network of drainage pipes installed prior to the deposit of waste, which carry the leachate to a topographic low from which it can be pumped to a tank or lagoon (Plate 8.4).

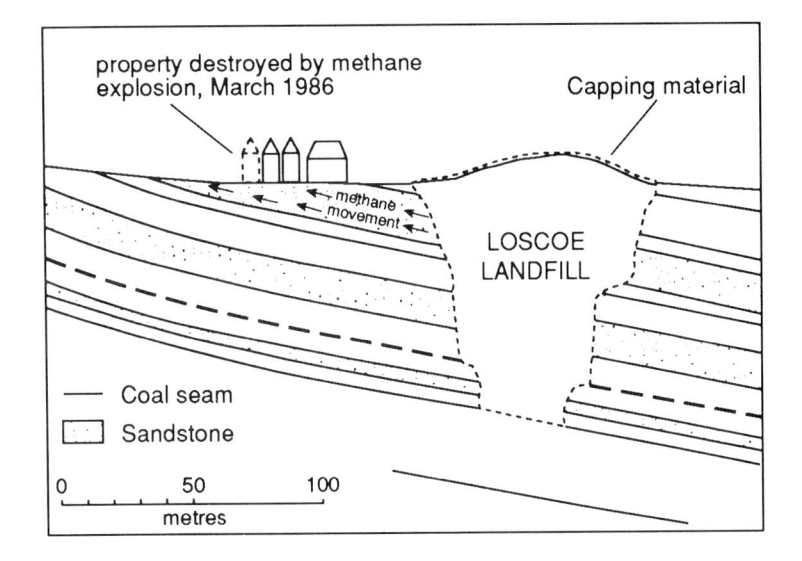

Figure 8.3 Geological cross-section through the Loscoe landfill to illustrate the gas escape. (Williams & Aitkenhead, 1991)

8.3.2 Methane and other landfill gases

Landfill gas is produced during the anaerobic decomposition of organic material by microbial activity. Such materials are present in several waste streams but notably in MSW, where the proportion by weight of biodegradable components can exceed 40 per cent and includes paper, wood and vegetable wastes (Table 8.3). Landfill gas is complex in its composition which changes with time and many other factors, but the principal gases present are methane and carbon dioxide, with the former commonly representing more than 50 per cent by volume. A hazard exists because methane in concentrations of only 10 per cent in air constitutes an explosive mixture as illustrated by the destruction of a bungalow at Loscoe, Derbyshire in 1986. The Loscoe example demonstrates the geological context in which methane was able to migrate laterally through the permeable sandstones towards the bungalow 70 m distant from the edge of the site (Figure 8.3). Prevention in such cases would have required effective barriers being installed around the landfill, gas collection systems within the landfill, and possibly monitoring boreholes as a precaution. As a result of this and other incidents, HMIP collated surveys by local authorities in the United Kingdom to assess the scale of the problem. Estimates in 1989 suggested that over 1000 sites were located within 250 m of residential areas and needed some gas control measures.

The migration of gas to the surface of a site is also connected with adverse effects on vegetation. This implies careful after-use of the site, although gas-tolerant grass species have been identified for some land uses such as parkland and golf courses (Plate 8.5). Experience with gas escapes from old landfills has demonstrated the need to design control measures at an early stage and to monitor closed landfills. For large domestic landfills, it is now increasingly common for gas to be collected and used to recover energy; on smaller sites it is either flared or vented to the atmosphere.

8.3.3 Incineration: principal emissions

MSW incineration emits a wide range of gaseous pollutants which may include dioxins and furans, heavy metals such as cadmium and lead, chlorine in the form of acidic gases as well as greenhouse gases such

Plate 8.5 Newly established golf course on old landfill, Charlotte, North Carolina, USA. Methane gas is vented to the atmosphere and some problems have occurred with the establishment and maintenance of a high quality grass cover

as carbon dioxide and nitrous oxides. The more toxic of these can be largely removed by pollution control devices. These include after-burners which are designed to burn up pollutants formed at low temperatures after the first incineration; electrostatic precipitators to trap ash particles; and scrubbers which further reduce emission levels. Some of these abatement measures result in toxic residues as the contaminated ash is removed before passing into the stack: these have to be landfilled. A recent study has shown that MSW incinerators are responsible for about 30 per cent of the total dioxin release to the UK atmosphere, with industrial chemical incinerators responsible for less than 1 per cent (Department of the Environment, 1989). This study also claims that the technology is available to construct new incinerators which will have very low levels of dioxin emissions and that the performance of existing plant could be improved by more stringent operating conditions. In contrast, Ontario, Canada, which has a massive 'garbage problem', has rejected MSW incineration for two reasons. Firstly, it is not convinced that the toxic emissions to the atmosphere or the toxic fly ash disposal to landfill can be properly controlled or monitored and, secondly, it feels that the technology is inconsistent with reduction, reuse and recycling which would attack the root of the problem.

Studies have been undertaken of emissions from Belgian MSW incinerators which have demonstrated the complexity of modelling concentrations in the plume due to the plethora of variables involved. These include the nature and moisture content of the waste feed; the control devices and operating

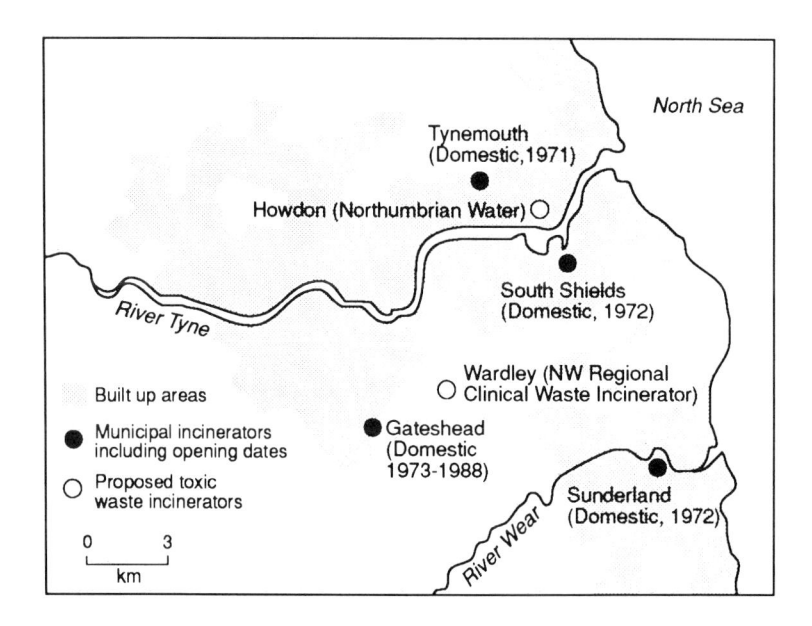

Figure 8.4 Map showing the distribution of incinerators and proposed incinerators in the Tyne and Wear region of North-East England

temperatures; the stack height and meteorological conditions (De Fre, 1986). In this debate many have placed their trust in the ability of technology to control air emissions. Yet regulation and enforcement are extremely difficult for some pollutants: with the continuous measurement techniques currently available it is not possible, for instance, to measure the emissions of dioxins and furans. The key technical requirements appear to be the achievement of very high combustion temperatures, a minimum residence time in the combustion chamber and a sufficient oxygen supply for total destruction (Argent, 1991).

Sweden incinerates about half of its MSW in 20 plants with linked district heating schemes. The technology is considered as established and well-operating. Nevertheless, much stricter controls were implemented during the 1980s when tests for dioxins in breastmilk showed that babies could have been ingesting between 50 and 200 times the daily 'safe' limit of dioxins. It is widely regarded that the source of many of the most toxic elements in MSW incinerator emissions is the presence of PVC plastic in the waste feed; so a possible management strategy would be to reduce or eliminate this source.

8.4 Case studies in a developing arena

8.4.1 Domestic waste, sewage sludge and toxic waste incineration on Tyneside

In the United Kingdom, the role of incineration in waste disposal has often been controversial. This is exemplified by the situation in Tyne and Wear where throughout the 1970s and 1980s the clear policy of the local authorities for the disposal of MSW was one of incineration. Four such incinerators were constructed in the early 1970s in response to the shortage of local landfill space. Together they handled nearly 70 per cent of the local domestic waste stream. Twenty years later the plants have closed or are facing closure. The incinerators are nearing the end of their design life and without the expensive fitment of modern pollution control devices will be unable to meet new emission standards being advocated by the European Community. The incinerators were only fitted with devices to control particulate emissions, not gaseous ones. The first to close, that serving Gateshead (Figure 8.4), shut in 1988 following a fair amount of media exposure resulting from concern over pollution emanating from it and the possible though unproven link with an identified leukaemia cluster (Openshaw, 1988). The conversion of the Gateshead incinerator to a baling plant and the likely phasing out of other incinerators at Sunderland, South Shields and Tynemouth is once more going to put pressure on landfill space on the urban periphery.

If in part the closure of the MSW incinerators is due to more stringent regulation of emission standards, then the plan to build a toxic waste incinerator at Howdon on Tyneside is largely the result of a government commitment to end the dumping at sea of liquid hazardous waste by the end of 1992 and the dumping of sewage sludge by 1998 (HMSO, 1990). The proposal by Northumbrian Water, who will be faced with a sewage disposal problem, is to build a high temperature incinerator jointly with a commercial firm who will use the facility for toxic waste disposal. This, together with similar proposals elsewhere in the United Kingdom, has prompted an urgent consideration of regional and national policy. For despite strong, well-argued opposition from campaigning groups against the proposal, it is clear that high temperature incineration is increasingly seen in Europe as the best available option. How many such incinerators are needed, where they should be sited and how efficiently they destroy the wastes will remain a subject of environmental controversy.

8.4.2 Toxic tips: the US experience at Love Canal

Love Canal, an old toxic waste site near Niagara Falls in New York State, United States, ranks alongside Chernobyl, Three Mile Island and Minimata in an inventory of environmental disasters. Although not involving the scale of human tragedy of some of those disasters, it became a symbol of environmental mismanagement recognized worldwide. In the United States the issue of pollution hurting people became vivid: it gave the environmental movement a shot in the arm by broadening its constituency and raising the national consciousness. Television pictures showed terrified and angry families forced to evacuate their homes as a result of the leaking chemical dump (Moyers, 1990).

About 21,800 tonnes of hazardous waste was dumped in an unfinished canal by the Hooker Chemical Company from 1942 to 1952 (Figures 8.5(a) and (b)). These wastes included chlorinated hydrocarbon residues, processed sludges, fly ash and some municipal solid waste. In 1953 the land was bought by the Niagara Falls Board of Education and a school and residential housing were constructed immediately adjacent to the closed dump site (Brown, 1981). However, it was not until 1978, following a succession of complaints from residents about chemical odours and reports of chemicals breaking through the soil cover, that the site received wide attention. An analysis of samples taken from the basements of homes adjoining the landfill revealed elevated levels of toluene, chlorobenzene, dichlorobenzene and other known carcinogens and the area was declared a threat to human health. Later that year, following the closure of the school and the recommended relocation of the residents most at risk — pregnant women and children under two — US President Carter declared a federal emergency, the first ever for a man-made disaster.

The questions posed as this scandal emerged included the formulation of the remedial measures necessary for the leaking landfill; the delimitation of the boundaries of the area most affected by the pollution; and the identification of the number of similar active and inactive hazardous landfills across the country which could pose equally lethal threats to health and well-being.

The remedial efforts at Love Canal took over ten years and cost $US250 million

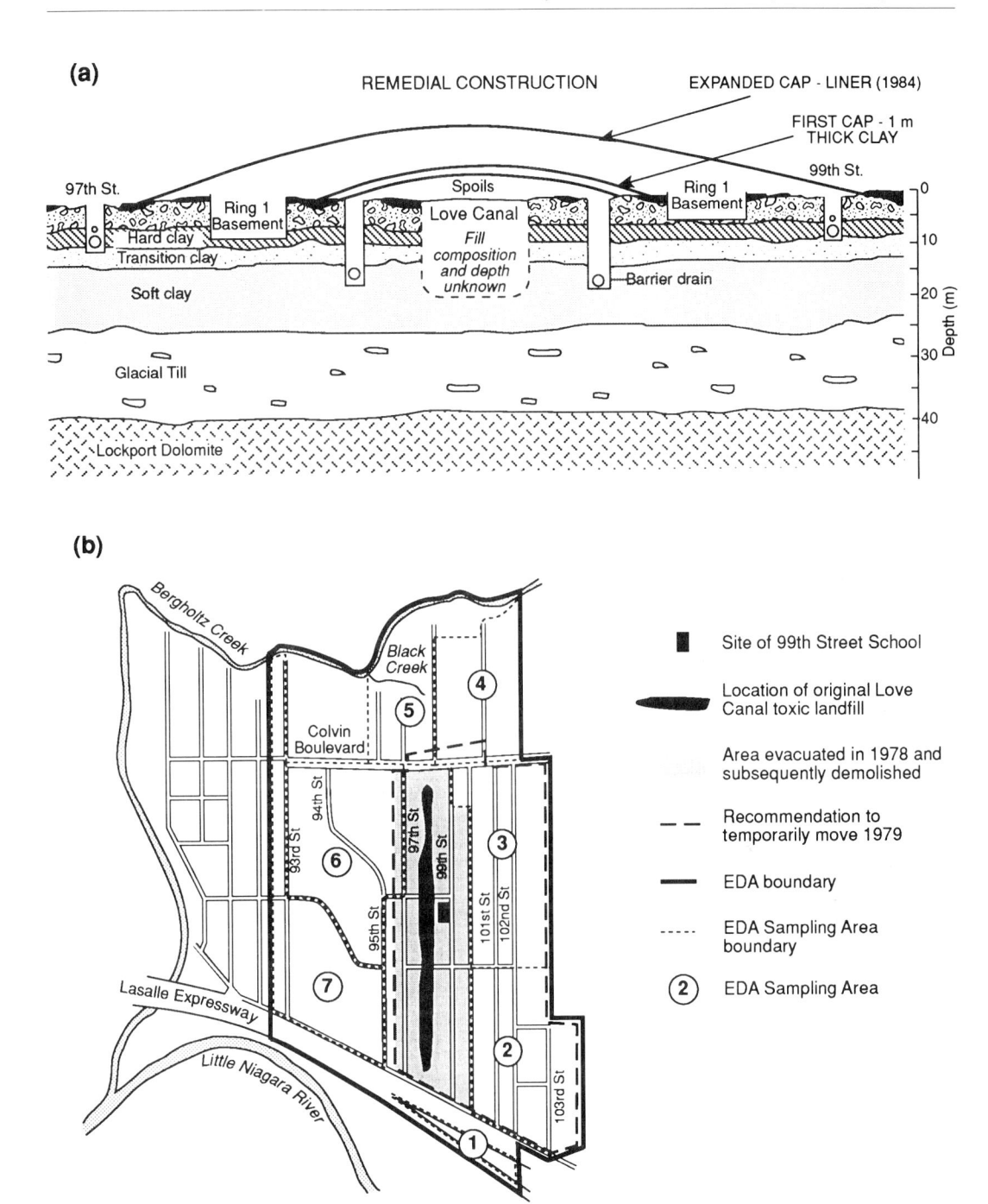

Patterns of land, water and air pollution by waste

Figure 8.5 Love Canal, New York State: (a) cross-section through the toxic landfill showing the remedial measures undertaken; (b) plan of landfill vicinity showing evacuation zones and limits of Habitability Study Sampling Areas. Areas 1–3 did not meet the habitability criteria, Areas 4–7 did and can continue to be used for residential purposes

(Silverman, 1989). The main thrust of the plan of action was to contain the leaking landfill and subsequently to clean up off-site contamination. The construction of the roads around the residential area had breached the walls of the old clay-lined canal with storm and sanitary sewers installed on a gravel bed, thus providing a migration route for escaping leachate. The original clay cap on the landfill had failed to prevent the ingress of rain and snowmelt, forcing many wastes to the surface and exposing them to the threats of volatization and surface runoff. The surface geology of the Love Canal area consists of silts and fissured clays, materials which had allowed the lateral migration of toxic leachates into the surrounding residential neighbourhood. Containment was achieved by encircling the dump site with barrier drains which collected migrating leachate, allowing it to be pumped to a storage facility at an on-site treatment plant. In 1979 the first clay cap was added to limit the ingress of water and to prevent the emission of contaminants to the air. Five years later, an expanded cap and plastic membrane was emplaced to cover the area of the inner ring of demolished houses and to limit further the ingress of water into the leachate collection system. These two caps and the leachate collection and treatment facility are part of a fenced-off area which isolates the inner area close to the landfill. The subsequent focus of the remedial plan was to clean up those wastes which had migrated off-site and this has entailed the cleaning or removal of 20,000 m of storm and sanitary sewers and the dredging of sediments in the Black and Bergholtz Creeks (Figure 8.5(b)) where dioxins had been identified on gravel samples and in fish.

It was the 'habitability' question, however, where the environmental imperatives and the social, political and health issues came together. Between 1982 and 1983, the US Environmental Protection Agency had made over 150,000 individual measurements of environmental contamination levels near Love Canal. The findings of this study indicated that outside the inner contaminated area where the houses had been demolished, levels of chemicals were comparable to those found in other cities and no environmental standards had been breached. These findings were not universally accepted and so a panel of independent experts was assembled to develop habitability criteria. Ten years after the Emergency Declaration Area (EDA) had been designated, the Habitability Study was published (New York State Department of Health, 1988) and concluded that about two thirds of the EDA (consisting of some 300 houses) was now suitable for unrestricted residential use, whilst a further 200 homes were declared uninhabitable (Figure 8.5(b); Plate 8.6). The criteria chosen for this lengthy and expensive benchmark study were comparisons of soil and air samples from the EDA and from other sites in the Niagara area with comparable soils and at a distance from known landfills. In all some 2500 samples were taken and 11 'indicator' chemicals were tested for. Such an approach was deemed necessary for, apart from the dioxin, the other Love Canal chemical pollutants were not associated with specific health standards in terms of their safe concentration in soil and air samples. The full health impact of the Love Canal disaster can only be established in time as many of the alleged effects of exposure to these toxins take time to reveal themselves. As is so often the case, establishing cause and effect in environmental health is fraught with difficulties. Medical studies of the Love Canal area have pointed to high rates of low birthweight infants in the area and children with significantly shorter stature for their age than expected (British Medical Association, 1991).

As the Love Canal story unfolded, attention was turned to the other possible dumps where extensive remediation might be necessary and where health hazards could exist. The US Environmental Protection Agency estimated that 2500 such sites could exist around the country, a problem which will take years and billions of dollars to deal with. The events at Love Canal expanded the focus of US environmental regulation from

Patterns of land, water and air pollution by waste

Plate 8.6 Abandoned house in the Love Canal Emergency Declaration Area, Niagara Falls, New York State. Over 200 homes were declared uninhabitable as a result of the leaking landfill. (Photo: T.D. Douglas)

water and air to include waste and also led to improvements in the legislative basis for dealing with toxic waste problems.

8.4.3 Integrated hazardous waste management: Alberta, Canada

The public mistrust of hazardous waste facilities following incidents such as that at Love Canal has resulted in numerous examples of vocal opposition to the siting of new facilities. This was the case in the Canadian province of Alberta in 1979 when a private company proposed to locate a hazardous waste treatment plant at Fort Saskatchewan near Edmonton. The provincial minister of environment placed a moratorium on all off-site development of waste-handling facilities until a management plan could be agreed (McQuaid-Cook and Simpson, 1986). The Alberta Special Waste Management Corporation was formed and invited proposals from private firms for the construction and management of an integrated waste facility which could treat and dispose of the annual provincial waste production as well as disposing of the stockpile of stored wastes which included PCBs from old electrical transformers. The siting strategy which emerged included extensive public involvement and dealt effectively with the sociological as well as the technological aspects of site selection. A number of constraints were identified (Table 8.6) and 120 community meetings were held throughout the province to explain the nature of the facility. Policy was that only constraint-free areas were to be considered for siting and that local authorities must

167

Table 8.6 Siting criteria for Alberta hazardous waste management facility

Physical constraints
   Geology
   Hydrogeology
   Surface water
   Topography
   Seismic activity potential
Biological constraints
   Forestry
   Soils
   Wildlife
   Birdlife
Land use constraints
   Agriculture
   Federal land
   Provincial Crown land
   Resource extraction
Human constraints
   Recreation areas
   Archaeological/historical sites

Source: McQuaid-Cook and Simpson (1986)

invite further investigations of those areas. The waste treatment plant was being planned to include waste storage and handling areas, a treatment plant, a stabilization facility, incinerators, a landfill, and deep-well disposal. Thus the geological constraints ruled out all areas with significant sand and gravel deposits as well as those bedrock areas where groundwater movement was through fractures or where groundwater recharge took place. Following a programme of community meetings throughout Alberta, 52 of the 70 local authorities requested assessments for plant suitability. As the siting process narrowed down possible contenders, some municipalities withdrew their bids following analysis of the constraints or public opposition. Eventually five candidate communities emerged in each of which detailed environmental assessments and test drilling were undertaken. In the final stage the two major contenders held plebiscites and with high voter turn-out, showed 79 per cent and 77 per cent in favour of hosting the facility (McQuaid-Cook and Simpson, 1986). In 1984 the site near Swan Hills was chosen and the plant was constructed and operating by 1987.

The site lies 200 km north west of Edmonton and is centrally located within the province. It occupies elevated ground covered by glacial tills and boreal forest. The plant is located 20 km from the town of Swan Hills itself which has a population of 2500. It is underlain by 10–15 m of dense clay till with measured hydraulic conductivities in the range of 10^{-7} cm/s, presenting a good natural containment barrier. Possible factors in the apparent enthusiasm of the community to host the plant were the employment opportunities it provided in the context of the existing dependence on the timber and oil industries and other local benefits such as improved services. Such factors clearly weighed more heavily than the perceived risks such as health fears, the threat of explosion and environmental degradation.

The hazardous waste plant represents state-of-the-art technology for such a regional facility. Environmental safeguards include comprehensive groundwater, air, soil, vegetation and small mammal monitoring and an 800-m buffer zone of forest surrounding the site. All surface water from the site is collected in retention ponds prior to deep-well disposal. The active landfill cell is covered with a steel shed to prevent the ingress of precipitation. One year of background environmental monitoring was undertaken prior to the opening of the plant to act as a reference datum for future comparisons.

The siting process can be seen as a remarkable achievement for environmental assessment and has been hailed as an example of YIMBY — Yes In My Back Yard! (Massam, 1988). Yet the participatory management which so clearly marked the siting process in Alberta is not necessarily always the key to success. In southern Ontario, an area which already has a number of 'toxic hotspots', the Ontario Waste Management Corporation has had fierce resistance from environmental groups and local inhabitants living near candidate sites for a large treatment facility. Here, however, many of the factors are radically different to those found in Alberta — population densities are greater with much less available land, the volume of wastes is much larger and the local memory of waste disposal includes Love Canal which is only a few kilometres away on the US side of the border. Although the clean-up of existing sites and the backlog of stored wastes necessitate new facilities being built and operated with the highest environmental safeguards, the existence of efficient integrated hazardous waste plants may encourage the future production of large volumes of waste when waste reduction policies would be the appropriate environmental option.

8.5 Towards environmental protection in the waste industry

8.5.1 Waste as a commercial entity

The waste industry is growing. This growth is in response to the larger volumes of waste being generated as well as to the greater efforts being made to regulate and manage it. In the United Kingdom, about 14,000 jobs are estimated to be associated with waste management with about 3500 of these as Waste Disposal Authority personnel in England and Wales.

The more stringent regulatory controls on many aspects of waste management and the shortage of facilities in some areas are giving rise to sharp increases in disposal costs. This trend will be most evident in landfilling costs, which many critics allege have been kept too low for too long and do not reflect the true environmental costs of disposal. The geographical expression of this new era of regulation will undoubtedly lead to fewer but larger companies operating in waste disposal in the 1990s and a reduction in the number of landfill sites licensed to operate. Table 8.7 shows the gross operational costs per tonne of MSW disposed using different methods. It is interesting to note that the disposal routes

Table 8.7 Average disposal costs for MSW in England and Wales, 1988–89

Disposal method	Gross operational costs per tonne (£)
Transfer: compaction/shredding/baling	4.34
Transfer: civic amenity sites	4.56
Incineration	11.71
Reclamation and other methods	6.25
Direct landfill by WDA	2.77
Direct landfill by contractor	3.46

Source: Waste Disposal Statistics 1988/9, CIPFA Statistical Information Service

involving technological input, such as incineration and reclamation, are still the most costly, although it has to be acknowledged that the data in Table 8.7 are averages for England and Wales and disguise considerable variation between London, where landfill by WDA costs are over £15 per tonne, and the non-metropolitan counties where costs are just £2 per tonne. The difference in costs between these alternative disposal methods can perhaps be explained with reference to environmental costs. The assessment of the economics of landfill disposal needs to include the costs of initial development, including engineering works, monitoring, restoration and post-restoration management. As landfill operators have to comply with a stricter regulatory framework, the costs of landfill relative to other methods of disposal will increase.

A good case can be made to support the assertion that the low prices paid for landfill in the past have been possible only through the degradation of the environment as a result of inadequate regulation and lax compliance. The adoption of a system of higher waste charges, more in line with the 'polluter pays' principle and the imposition of waste taxes, has clear commercial implications: clean technologies, waste minimization schemes and recycling initiatives may be given a much-needed impetus.

8.5.2 Controls on operation

There is good evidence that the way in which a waste disposal facility is designed and operated, even when it is well sited, can greatly influence its potential to cause pollution. The control instrument used in the United Kingdom is the site licence and the conditions attached to it. Under the Control of Pollution Act (1974), the WDA awarding the site licence could impose conditions which require the monitoring of leachate on-site and the monitoring of groundwater by off-site boreholes. Critics of the site licence system have pointed to the disparity of practice between the 195 separate WDAs within the United Kingdom as well as the need to design controls which are specific to the requirements of an individual site.

At the start of the 1990s, it was clear that the United Kingdom needed larger regional control bodies in which sufficient expertise could be gathered; as a result a new administrative tier of Waste Regulation Authorities (WRAs) was provided for in the 1990 Environmental Protection Act. A fundamental change in controlling the post-restoration phase of a waste site will be that a licence will only be able to be surrendered when the WRA approves the 'residual environmental liability' as being acceptable.

8.5.3 Integrated pollution control and waste

It is abundantly clear from the examples in this chapter that waste discharges have a potential impact on all three media (air, water and land) and that effective management of the waste industry can only be undertaken through integrated pollution control as opposed to fragmented, piecemeal regulation. Integrated systems have emerged under the US Environmental Protection Agency and are a fundamental ingredient of the Environmental Protection Act (1990) in England and Wales.

The components of an integrated waste management system embrace:

Waste avoidance: clean technology; waste minimization; reuse.
Material recovery: separate collection; recycling and recovery.
Energy recovery: RDF production; waste to energy incineration.
Waste transport and handling: implementation of 'Duty of Care'.
Final disposal: secure, properly engineered landfill.

Such a pattern of 'systems thinking', where waste is viewed in its true environmental context, is going to require a fundamental rethink of attitudes towards waste management (Turner and Powell, 1991). It must be the hoped that, at the end of the twentieth century, waste management systems no longer bequeath future generations a legacy of expensive pollution clean-ups.

References

Argent, F. (1991), 'Toxic and hazardous waste', *Environmental Policy and Practice*, 1(3), 10–18.
Armson, R. (1985), 'Turning rubbish to heat and electricity', *New Scientist*, 7 November, 33–6.
Baccini, P. (ed.) (1989), *The Landfill: Reactor and Final Storage*, Springer-Verlag, Berlin, 438pp.
Baillie, A.D. (1984), 'An examination of the geological controls, practice and legislation involved in the restoration of sites of mineral extraction by the use of domestic refuse as fill material', unpublished M.Sc. thesis, Imperial College of Science and Technology, University of London, 137 pp.
Barron, J.M. (1989), *An Introduction to Wastes Management*, Institution of Water and Environmental Management, London, 62 pp.
British Medical Association (1991), *Hazardous Waste and Human Health*, Oxford University Press, Oxford, 242 pp.
Brown, M. (1981), *Laying Waste: The Poisoning of America by Toxic Chemicals*, Washington Square Press, New York.
De Fre, R. (1986), 'Dioxin levels in the emissions of Belgian municipal incinerators', *Chemosphere*, 15 (9–12), 1255–60.
Department of the Environment (1986), *Waste Management Paper No. 26: Landfilling Wastes*, HMSO, London, 205 pp.

Department of the Environment (1989), *Pollution Paper No. 27: Dioxins in the Environment*, report of an interdisciplinary working group on polychlorinated dibenzo-para-dioxins (PCDDs) and polychlorinated dibenzofurans (PCDFs), HMSO, London.

Department of the Environment (1990), *Digest of Environmental Protection and Water Statistics No. 12*, HMSO, London.

Douglas, T.D. (1988), 'Waste disposal and landfill in the Northern Region', *Northern Economic Review*, 16, 47–51.

Environmental Protection Agency (1990a), *The Nation's Hazardous Waste Management Program at a Crossroads*, EPA, Washington, DC, 114 pp.

Environmental Protection Agency (1990b), *Characterization of Municipal Solid Waste in the United States: 1990 Update*, EPA, Washington, DC, 15 pp.

Furuseth, O.J. and Johnson, M.S. (1988), 'Neighbourhood attitudes towards a sanitary landfill: a North Carolina study', *Applied Geography*, 8, 135–45.

Gatrell, A.C. and Lovett, A.A. (1986), 'The geography of hazardous waste disposal in England and Wales', *Area*, 18(4), 275–84.

Gray, D.A., Mather, J.D. and Harrison, I.B. (1974), 'Review of groundwater pollution from waste disposal sites in England and Wales, with provisional guidelines for future site selection', *Quarterly Journal of Engineering Geology*, 7, 181–96.

Hershkowitz, A. (1987), 'Burning trash: how it could work', *Technology Review*, 90(5), 26–34.

HMSO (1990), *This Common Inheritance: Britain's Environmental Strategy*, HMSO, London, 291 pp.

House of Commons (1988/9), *Environment Committee, Second Report, Toxic Waste*, 3 vols., HMSO, London.

House of Lords (1988/9), *Select Committee on the European Communities: Air Pollution from Muncipal Waste Incineration Plants*, HMSO, London.

Institution of Civil Engineers (1991), *Recycling Household Waste: The Way Ahead*, The Institution of Civil Engineers, London, 96 pp.

Johnson, L. (1990), *Green Future: How to Make a World of Difference*, Penguin Books Canada, Toronto, 231 pp.

Knox, K. (1985), 'Leachate treatment with nitrification of ammonia', *Water Resources*, 19(7), 895–904.

Massam, B.H. (1988), *Environmental Assessment in Canada: Theory and Practice*, Canada House Lecture Series No. 39, 32 pp.

McKechnie, R., Simpson-Lewis, W. and Niemanis, V. (1983), 'Sanitary landfills and their impact on land', in W. Simpson-Lewis, R. McKechnie and V. Niemanis (eds.), *Stress on Land*, Environment Canada, Ottawa, pp. 34–91.

McQuaid-Cook, J. and Simpson, K.J. (1986), 'Siting a fully Integrated Waste Management Facility', *Journal of the Air Pollution Control Association*, 36(9), 1031–6.

Moyers, B. (1990), *Global Dumping Ground: The International Traffic in Hazardous Waste*, Seven Locks Press, Washington, DC, 152 pp.

Nagelhout, D., Joosten, M. and Keimpe, W. (1990), 'Future waste disposal in the Netherlands', *Resources, Conservation and Recycling*, 4, 283–95.

New York State Department of Health (1988), *Love Canal Emergency Declaration Area Habitability Study, Final Report of the Technical Review Committee*, 5 vols.

Openshaw, S. (1988), 'Leukaemia patterns in Northern England: a new method for finding cancer clusters', *Northern Economic Review*, 16, 52–9.

Silverman, G.B. (1989), 'Love Canal: a retrospective', *Environment Reporter*, 20(20), 835–50.

Simpson, S. (1990), *The Times Guide to the Environment*, Times Books, London, 224 pp.

Turner, R.K. and Powell, J. (1991), 'Towards an integrated waste management strategy', *Environmental Management and Health*, 2(1), 6–12.

United Nations Environment Programme (1990), *Environmental Data Report*, Blackwell, Oxford, 547 pp.

Williams, G.M. and Aitkenhead, N. (1991), 'Lessons from Loscoe: the uncontrolled migration of landfill gas', *Quarterly Journal of Engineering Geology*, 24, 191–207.

9

Metal pollution of soils and sediments: a geographical perspective

Mark Macklin

Environmental contamination and degradation arising from the mining, production and utilization of metal constitutes the earliest perturbation of geochemical cycles by human activity. From the time that copper was first shaped and smelted in the Near East around 7000 BC (Renfrew and Bahn, 1991), whenever metals and their compounds are heated, pulverized or dissolved they become labile or unstable and susceptible to redistribution by air, water or dumping until reaching a 'sink' such as soil or sediment. Here they are stored usually for extended periods, frequently approaching centuries or millennia. The prolonged residence time of heavy metals in environmental media, as well as their persistence and bioaccumulatory nature in plants and animals, is one of the most significant aspects of metal contamination and engenders them with a greater potential for global distribution and global effect.

9.1 Patterns of pollution in time and space

Following the recognition that soils are the ultimate sink for metals in the terrestrial environment (Nriagu, 1990), and that fluvial processes are the primary mechanism for their transportation and redistribution at the earth's surface (e.g. Graf, 1990), has come the necessity of developing sound generalizations to explain the physical dynamics of particulate pollutants. This has drawn physical geographers, and especially river geomorphologists, from the periphery of metal pollution studies to become key players in investigations of the global cycling of trace metals.

Three areas in the field have recently seen notable geographical and geomorphological contributions. First, there is a growing body of literature (summaries can be found in Horowitz, 1985; Lewin and Macklin, 1987; and Graf, 1990), emanating from geomorphological studies in Australia, Europe and North America, on river systems, alluvial soils and sediments that have been affected by the discharge of metal mining wastes. Second, lake sediment studies (e.g. Salomons and Forstner, 1984) have been used as a historical record of changing metal concentrations through time. Third, heavy metal concentrations have been examined in urban and amenity soils (e.g. Aspinall, Macklin and Openshaw, 1988). Many of these studies have focused on estimating spatial and temporal metal pollution fluxes and their controls, which is also a central theme of this book and chapter. There has also been a move away from using sediment and soils as media solely for the purposes of pollution monitoring, as it has become apparent that soil and sediment themselves can be very important secondary sources of metal pollutants. Thus, in the light of these recent developments, it would seem timely to review metal pollution in terrestrial environments from a geographical viewpoint.

This chapter is subdivided into two sections. The first provides a general background to heavy metals in the terrestrial environment and considers:

(i) the nature and history of metal pollution (global and local) since the Industrial Revolution;
(ii) sources of metal contaminants;
(iii) biogeochemical pathways that metals follow in the environment, including the principal dispersal agencies and environmental compartments where longer-term metal accumulation occurs;
(iv) approaches to identifying metal contaminated/polluted soils and sediment;
(v) links between heavy metal pollution and plant, animal and, in particular, human health.

In the second section geographical approaches to environmental auditing of heavy metals are illustrated using two case-studies from northeast England. The first examines metal levels in urban soils within the intensely industrialized and highly populated area of Tyneside. The second study considers the pollution legacy of historic metal mining within the Tyne River and its catchment.

9.2 Nature and history of metal pollution

Following Davies (1980), heavy metals are defined here as metallic elements of a density greater than 6 g/cm^3 and some of the more important heavy metals are listed in Table 9.1. Although the term heavy metal is well established in the literature its definition is based on a rather arbitrary chosen parameter (e.g. Jarvis, 1983, sets the density threshold for heavy metals at 4.5 g/cm^3) and consequently includes elements with widely different chemical properties. Some metals (e.g. Cr, Co, Cu, Ni, Sn, Zn) are essential for the life functions of plants and animals, and only pose a problem when their concentrations in air, soil or water are significantly elevated above natural levels by human activities. In this way heavy metal investigations differ from studies of organic compounds (e.g. pesticides) or man-made radionuclides where detection is itself an indication of contamination (Davies, 1980).

Other heavy metals, including the most common pollutants (e.g. Cd, Pb, Hg) not yet identified as serving a beneficial function, are termed non-essential and are toxic even at low concentrations. In relation to environmental pollution heavy metals which are very toxic, relatively abundant in nature and readily available as a soluble species pose the greatest threat to biological systems (Wittmann, 1981). Heavy metals which fall within this category (as defined by Wood, 1974) are indicated in Table 9.1.

From the advent of metallurgy in prehistoric times up until the beginning of the Industrial Revolution, metal contamination was a local problem generally restricted to larger towns and cities, and areas around non-ferrous mine workings. In the past three hundred years, particularly since the Second World War, mining and the use of metals have grown rapidly as have the extent and severity of metal contamination in the environment. The scale of the problem has changed from local or regional in the late nineteenth century (e.g. highlighted by the Rivers Pollution Commission (1874) report on the base-metal mining areas of Britain), to continental and global in the 1990s as shown by trace metal contamination in remote polar regions (Pacyna and Winchester, 1990). Sediments and soils, similar to ice bodies, can uniquely preserve the historical sequence of pollution intensities as well as indicating current environmental contamination. Figure 9.1 illustrates historical changes in metal loading recorded in a variety of environmental media, reflecting anthropogenic perturbation of geochemical cycles over the last 100–300 years. Progressive increases in soil

Table 9.1 The more important heavy metals (d > 6 g/cm³) with their commonly accepted densities and crustal abundances

Element	Density (g/cm³)	Mean content in crustal rocks (ppm)	Essential (plants/animals)	Known pollutant	Very toxic and relatively available (Wood, 1974)
Ag	10.5	0.07		P	✓
Au	19.3	0.05			✓
Bi	9.8	0.17		P	✓
Cd	8.7	0.2		P	✓
Cr	7.2	100	E	P	
Co	8.9	25	E	P	✓
Cu	8.9	55	E	P	✓
Fe	7.9	6×10^4	E	P	
Hg	13.6	0.08		P	✓
La	6.2	25			
Mn	7.4	950	E		
Pb	11.3	13		P	✓
Mo	10.2	1.5	E	P	
Ni	8.9	75	E	P	✓
Pt	21.5	0.05			
Tl	11.9	0.45		P	✓
Th	11.5	9.6		P	
Sn	7.3	2	E	P	✓
U	19.1	2.7		P	✓
V	6.1	135	E		
W	19.3	1.5	E	P	
Zn	7.1	70	E	P	✓
Zr	6.5	165			

Source: After Davies (1980)

Cd (Figure 9.1(a)) at Rothamsted Experimental Station (UK) (Jones, Symon and Johnston, 1987) and Pb concentrations in Greenland snow (Figure 9.1(b)) since the mid-1800s (Wolff and Peel, 1985) both reflect longer-term atmospheric contamination. In the Mississippi (Figure 9.1(c), Trefry, Metz, Trocine and Nelsen, 1985) and Rhine (Figure 9.1(d), Salomons and Forstner, 1984) rivers, maximum concentrations of most metals occurred in the 1960s and 1970s, a time when in both North America and Europe aquatic metal pollution of surface waters reached its peak. Within the last decade inputs of pollutant Pb to the Gulf of Mexico from the Mississippi River have declined by about 40 per cent following partial regulation of Pb additions in gasolene. In the lower reaches of the Rhine, sediment metal concentrations are also now decreasing as the result of increasing efforts to reduce waste-water input. Despite these recent reductions metal levels in the Rhine and Mississippi are still between one to two orders of magnitude higher than those of the pretechnological era. Although the Mississippi and Rhine Deltas presently act as efficient filters and sinks for heavy metals, given the large quantities of metals currently stored in these deposits, their long-term reactivity and biological availability need to be assessed.

9.3 Sources, pathways and targets

9.3.1 Sources of heavy metals in the environment

Both natural and human activities introduce trace metals into the environment. The

Metal pollution of soils and sediments

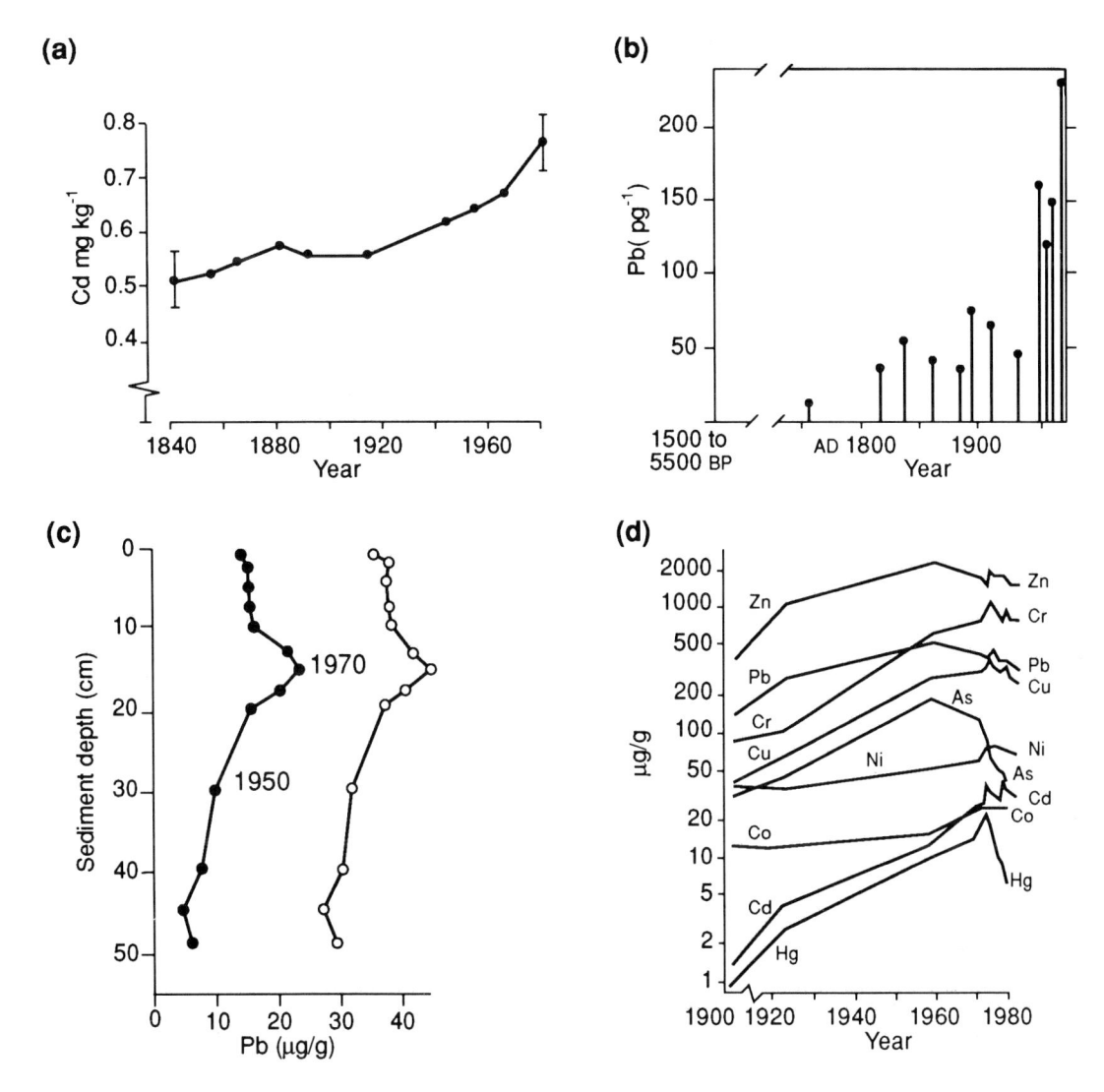

Figure 9.1 Metal accumulation in the environment. (a) Changes in Cd levels at Rothamsted Experimental Station. Error bars relate to analyses of eight separately digested samples for both 1846 and 1980. All other data points are the mean of duplicate analyses (after Jones *et al.*, 1987). (b) Lead concentrations in Greenland snow (after Wolff and Peel 1985). (c) Profiles of total (○) and pollutant (●) lead plotted against depth in the sediments of the Mississippi delta (after Trefry *et al.*, 1985). (d) History of trace metal pollution in the River Rhine (Netherlands) as reflected in its sediments (after Salomons and Forstner, 1984)

Table 9.2 Global annual average contributions of trace elements from natural and anthropogenic high-temperature processes, 1983

	Arsenic	Cadmium	Lead (thousands of metric ions)	Selenium	Mercury
Natural					
Dust	0.24	0.25	10	0.3	0.03
Volcanoes	7	0.5	6.4	0.1	0.03
Forest fires	0.16	0.01	0.05		0.01
Vegetation	0.26	0.2	1.6		
Sea salt	0.14	0.002	0.1		0.003
Total	7.8	0.96	18.6	0.4	0.16
Anthropogenic					
Mining	—	—	8.1	0.005	—
Smelting of non-ferrous metals	15.2	5.3	77.2	0.28	0.29
Iron production	4.3	0.1	50.0	0.01	0.45
Other industrial activities	—	0.04	7.2	0.05	—
Waste incineration	0.4	1.4	9.0	—	—
Phosphorous fertilizer production	2.6	0.2	—	—	—
Coal combustion	0.5	0.05	13.9	0.68	0.63
Wood combustion	0.5	0.2	1.0	—	—
Petroleum combustion	—	—	273.0	0.06	0.27
Total	23.6	7.3	449	1.1	1.8

Source: After Brown, Kasperson and Raymond (1990)

principal natural sources of metals in the atmosphere are wind-blown soil particles, volcanoes, seasalt spray, wild forest fires and aerosols of biogenic origin (Nriagu, 1989). Under pristine conditions metal levels within aquatic and terrestrial environments are determined by continental weathering and erosion processes, in addition to atmospheric inputs. Anthropogenic metal sources fall into two main categories: *production-related* activities including mining, smelting, energy generation and agriculture and second, *consumption-related* activities such as use, wear and disposal of consumer and commercial products (Brown, Kasperson and Raymond, 1990). Among the production-related acivities high temperature processes are significant contributors to metal pollution, mainly through direct emissions of fine particles and gases into the atmosphere.

In Table 9.2 the contribution of Cd, Pb and Hg to the atmosphere from natural and anthropogenic high-temperature processes are compared. Natural sources of heavy metals are an order of magnitude smaller than those from high-temperature industrial activities and it is clearly evident that mankind has become the most important element in the global biogeochemical cycling of trace metals. Combustion of hard coal, lignite and brown coal in electric power plants and in industrial, commercial and residential burners is the major source of airborne Hg (Nriagu and Pacyna, 1988). The bulk of the Cd emissions comes from metal smelter and waste incineration, and Pb is introduced mainly by the combustion of leaded gasolene with non-ferrous and ferrous metal industries also being important contributors.

Tables 9.3 and 9.4 (Nriagu and Pacyna, 1988) list the principal sources of anthropogenic inputs of heavy metals into aquatic ecosystems and onto land. The major sources of Cd, Pb and Hg pollution in aquatic ecosystems, including the ocean, are coal-burning power plants (Hg in particular), non-ferrous metal smelters (Cd and Pb) and

Table 9.3 Anthropogenic inputs of trace metals into the aquatic ecosystems (10^6 kg yr^{-1})

Source category	Cd	Hg	Pb
Domestic wastewater			
Central	0.18–1.8	0–0.18	0.9–7.2
Non-Central	0.3–1.2	0–0.42	0.6–4.8
Steam electric	0.01–0.24	0–3.6	0.24–1.2
Base metal mining and dressing	0–0.3	0–0.15	0.25–2.5
Smelting and refining			
Iron and steel			1.4–2.8
Non-ferrous metals	0.01–3.6	0–0.4	1.0–6.0
Manufacturing processes			
Metals	0.5–1.8	0–0.75	2.5–22
Chemicals	0.1–2.5	0.02–1.5	0.4–3.0
Pulp and paper	–	–	0.01–0.9
Petroleum products	–	0–0.02	0–0.12
Atmospheric fallout	0.9–3.6	0.22–1.8	87–113
Dumping of sewage sludge	0.08–1.3	0.01–0.31	2.9–16
Total input, water	2.1–17	0.3–8.8	97–180
Median value	9.4	4.6	138

Source: After Nriagu and Pacyna (1988)

Table 9.4 Worldwide emissions of trace metals into soils (10^6 kg yr^{-1})

Source category	Cd	Hg	Pb
Agric. and food wastes	0–3.0	0–1.5	1.5–27
Animal wastes, manure	0.2–1.2	0–0.2	3.2–20
Logging and other wood wastes	0–2.2	0–2.2	6.6–8.2
Urban refuse	0.88–7.5	0–0.26	18–62
Municipal sewage sludge	0.02–0.34	0.01–0.8	2.8–9.7
Miscellaneous organic wastes including excreta	0–0.01	–	0.02–1.6
Solid wastes, metal mfg.	0–0.08	0–0.08	4.1–11
Coal fly ash and bottom fly ash	1.5–13	0.37–4.8	45–242
Fertilizer	0.03–0.25	–	0.42–2.3
Peat (agricultural land fuel uses)	0–0.11	0–0.02	0.45–2.6
Wastage of commercial products	0.78–1.6	0.55–0.82	195–390
Atmospheric fallout	2.2–8.4	0.63–4.3	202–263
Total input, soils	5.6–38	1.6–15	479–1,113
Median value	22	8.3	796
Mine tailings	2.7–4.1	0.55–2.8	130–390
Smelter slags and wastes	1.6–3.2	0.05–0.28	0.1–0.2
Total discharge on land	9.9–45	2.2–18	6.4–77

Source: After Nriagu and Pacyna (1988)

the dumping of sewage sludge (Pb). On a worldlink basis the atmosphere is the major route of Pb into continental and marine waters. Total discharges of Pb, Cd and Hg on to land surpass emissions into atmosphere and aquatic ecosystem combined by a factor of between 3.4 (Pb) and 1.2 (Hg). Sources are diverse as can be seen from the list (Table 9.4), though the main ones appear to be ash residues from coal combustion (Hg, Cd, Pb), mine tailings and smelter slag wastes (Hg, Cd, Pb), urban refuse (Cd and Pb), wastage of commercial products (Hg, Pb, lost due to corrosion, dispersed in soils from usage as chemical, pesticides, crop preservatives), logging and other wood wastes (Hg).

Emission inventories for Pb, Hg and Cd presented in Tables 9.2 to 9.4 tend to focus primarily on industrial and energy-generating sources and, with the exception of lead emissions from gasoline combustion, rates and amounts of metals released into the environment through the use and wear of consumer and commercial products are largely unknown. Tarr and Ayres (1990), however, have recently documented metal emissions arising from consumption uses into the Hudson-Raritan river basin, USA, over the last 100 years. They recognized four categories of dissipative consumption:

1. weathering of paints and pigments;
2. incineration of discarded pharmaceuticals, batteries, electronic tubes, plastics and photographic film;
3. wear and weathering of electroplated surfaces, leather and plastics;
4. decomposition or combustion of treated wood.

Whereas the air is the principal target of metal emissions from some of these sources (e.g. incineration of discarded material), it is the soil and surface water that receive the bulk of the emissions from consumption-related dissipated processes, mainly via surface runoff and sewage treatment plants (Brown, Kasperson and Raymond, 1990). During the past 30 years or so the principal source of heavy metal emissions in the Hudson-Raritan river estuary has shifted from productive- to consumption-related processes (Table 9.5). It is very likely that this pattern is repeated in many other industrialized river basins around the world, and suggests that in the future pollution control technology and monitoring will need to focus not just on industrial and energy-generating sources but also on diffuse and diverse metal inputs from dissipative consumption. This will not be an easy task.

9.3.2 Pathways: dispersal and storage of metals in the environment

The route that metal contaminants take through the environment, and their rate and pattern of transfer, depends on whether they are discharged into air or water, or on to land, and also whether they are emitted in a gaseous, liquid or particulate form. Solid material such as mine waste, dredging, urban refuse, animal slurries or sewage dumped on land is usually slowly remobilized and redistributed to other environmental compartments, unless it is subject to enhanced rates of leaching (arising from acidification), oxidation or bacterial activity (e.g. acid mine drainage), erosion or uptake by biota. Metal contaminants emitted into the atmosphere are carried different distances by wind depending on their state (gaseous, vapour or particulate) before they fall or are washed out of the air. In the case of metals associated with particulate matter, particle size is the decisive factor with most metals deposited within 3 km or 50 per cent of the point of emission with concentrations declining exponentially until background levels are reached between 10 and 15 km (Davies and Roberts, 1978). The ground pattern of contamination is usually an ellipse with the major axis aligned, with the direction of the prevailing wind, as shown (Davies, 1983a) around a metal refinery in South Wales (Figure 9.2). Lead concentrations in roadside soils resulting from car exhaust emissions also decline exponentially with distance to almost background concentrations within

Table 9.5 Emissions of selected metals (over 100 yrs) in the Hudson-Raritan river basin (in tons)

Year	Metallurgical operations Low	Metallurgical operations High	Fossil-fuel combustion	Consumptive uses	Total emissions Low	Total emissions High
Average annual cadmium emissions						
1980	0.0	0.0	31.5	20.8	52.3	52.3
1970	0.1	0.4	46.6	71.5	118.3	118.6
1960	0.4	1.1	22.8	80.9	104.1	104.8
1950	0.6	1.6	17.2	71.7	89.4	90.4
1940	0.7	1.8	11.9	69.0	81.5	82.6
1930	14.2	24.9	10.5	20.7	45.4	56.1
1920	20.2	49.3	24.6	3.0	47.8	76.9
1900	1.6	3.7	23.2	0.0	24.8	26.9
1880	0.4	0.6	7.1	0.0	7.5	7.6
Average annual mercury emissions						
1980	0.00	0.00	0.03	64.1	64.2	64.2
1970	0.01	0.02	0.10	106.3	106.4	106.4
1960	0.03	0.04	0.16	56.6	56.8	56.8
1950	0.04	0.06	0.32	66.4	66.7	66.8
1940	0.04	0.05	0.71	66.0	66.7	66.8
1930	0.11	0.16	1.04	57.8	58.9	58.9
1920	0.16	0.31	3.54	48.9	52.6	52.7
1900	0.05	0.08	4.21	27.7	31.9	31.9
1880	0.01	0.01	1.28	27.9	29.2	29.2
Average annual lead emissions						
1980	43	51	5,312	95	5,451	5,458
1970	112	261	9,684	199	9,994	10,143
1960	295	688	7,203	126	7,625	8,017
1950	564	2,173	4,884	81	5,529	7,138
1940	957	3,163	2,504	18	3,480	5,685
1930	864	2,306	266	13	1,144	2,586
1920	1,254	2,902	74	14	1,342	2,991
1900	1,470	1,708	92	0	1,562	1,800
1800	16	23	29	0	45	53

Source: After Tarr and Ayres (1990)

100 m of the road (Figure 9.3). For the gaseous phase deposition can take place 200–2000 km from a source and is important for the aerial transport of Hg, Cd, Zn and Pb particularly to the oceans.

Metal pollutants which enter the fluvial system, in solution or as particle material, follow the same environmental pathways as any other ion or sediment-associated element. Prediction of the fate of metals in the river environment therefore requires a full understanding of short- and long-term sediment dynamics (erosion, transportation and storage) which can only come from fluvial geomorphology or sedimentology. Although metallic pollutants may enter rivers initially as simple ions, they react with and are complexed by dissolved organic material, suspended and fine-grained bottom sediments. As a result particulate metals generally predominate in river metal fluxes (Table 9.6), although dissolved metal discharges are significantly higher in polluted rivers particularly for Cd, Zn and Cu.

Metal concentrations in channel and floodplain sediments, as well as in solutions,

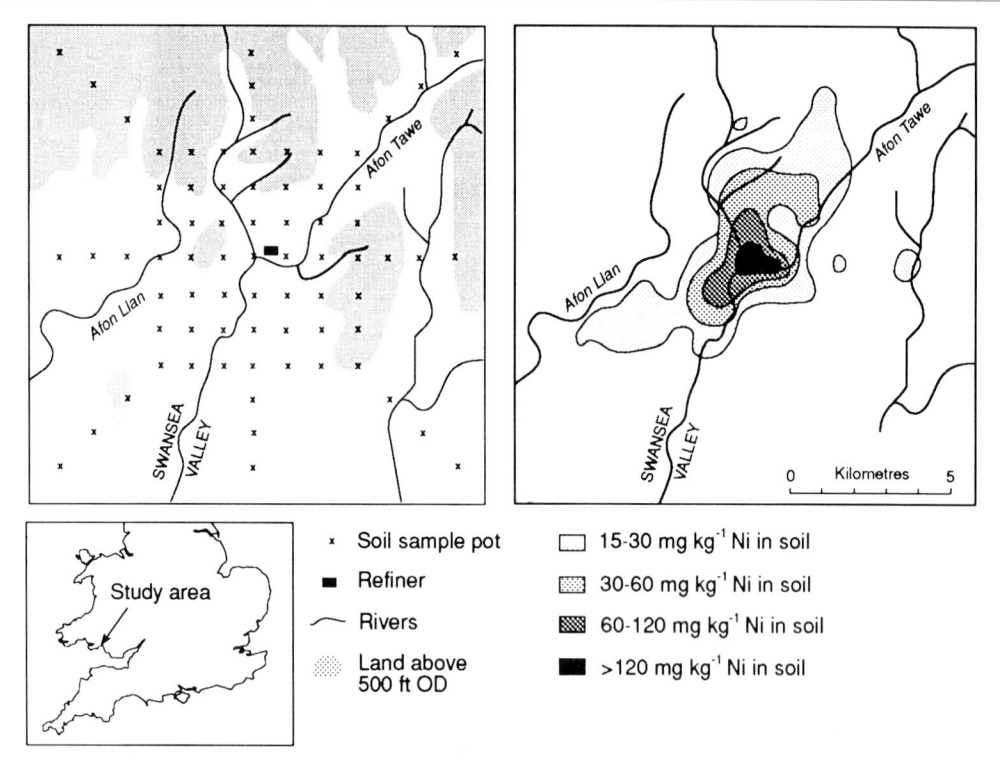

Figure 9.2 The pattern of accumulation of Ni in surface soils around a refinery in South Wales. (After Davies, 1983a)

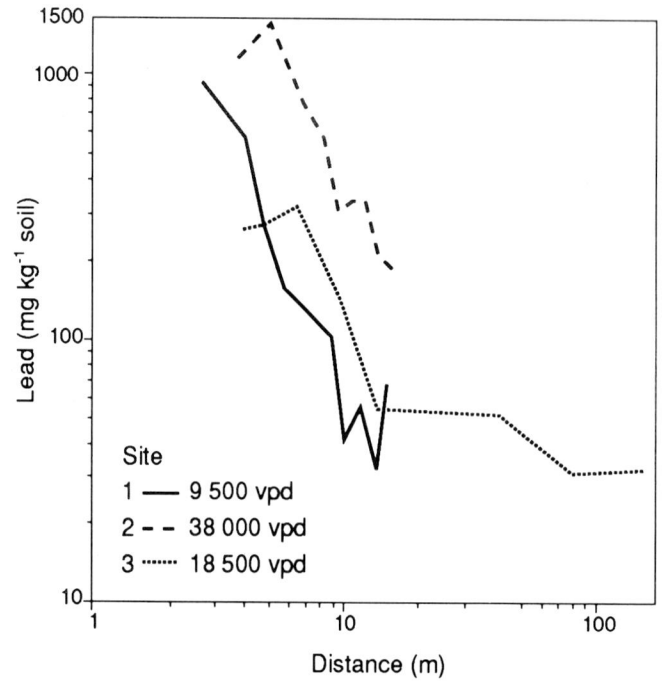

Figure 9.3 Lead concentration in surface soil at different traffic volume locations as a function of distance from a highway (vpd = vehicles per day). (Welch and Dick, 1975)

Table 9.6 Percentage particulate-bound metal of total discharge

	Amazon River %	Mississippi River %	Polluted rivers in USA %	FRG %
Cd	—	88.9	—	30
Zn	—	90.1	40	45
Mn	83	98.5	—	8–97
Cr	83	98.5	76	72
Cu	93	91.6	63	55
Pb	—	99.2	84	79
Fe	99.4	99.9	98	98

Source: After Salomons and Forstner (1984)

tend to decrease downstream from a point or area of emission. Dissolved metal species under normal physiochemical conditions are usually immobilized as the result of being rapidly adsorbed onto particulate material. Downstream dispersal rates and patterns of sediment-associated metals, however, reflect: (i) hydraulic sorting according to differential particle density and size, (ii) sediment-mixing processes with addition of uncontaminated material from tributaries or bank erosion resulting in dilution of heavy metals, (iii) floodplain deposition and storage (Lewin and Macklin, 1987). In many river systems downstream reductions in sediment metal concentration can be approximated (using regression analysis, e.g. Wolfenden and Lewin, 1987, or simple mixing models, e.g. Marcus, 1987) by a simple decay model of a negative linear exponential or power form. Departures from modelled dispersion patterns can be used as an indication of the relative importance of hydraulic, chemical and dilution processes (Marcus, 1987; Macklin and Dowsett, 1989). In some ephemeral, semi-arid streams, however, where solution-dissolution processes are not very active and organic matter contents are low, heavy metal concentrations are not related to distance from source by a simple exponential function. Graf (1990) found that with concentrations of heavy metals in the Puerco River, New Mexico, following the failure of a uranium tailings pond, alternatively increased and decreased in a downstream direction. Metal levels in channel sediments were shown to be inversely related to unit stream power and to the length of time that shear stress exceeded critical values during the passage of the flood wave.

Heavy metals which are incorporated within soils and sediment in the floodplain domain have comparatively long environmental residence times. Although metals are slowly depleted from these environmental compartments by leaching (e.g. Cd 75–380 years, Hg 500–1000 years, Pb and Zn 1000–3000 years in temperate soils, Salomons and Forstner, 1984) and plant uptake, where their cycling is accelerated by high rates of erosion or deflation, metal-rich sediment or soil constitutes a very important secondary source of contamination (Marron, 1989; Macklin and Klimek, 1992). Chemical budgets of the Geul river, the Netherlands (Figure 9.4) calculated by Leenaers (1989) show that contaminated overbank deposits are now functioning as a major non-point source of heavy metals with 66 per cent of Pb entering the channel being supplied through river bank erosion.

Metals that are not abstracted into fluvial storage are either retained in estuaries or pass out into the ocean which constitutes the ultimate metal sink. Data recently compiled by Mance (1987) evaluating the input routes of heavy metals into the North Sea show that land-based contributions exceed atmospheric sources for all metals listed except Cd (Table 9.7).

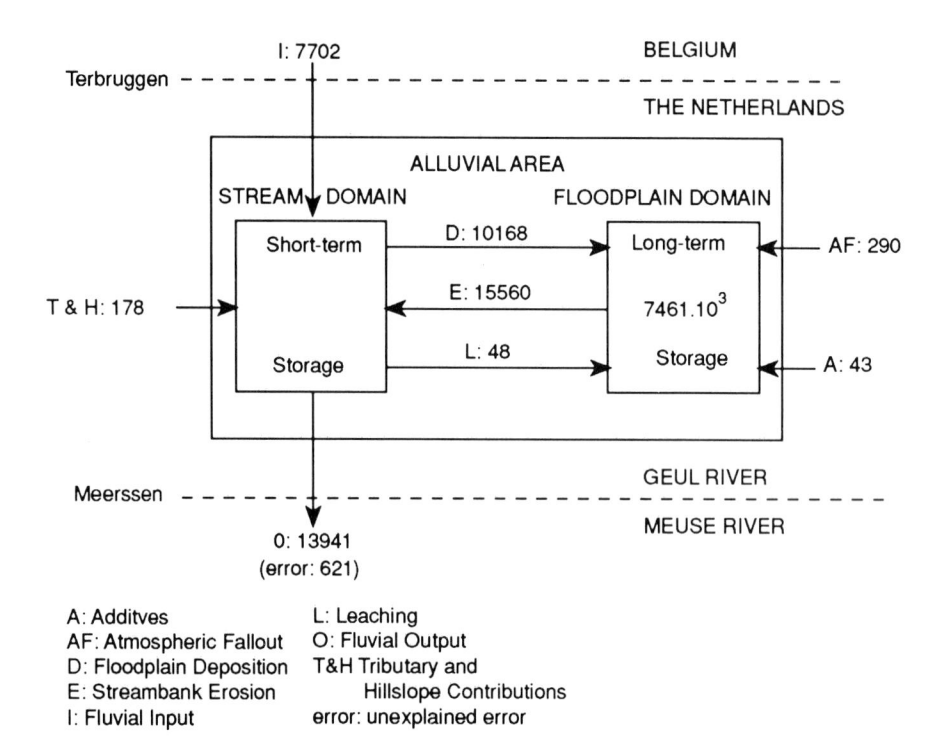

```
                    I: 7702                BELGIUM
    Terbruggen ----+------------------------------
                   |                       THE NETHERLANDS
          ┌────────┼──────────────────────────────┐
          │           ALLUVIAL AREA               │
          │  STREAM↓ DOMAIN      FLOODPLAIN DOMAIN│
          │  ┌──────────┐  D: 10168  ┌──────────┐ │
          │  │Short-term│───────────▶│Long-term │◀── AF: 290
          │  │          │  E: 15560  │          │ │
   T & H: 178→│          │◀───────────│ 7461.10³ │ │
          │  │          │  L: 48     │          │ │
          │  │ Storage  │───────────▶│ Storage  │◀── A: 43
          │  └──────────┘            └──────────┘ │
          │                                       │
          └────────┼──────────────────────────────┘
                   │                       GEUL RIVER
    Meerssen ----+------------------------------
                   ▼                       MEUSE RIVER
                 O: 13941
                 (error: 621)

    A: Additves              L: Leaching
    AF: Atmospheric Fallout  O: Fluvial Output
    D: Floodplain Deposition T&H Tributary and
    E: Streambank Erosion        Hillslope Contributions
    I: Fluvial Input         error: unexplained error
```

Figure 9.4 Tentative lead budget of the alluvial area of the Geul in the Netherlands. Values in kg. (Leenaers, 1989)

Table 9.7 Land-based and atmospheric contributions to heavy metals in the waters of the North Sea

Metal	Land-based (t/day)	Atmospheric (t/day)	Relative contribution (land atmospheric)
Arsenic	2.2	0.63	3.5
Cadmium	0.68	1.56	0.4
Chromium	18.3	1.83	10.0
Copper	12.9	10.8	1.2
Lead	17.9	8.0	2.2
Mercury	0.18	0.14	1.3
Nickel	11.1	4.3	2.6
Zinc	86.3	19.2	4.5

Source: After Mance (1987) and Brown *et al.* (1990)

Table 9.8 Background concentrations (mg kg^{-1}) of heavy metals in sediments and soils

Metal	Shale and clays	Sub-recent Rhine sediments	Lacustrine sediments	Soils
Fe (%)	4.72	3.23	4.34	3.2
Mn	600	960	760	760
Zn	95	115	118	59.8
Cr	83 (*60)	47	62	84
Ni	68 (*32)	46	66	33.7
Cu	45 (*31)	51	45	25.8
Pb	20	30	34	29.2
Co	19 (*13)	16	16	12
Hg	0.2	0.2	0.35	0.098
Cd	0.2	0.3	0.40	0.62

*Recent freshwater sediments.
Source: Salomons and Forstner (1984)

9.3.3 Likely targets: identification of contaminated and polluted soil and sediment

Although there are clear guidelines of limit values for maximum metal concentrations in water, air and food, as yet there is no equivalent consensus of permissible levels of metals in soils and sediment. This arises primarily because of great uncertainties of metal dose–response relationships in soil and sediment media and has resulted in bewildering and sometimes contradictory guidance from government environmental protection agencies. However, approaches for identifying metal-contaminated sediment or soil, aside from attempting to quantify environmental impact or risk to health, are better established. The principal task is to determine whether metal concentrations fall within the range 'normally' found in soils and sediments unaffected and uncontaminated by metal utilization and mining. This can be assessed in a number of ways. Metal concentrations can be compared with levels in sediments (fluvial, lacustrine, aeolian) deposited before the industrial era, in recent deposits from unpolluted areas or with a rock standard (e.g. average shale data compiled by Turekian and Wedepohl, 1961). 'Pre-civilization' metal levels in soils are more difficult to establish because they are susceptible to continuous overprinting by metal contaminants. This is even a problem in areas remote from industry, as shown by elevated Pb concentrations in polar regions. In Table 9.8 background concentrations of some of the more common metal contaminants in soils and sediments are listed.

Contaminated and uncontaminated soils (and sediment) can also be distinguished using statistical techniques such as the method outlined by Davies (1983b). Percentage cumulative frequency distributions of the log metal content of soil or sediment are plotted on probability graph paper and different geochemical populations are identified by sharp changes in the slope of the curve. The lower, linear portion of the probability curve is interpreted as a lognormal population derived from non-contaminated soils. This can be separated and replotted, and from this geometric means and deviations calculated allowing the highest probable (threshold) metal value for an uncontaminated soil to be estimated. Where statistical analyses are carried out on pedogenically similar soils, or sediments of a similar origin within a defined region (e.g. a river basin), they can provide an objective contamination threshold. Some statistically

Table 9.9 Statistically based threshold values for lead in soils

Area	Threshold (mg kg^{-1})
Agricultural soils in England and Wales	108
Hamps and Manifold valleys, North Staffordshire	53
Ceredigion	116
Tamar Valley	107
Halkyn Mountain	90
North Somerset	110
Ystwyth Valley, mid-Wales	125
Lox Yeo Valley, Somerset	67
Tyneside	80

Source: After Aspinall, Machlin and Openshaw (1988)

based, highest probable Pb values in uncontaminated soils, collated by Aspinall *et al.* (1988), are presented in Table 9.9. Although they exceed background Pb concentrations (Table 9.8 suggests these should be a conservative estimate of the threshold between contaminated and uncontaminated material), they all fall within the normal range (defined as that between two standard deviations above and below the mean) of Pb content of agricultural soils in England and Wales (Archer and Hodgson, 1987).

If metal concentrations are found to exceed natural levels, the next step is to assess if they constitute an environmental problem. To do this means deciding at what level metal contamination becomes metal pollution and poses a measurable risk to plants, animals and humans. This is not straightforward because the relationship between soil (and sediment), metal concentrations and human health is not well established, and consequently it has been difficult to set precise protection standards. The toxic effect of metals on plants is clearer because for obvious ethical reasons it has been possible to conduct more scientific experiments on plants than people. In Britain, with no statutory limits for metals in contaminated land, the Inter-Departmental Committee on the Redevelopment of Contaminated Land (ICRCL) has put forward the concept of 'trigger concentrations' (ICRCL, 1983; 1987). Trigger concentrations are defined as those values below which a site could be regarded as uncontaminated — that is, it will present no additional risk to a critical group. Tentative trigger concentrations for selected metals that either pose a hazard to health (Cd, Cr, Pb, Hg) or are phytotoxic (Cu, Ni, Zn) but are not normally hazards to health, are listed in Table 9.10. Trigger concentrations vary with the proposed use of the site, thus for domestic gardens and allotments they are set much lower than in amenity soils in public open spaces. Similar guidelines are also available for agricultural (ICRCL, 1990) restoration and aftercare of metalliferous mining sites. The threshold trigger concentration (Table 9.11) is defined as metal levels below which the soil is considered to be safe and that would not give rise to phytotoxic or zootoxic effects. The maximum or 'action trigger' concentration is that above which there is a very high probability of damage to stock or reduced crop yields if these thresholds are continuously exceeded. It was acknowledged that between these two categories there could be sub-clinical effects in animals, although this will depend on soil pH (generally solubility, toxicity and plant availability increases with decreasing pH), sward cover (animals are not likely to ingest soil metals where vegetation is damaged or incomplete), type of stock management and length of exposure.

The Netherlands government, like the British authorities, has also introduced a set of signal or indicator values to evaluate the extent of metal pollution and to provide guidelines for soil sanitation (Anonymous, 1983). The A-value (Table 9.12) refers to an acceptable or natural metal concentration, while soils that exceed the B- (warning) value are considered contaminated enough to warrant further study to evaluate the precise nature and degree to which metal levels may be unacceptably enhanced. In cases where the C- (danger) value is surpassed the feasibility of removal or burial of

Metal pollution of soils and sediments

Table 9.10 Tentative trigger concentrations for selected metal contaminants

Contaminants	Planned uses	Trigger concentrations (mg/kg air-dried soil)
Group A: Contaminants which may pose hazards to health		
Cadmium	Domestic gardens, allotments, parks, playing fields,	3
	open space	15
Chromium (hexavalent)	All uses	25
Chromium (total)	Domestic gardens, allotments, parks, playing fields,	600
	open space	1000
Lead	Domestic gardens, allotments, parks, playing fields,	500
	open space	2000
Mercury	Domestic gardens, allotments, parks, playing fields,	1
	open space	20
Group B: Contaminants which are phytotoxic but not normally hazards to health		
Copper ('available')	Any uses where plants are to be grown	50
Nickel ('available')	Any uses where plants are to be grown	20
Zinc ('available')	Any uses where plants are to be grown	130

Source: ICRCL (1987)

Table 9.11 Guidelines for toxic element trigger concentrations in minespoil-contaminated 'soils'

Element	Threshold trigger concentrations	Maximum (action trigger) concentrations (values not to be exceeded for use as specified)	
		For grazing livestock	For crop growth (risk of phytotoxicity)
Zinc	1000	3000	1000
Cadmium	3	30	50
Copper	250	500	250
Lead	300	1000	—

Source: ICRCL (1990)

Table 9.12 Signal values of soil pollution according to Dutch guidelines (mg kg^{-1})

Metal	A-value	B-value	C-value
Zn	200	500	3000
Cd	1	5	20
Cu	50	100	500
Pb	50	150	600
Co	20	50	300
Ni	50	100	500
Cr	100	250	800

Source: Anonymous (1983)

Table 9.13 Signal values for metal concentrations in the soil (mg kg^{-1}) for various types of agricultural land-use/products

Land use	Pb	Zn	Cd	Cu
Consumer crops	200	350	1.0	200
Pasture	200	350	3.0	—
Grazing sheep	—	—	—	30
Grazing cattle	—	—	—	80

Source: Leenaers (1989)

contaminated material needs to be investigated. The Dutch Ministry of Agriculture (Leenaers, 1989) also provides signal values for metal concentrations in soil for various types of agricultural land-use/products (Table 9.13).

There is a clear and somewhat worrying disparity between what the Dutch and British authorities view as acceptable and unacceptable levels of toxic metals in soils, with the British government setting significantly higher, and less stringent thresholds. For example, in Britain Pb levels up to 500 mg kg^{-1} are considered acceptable in garden soils and land used for grazing livestock; however, in the Netherlands it is considered unwise to grow crops for human consumption, or utilize land for pasture, when Pb concentrations exceed 200 mg kg^{-1}. Indeed, the ICRCL trigger concentration for Pb in domestic gardens and allotments approaches the Dutch C signal value where levels are believed to be sufficiently high to require removal or some other form of soil sanitation! This raises major questions not only on the definition of 'acceptable' metal concentrations but also more generally on how risk is assessed. Until more is known of metal dose–response relationships in soil and sediment systems, setting and implementing standards will remain an economic and therefore ultimately a political decision. This is the 'bottom line' in many aspects of environmental protection and management, and is explored in greater detail towards the end of this chapter and in Chapter 14.

9.4 The toxicology of heavy metals

Toxicology is the study of the harmful interactions between chemicals and biological systems (Timbrell, 1989). The toxicity of heavy metals such as Pb has been recognized since the time of Classical Greece, although widespread, inadvertent subclinical poisoning resulting from environmental pollution is a relatively recent post-Industrial Revolution phenomenon. This is very clearly illustrated by the estimated hundredfold increase of daily amounts of lead absorbed into the blood of present-day North Americans compared with prehistoric humans (Settle and Patterson, 1980, Table 9.14). The toxicology of heavy metals is unique for each element and for each route of exposure (e.g. ingestion versus inhalation)

Table 9.14 Inventory of estimated average daily amounts of lead (in nanograms) absorbed into blood of adult humans in prehistoric times

Source	Prehistoric (natural)*	Contemporary (urban American)**
Air	0.3	6,400
Water	<2	1,500
Food	<210	21,000
Total	<210	29,000

*Lead in air, 0.04 ng/m^3 at 20 m^3/day × 0.4; in water, <20 ng/kg at 1 kg/day × 0.1; in food, <2 ng/g at 1.5 kg/day × 0.7.
**Lead in air, 800 ng/m^3 at 20 m^3/day × 0.4; in water, 15,000 ng/kg at 1 kg/day × 0.1; in food, 200 ng/g at 1.5 kg/day × 0.07.

Source: Settle and Patterson (1980)

but all have been shown to produce adverse effects in humans at very low levels of exposure, possibly without a meaningful threshold.

Compared with the well-documented cases of catastrophic metal poisoning (e.g. consumption of wheat seed treated with mercurial fungicides, Bakir *et al.*, 1973; pollution of irrigation waters by Cd in the Jinktu River area, southern Japan,

Kobayashi, 1971; Hg discharge into coastal waters around Minamata, Japan, Takenchi, 1972), it has been very difficult to diagnose and link adverse health effects resulting from continuous exposure to lower metal levels more usually found in the environment. Studies both in the United States (Needleman et al., 1979) and in the United Kingdom (Royal Commission on Environmental Pollution, 1983), however, have shown a strong association between increased body Pb burdens (estimated from either blood or tooth Pb levels) and IQ decreases in children living in urban areas. In Needleman et al.'s (1979) study of children in Massachusetts, USA, after allowing for socio-economic factors, there was a consistent and statistically significant (at the 3 per cent level) metal dose–response relationship with children with high Pb levels performing less well in the classroom. The significance of this study is that it took 'normal' children from an urban setting and found what would appear to be strong evidence of a decrease in IQ, and other functioning, due to integrated exposure to Pb. There is little doubt that most people in the industrialized world are exposed to Pb levels today which are between two to three orders of magnitude greater than those of the prehistoric period. Overtly poisonous Pb doses are only two to five times greater than the US typical daily intake and would indicate that an unrecognized subclinical form of poisoning caused by excessive exposure to Pb may affect most North Americans (Settle and Patterson, 1980).

9.5 Geographical audits of metal-contaminated soil and sediment systems

9.5.1 Heavy metal contamination in amenity soil, Tyneside

In comparison with agricultural areas in Britain, soils in the urban environment, in which the majority of the population live and come into contact with soil, have been almost totally neglected (Aspinall et al., 1988; Thornton, 1990). A wider appreciation of the nature of this problem has been hampered by the general absence in Britain (and in many other countries also) of systematic data on the distribution of metal contaminants in the urban environment mapped at an appropriately fine level of geographical detail. Difficulties arise most often in urban areas because of problems in ensuring consistent sampling procedures, through soil disturbance and also marked local variations in soil-metal concentrations. In addition, interpretation of soil-metal levels and patterns is frequently problematic because the long industrial history of Britain (with a continuous overprinting by metal contaminants) makes it difficult to establish unequivocally how far present metal levels in urban soils can be attributed to current or past industrial activities. These problems have been addressed by the author and colleagues at Newcastle University in an investigation of metal levels in amenity soil on Tyneside. This study focused on mapping metal concentrations in urban soils collected from public parks and grassed areas, and attempted to estimate the number of people living in contaminated areas. Details of sampling and analytical methods can be found in Aspinall et al., 1988.

The class intervals for the chloropleth map of Pb concentrations in amenity soils on Tyneside (Figure 9.5) are based on published British guidelines of acceptable soil-metal concentrations (ICRCL, 1987), discussed in section 9.5, and our own and Royal Commission's (1983) Pb thresholds for contaminated soil (80 and 150 mg kg^{-1} respectively). Sixty-nine per cent of the soil samples analysed in Tyneside (Table 9.15) were identified as being contaminated by Pb (concentrations greater than 80 mg kg^{-1}) and were found in most parts of the city. It is estimated that more than 80 per cent of the population in Tyneside live in wards where public open spaces and amenity land is contaminated by Pb. Pb concentrations between 150 and 500 mg kg^{-1} (34 per cent of samples analysed) tended to be restricted

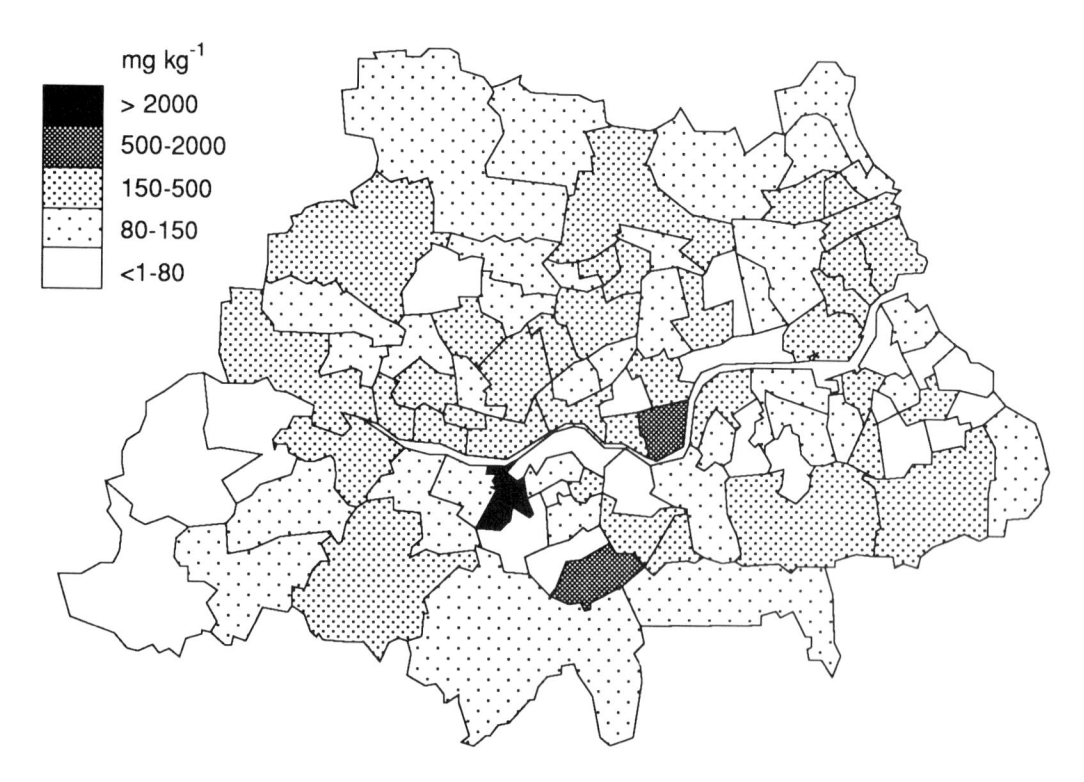

Figure 9.5 Lead concentrations in amenity soils, Tyneside (Aspinall, Macklin & Openshaw, 1988)

Table 9.15 Percentage of Tyneside population living in wards where measured values exceed Pb thresholds, and percentages of soils sampled exceeding thresholds

Metal	Class Intervals (mg/kg)	Source of threshold value	Percentage of Tyneside soils sampled exceeding class interval	Percentage of Tyneside population living in contaminated wards
Lead	80	Aspinall, Macklin and Openshaw (1988)	69.1	84.3
	150	Royal Commission 9th Report (1983)	35.4	44.0
	500	ICRCL (1987)	6.5	4.8
	2000	ICRCL (1987)	1.3	0.7

Source: After Aspinall, Macklin and Openshaw (1988)

to soils within central Newcastle and along the River Tyne where 44 per cent of the population of Tyneside live. Four areas (6.5 per cent of samples) have Pb concentrations over 500 mg kg^{-1}: the north bank of the Tyne around Elswick and Walker and south of the river in the Harlow Green-Wrekenton and Team Valley areas. The highest mean Pb concentrations by ward (>2000 mg kg^{-1}) were recorded in the industrialized Team Valley. Although the proportion of population living in this area is quite small (0.7 per cent), somewhere around 7000 people are exposed to significantly elevated amenity soil metal levels.

The results of the Tyneside survey are consistent with many previous studies (e.g. Bridges, 1989) which have shown that wherever people live or work metal concentrations in nearby soil generally rise. This maxim holds true as much for ancient settlements in Greece (Bintliff et al., 1990) as it does for modern Tyneside. Metal contamination of soil is a cumulative process, which in Tyneside must have begun when it became a significant urban centre in the Middle Ages, and has continued at an accelerating rate over the last 100–150 years. The degree to which elevated soil metals impinge directly on human health in Tyneside is a highly contentious issue. Recent studies of young children in inner-city Birmingham (Davies et al., 1987), however, have shown that ingestion (diet 34 per cent, dust 63 per cent) accounts for 97 per cent of total Pb uptake, and that contaminated soils in parks and public playing areas may constitute a significant source of metals to this vulnerable group. It is very likely with the move towards a more federal Europe that the British government may not only face growing public pressure to come into line with more stringent European guidelines for 'acceptable' levels of metals in soils but also a considerable bill for soil sanitation particularly in urban and former metal-mining areas.

9.5.2 Contamination of river systems by historic metal mining: the Tyne River, northern England

Physical geographers at Newcastle University have since 1983 been investigating the impact of historic Pb and Zn mining on river channels and floodplains in the Tyne basin (Macklin and Rose, 1986; Macklin and Lewin, 1989; Macklin, Rumsby and Newson, 1992), and especially pollution arising from the release of toxic metals during the processing of metal ores (Macklin and Smith, 1990). The South Tyne river, one of the two major tributaries of the Tyne, drains the northern part of the Northern Pennine orefield formerly the most productive base-metal mining area in Britain (Schnellmann and Scott, 1970). The principal metal ores exploited were galena (Pb S) and sphalerite (Zn S), the latter ore containing significant trace concentrations of Cd and Hg. The peak of Pb-mining in the Northern Pennines occurred between the end of the Napoleonic wars and 1880, while the heyday of Zn extraction (1880–1920) was somewhat later and more short-lived. Many spoil heaps, however, were later reworked for metals during the Second World War with the last major metal mine in the Tyne basin closing in the 1950s.

As in many other base-metal mining areas in Britain (Lewin, Davies and Wolfenden, 1977; Bradley and Cox, 1986; Lewin and Macklin, 1987), mining activity increased the quantity of fine (sand-size and smaller) and coarse (gravel) sediment delivered to river channels in the Tyne catchment. The input of coarse material took place mainly in the seventeenth and eighteenth centuries, primarily through a primitive, but very effective form of hydraulic mining called hushing. Vast amounts of finely divided metal waste were discharged into local rivers during the sorting and preparation of ores. Up until the end of the nineteenth century large quantities of fine-grained wastes were indiscriminately fed into local streams before this was significantly reduced by preventive legislation enacted in 1876 following the report of the

Rivers Pollution Commission (1874). Environmental protection agencies in the region (e.g. NRA, ADAS) have attempted to control the pollution legacy of metal-mining in the Tyne system *at source* through the rehabilitation of mine workings and spoil heaps. These are the most visible reminder of the mining era and are considered by most to be the most direct threat to water quality and farming. Fortunately (as the result of neutral to alkaline river water), present 'dissolved' metal concentrations in the Tyne basin generally do not exceed EEC or WHO guidelines, and monitoring indicates that water pollution from historic mining is localised and restricted to streams immediately downstream of old mine workings (Say and Whitton, 1981).

Prior to 1983, with the exception of Harding *et al.*'s (1981) survey of Derwent Reservoir and Wehr *et al.*'s (1981) study of the River Team, little attention had been given to the sediment phase of heavy metal dispersal and storage, or the extent to which floodplains in the region had been contaminated by metal mining. A basin-wide survey (150 sites) of Pb concentrations in contemporary channel fines in the Tyne catchment, carried out by the author in 1986 (Figure 9.6), however, revealed that fine metal-rich mine waste had been extensively dispersed many tens of km beyond the orefield. This had resulted in significant contamination of the South Tyne and Allen Systems, and much of the Tyne River downstream as far as Newcastle. Thus, the current distribution of metals in Tyne basin river sediments largely reflects a historical pollution event, as shown by metal concentrations in the rural South Tyne catchment (formerly an important Pb- and Zn-mining area) greatly exceeding those in the industrialized Team Valley (Figure 9.6). Subsequent investigations (Macklin, 1988) showed that at present the principal source of metalliferous fines was not spoil heaps or mine tailings but erosion of metal-contaminated alluvium, deposited in the late nineteenth century and first half of the twentieth century during the peak period of mining. It is estimated that around 18 km^2 of the alluvial valley floor of the Tyne and its tributaries have soil Pb, Zn and Cd concentrations above those considered acceptable by ICRCL and MAFF for agricultural land. This is of particular concern with respect to the uptake of metals by plants growing on contaminated alluvium and also as a source of pollution to adjacent water courses.

A number of riverside farms in the South Tyne and Tyne valleys, where analysis has shown severe soil-metal contamination, commonly report poor performance of crops (stunting, chlorosis of leaves). Of greater concern, however, are riparian areas where metal contamination has gone unrecognized because metal burdens (though high) do not obviously inhibit plant growth. This is particularly a problem with Cd which is taken up by plants more easily than most other metals, and can accumulate toxic levels for animals before the vegetation shows any sign of damage. Cd levels in alluvial soils within both the South Tyne catchment and Tyne valley frequently exceed ICRCL and Netherlands government acceptable levels (>3.5 and 3 mg kg^{-1} respectively).

Significant quantities of sediment-borne metals still enter local river channels through the erosion of mining-age alluvium during major floods. This occurred most recently between 25 and 26 August 1986, when extensive sheets of fine-grained sediment containing high levels of Cd, Pb and Zn were deposited across riverside agricultural land along more than 80 km of the Tyne River (Macklin and Dowsett, 1989). Selective extractions, used to assess the chemical form of the sediment-associated metals, showed that only a tiny proportion of the deposited Pb and Zn was present as finely divided ore and highly toxic meals such as Cd were deposited in a chemically active and biologically available form (Macklin and Dowsett, 1989).

Given the scale and diffuse nature of metal contamination in the Tyne catchment, the only viable remedial option at present would appear to be a programme of environmental management which seeks to minimize and

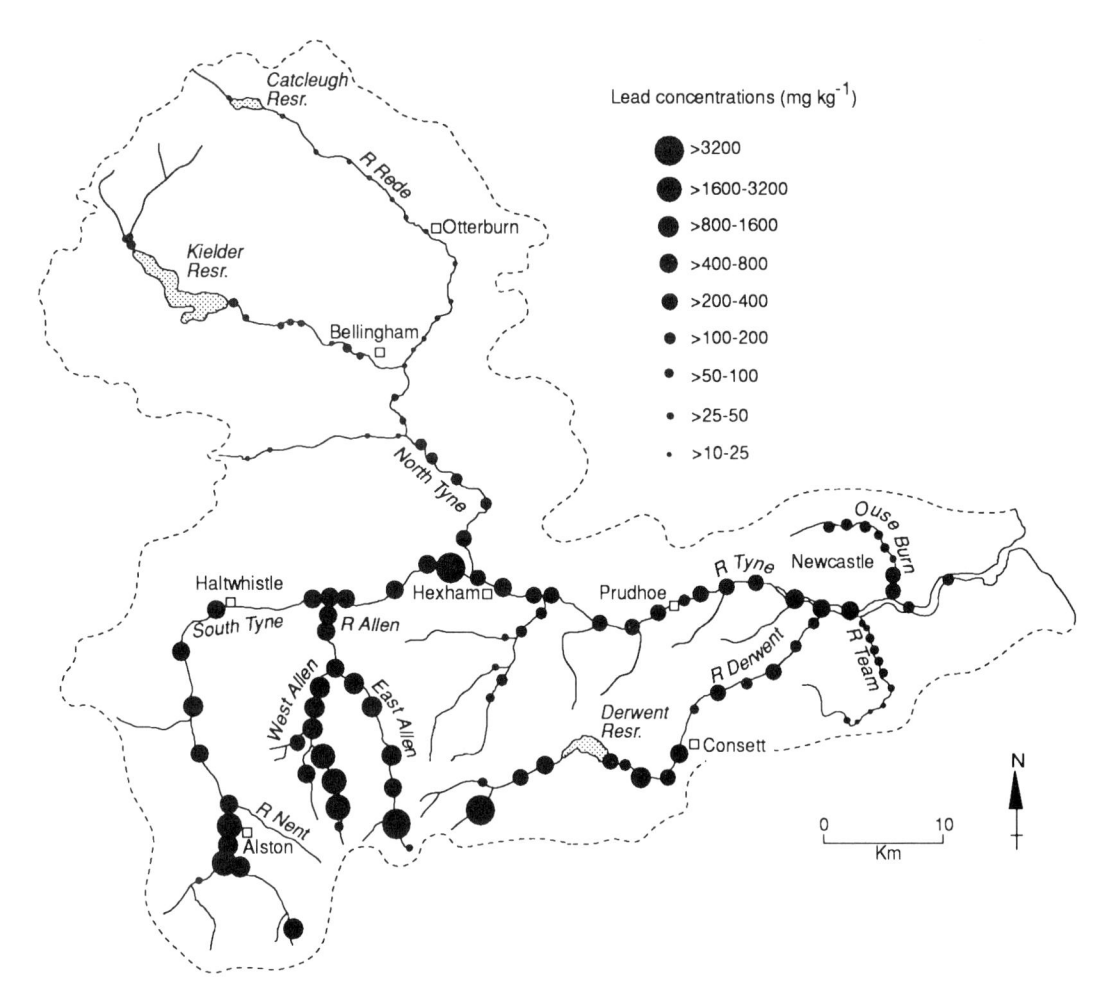

Figure 9.6 Lead concentrations in fine-grained channel bed sediment in the Tyne basin

regulate plant, animal and human exposure to toxic metals. This is a more difficult exercise than controlling metal input from extant industry or mines where emission sites are known and many are monitored by the NRA. The first corrective step in the Tyne, or for that matter in any river basin affected by base-metal mining, is to identify areas of the floodplain and alluvial valley floor contaminated by metals. This is obviously a considerable task, but, in the knowledge that alluvium deposited during the peak mining period (*c.* 1860–1950) is severely contaminated by heavy metals, identifying sediments of the age using historic maps and aerial photographs would suffice in most instances as a preliminary framework for environmental management and pollution. If a site was then suspected of being contaminated it would be important to restrict the dispersal of metals by chemical, biological or physical processes. Toxic metals should be prevented from re-entering the river systems through bank erosion, dredging or alluvial gravel extraction, or from finding their way into the food chain by growing crops or raising livestock on contaminated land. Controlling chemical remobilization of metals stored in alluvial sediments or soils, caused by acid precipitation or redox changes following channel incision and lowering of water tables, is more difficult. Oxidation of floodplain sediment and transformation of metals into a toxic ionic form is a particular threat, although soil acidification can be counteracted (at least in the short term) by the addition of lime and organic material.

Alluvial sediments contaminated by historic Pb- and Zn-mining constitute the most important *single* source of metal contaminants in the Tyne river system today. This is also very likely to be true in many other river basins in Britain and elsewhere that have experienced base-metal mining. Local farmers, environmental protection agencies and scientists in the region must therefore continue to be vigilant to prevent the inadvertent unlocking of a chemical 'time bomb' that could be the Tyne's environmental equivalent of 'Pandora's box'. With some estimated 4000 km^2 of agricultural land in England and Wales affected by base-metal mining alone (Thornton, 1977), this is a national problem which could be significantly exacerbated by the hydrological changes expected as the result of global warming.

9.6 Conclusions

The release of metal contaminants to the environment seems to be an unavoidable concomitant of metal extraction, production and use. Metal contamination of the hydrosphere probably started with Roman mining operations and appears to have reached its peak in surface waters in Western Europe, Japan and North America during the 1970s. Subsequent improvements in waste water treatment and replacement of metals in several applications (e.g. Pb in gasoline) has seen a significant reduction in 'dissolved' metal concentrations in many rivers. Ultimately, though, all metals discharged into the environment by human activities (past and present) are eventually incorporated into soil or sediment. Although metals usually remain stored in these environmental compartments for considerable periods of time, these are not static or sealed reservoirs and metals can be remobilized if sites are disturbed or environmental conditions change. The protracted environmental residence times of metals in soils and sediment make metal pollution in these media a more serious and insidious problem than in air or water, and will continue to have significant ecological impacts into the twenty-first century and beyond. If sustainable and effective strategies in metal pollution control are to emerge, greater attention has to be paid to the longer-term dynamics and cycling of soil and sediment-borne metals over 10^1–10^3 year time scales and, in particular, to the relative stability of metal contaminants within interim storage stages (e.g. floodplain soils and sediments). Physical geographers and geomorphologists therefore have a key role to play in managing and monitoring metal contamination and in establishing time and space scales of metal transfer, storage, transformation and uptake in soil and sediment systems.

Acknowledgements

This chapter is dedicated to Watts Stelling without whom there could never have been heavy metal studies at the Department of Geography, Newcastle University!

References

Anonymous (1983), *Leidraad Bodemsanering, Dutch Ministry of Housing, Physical Planning and Environment*, Leidschendam, the Netherlands, 133 pp.

Archer, F.C. and Hodgson, I.H. (1987), 'Total and extractable trace element contents of soils in England and Wales', *Journal of Soil Science*, 38, 421–31.

Aspinall, R.J., Macklin, M.G. and Openshaw, S. (1988), 'Heavy metal contamination in soils of Tyneside: a geographically-based assessment of environmental quality in an urban area', in J.M. Hooke (ed.), *Geomorphology in Environmental Planning*, John Wiley and Sons, Chichester, pp. 87–102.

Bakir, F., Damluji, S.F., Amin-Zaki, L.,

Murtada, M., Khalidi, A., Al-Rawi, N.Y., Tikriti, S., Dhalil, H.I. Clarkson, T.W., Smith, J.C. and Doherty, R.A. (1973), 'Methylmercury poisoning in Iraq', *Science*, 181, 230-41.

Bintliff, J.L., Gaffney, C., Waters, A., Davies, B. and Snodgrass, A. (1990), 'Trace metal accumulation in soils on and around ancient settlements in Greece', in S. Bottema, G. Entjes-Nieborg and W. van Zeist (eds), *Man's Role in Shaping the Eastern Mediterranean Landscape*, Balkema, Rotterdam, pp. 159-72.

Bradley, S.B. and Cox, J.J. (1986), 'Heavy metals in the Hamps and Manifold Valleys, north Staffordshire, U.K.: distribution in floodplain soils', *Science of the Total Environment*, 50, 103-28.

Bridges, E.M. (1989), 'Toxic metals in amenity soil', *Soil Use and Management*, 5(3), 91-100.

Brown, H.S., Kasperson, R.E. and Raymond, S. (1990), 'Trace pollutants', in B.L. Turner, W.C. Clark, R.W. Kates, J.F. Richards, J.T. Mathews and W.B. Meyer (eds), *The Earth as Transformed by Human Action*, Cambridge University Press with Clark University, Cambridge, pp. 437-55.

Davies, B.E. (1980), 'Trace element pollution', in B.E. Davies (ed.), *Applied Soil Trace Elements*, John Wiley and Sons, Chichester, pp. 287-351.

Davies, B.E. (1983a), 'Heavy metal contamination from base metal mining and smelting: implications for man and his environment', in I. Thornton (ed.), *Applied Environmental Geochemistry*, Academic Press, London, pp. 425-62.

Davies, B.E. (1983b), 'A graphical estimation of the normal lead content of some British soils', *Geoderma*, 29, 67-75.

Davies, B.E. and Roberts, L.J. (1978), 'The distribution of heavy metal contaminated soils in NW Clwyd, Wales', *Water, Air and Soil Pollution*, 9, 507-18.

Davies, D.J.A., Thornton, I., Watt, J.M., Culbard, E.B., Harvey, P.G., Delves, H.T., Sherlock, J.C., Smart, G.A., Thomas, J.F.A. and Quinn, M.J. (1987), 'Relationship between blood lead and lead intake in two-year old urban children in the UK', in S.E. Lundberg and T.C. Hutchinson (eds), *Heavy Metals in the Environment*, CEP Consultants, New Orleans, Edinburgh.

Graf, W.L. (1990), 'Fluvial dynamics of thorium-230 in the church rock event, Puerco River, New Mexico', *Annals of the Association of American Geographers*, 80 (3), 327-42.

Harding, J.P.C., Burrows, I.G. and Whitton, B.A. (1981), 'Heavy metals in the Derwent Reservoir catchment', in P.J. Say and B.A. Whitton (eds), *Heavy Metals in Northern England: Environmental and Biological Aspects*, University of Durham, Department of Botany, pp. 73-86.

Horowitz, A.J. (1985), 'A primer on trace metal-sediment chemistry', *US Geological Survey Water-Supply Paper*, 2277, 67 pp.

Inter-Departmental Committee on the Redevelopment of Contaminated Land (1983), 'Guidance on the assessment and redevelopment of contaminated land', *ICRCL Paper 59/83*, Department of the Environment, London.

Inter-Departmental Committee on the Redevelopment of Contaminated Land (1987), 'Guidance on the assessment and redevelopment of contaminated land', *ICRCL Paper 59/83*, Department of the Environment, 2nd edn, London.

Inter-Departmental Committee on the Redevelopment of Contaminated Land (1990), 'Notes on the after care of metalliferous mining sites for pasture and grazing', *Guidance Note 70/90*, Department of the Environment, London.

Jarvis, P.J. (1983), *Heavy Metal Pollution: An Annotated Bibliography: 1976-1980*, Geo-Books, East Anglia, UK.

Jones, K.C., Symon, C.J. and Johnston, A.E. (1987), 'Retrospective analysis of an archived soil collection 11: Cadmium', *Science of the Total Environment*, 67, 75-89.

Kobayashi, J. (1971), 'Relation between the "Itai-Itai" disease and the pollution of river water by cadmium from a mine', Proceedings 5th International Conference San Francisco and Hawai, *Advances in Water Pollution Research*, 1-25, 1-7.

Leenaers, H. (1989), *The Dispersal of Metal Mining Wastes in the Catchment of the River Geul (Belgium — The Netherlands)*, Geografische Instituut, Rijksumi-versitat Utrecht, Amsterdam.

Lewin, J., Davies, B.E. and Wolfenden, P.J. (1977), 'Interactions between channel changes and historic mining sediments', in K.J. Gregory (ed.), *River Channel Change*, John Wiley and Sons, Chichester, pp. 353-67.

Lewin, J. and Macklin, M.G. (1987), 'Metal mining and floodplain sedimentation in Britain', in V. Gardiner (ed.), *International Geomorphology 1986, Part 1*, John Wiley and Sons, Chichester, pp. 1009-27.

Macklin, M.G. (1988), 'A fluvial geomorphological based evaluation of contamination of the Tyne basin, north-east England by sediment-borne heavy metals', Report to the Natural Environment Research Council, 29 pp.

Macklin, M.G. and Dowsett, R.B. (1989), 'The chemical and physical speciation of trace metals

193

in fine grained overbank flood sediments in the Tyne basin, north-east England', *Catena*, 16, 135–51.

Macklin, M.G. and Klimek, K. (1992), 'Dispersal, storage and transformation of metal-contaminated alluvium in the upper Vistula basin, southwest Poland', *Applied Geography*, 12, 7–30.

Macklin, M.G. and Lewin, J. (1989), 'Sediment transfer and transformation of an alluvial valley floor, the River South Tyne, Northumbria, U.K.', *Earth Surface Processes and Landforms*, 14, 233–46.

Macklin, M.G. and Rose, J. (1986), *Quaternary River Landforms and Sediments in the Northern Pennines, England*, Field Guide, British Geomorphological Research Group/Quaternary Research Association, 88 pp.

Macklin, M.G., Rumsby, B.T. and Newson, M.D. (1992), 'Historic overbank floods and vertical accretion of fine-grained alluvium in the lower Tyne Valley, northeast England', in P. Billi, R.D. Hey, P. Tacconi and C. Thorne (eds), *Dynamics of Gravel-bed Rivers*, Proceedings of the Third International Workshop on Gravel-Bed Rivers, John Wiley and Sons, Chichester, pp. 564–80.

Macklin, M.G. and Smith, R.S. (1990), 'Historic riparian vegetaion development and alluvial metallophyte plant communities in the Tyne basin, north-east England' in J.B. Thornes (ed.), *Vegetation and Erosion*, John Wiley and Sons, Chichester, pp. 239–56.

Mance, G. (1987), *Pollution Threat of Heavy Metals in Aquatic Environments*, Elsevier Applied Science, London and New York.

Marcus, W.A. (1987), 'Copper dispersion in ephemeral stream sediments', *Earth Surface Processes and Landforms*, 12, 217–28.

Marron, D.C. (1989), 'Physical and chemical characteristics of a metal-contaminated overbank deposit, west-central south Dakota, U.S.A.', *Earth Surface Processes and Landforms*, 14, 419–32.

Needleman, H.L., Gunnoe, C., Leviton, A., Reed, R., Peresie, H., Maker, C. and Barrett, P. (1979), 'Deficits in psychologic and classroom performance of children with elevated dentine lead levels', *New England Journal of Medicine*, 300, 689–95.

Nriagu, J.O. (1989), 'A global assessment of natural sources of atmospheric trace metals', *Nature*, 338, 47–9.

Nriagu, J.O. (1990), 'Human influence on the global cycling of trace metals', *Palaeogeography, Palaeoclimatology, Palaeoecology (Global and Planetary Change Section)*, 82, 113–20.

Nriagu, J.O. and Pacyna, J.M. (1988), 'Quantitative assessment of worldwide contamination of air, water and soils by trace metals', *Nature*, 333, 134–9.

Pacyna, J.M. and Winchester, J.W. (1990), 'Contamination of the global environment as observed in the Arctic', *Palaeogeography, Palaeoclimatology, Palaeoecology (Global and Planetary Change Section)*, 82, 149–57.

Renfrew, C. and Bahn, P. (1991), *Archaeology, Theories, Methods and Practice*, Thames and Hudson, London, 543 pp.

Rivers Pollution Commission 1868 (1874), *Fifth Report of the Commissioners Appointed in 1868 to Inquire into the Best Means of Preventing the Pollution of Rivers*.

Royal Commission on Environmental Pollution (1983), *Lead in the Environment*, 9th Report CMDN 8852, HMSO, London.

Salomons, W. and Forstner, U. (1984), *Metals in the Hydrocycle*, Springer-Verlag, Berlin, 349 pp.

Say, P.J. and Whitton, B.A. (eds) (1981), *Heavy Metals in Northern England: Environmental and Biological Aspects*, University of Durham, Department of Botany.

Schnellmann, G.A. and Scott, B. (1970), 'Lead-zinc mining areas of Great Britain', in R. Jones (ed.), *Proceedings of the Ninth Commonwealth Mining and Metallurgical Congress 1969, Vol. 2*, Institute of Mining and Metallurgy, London, pp. 325–56.

Settle, D.M. and Patterson, C.C. (1980), 'Lead in Albacore: guide to lead pollution in Americans', *Science*, 207, 1167–76.

Takenchi, T. (1972), 'Distribution of mercury in the environment of Minamata Bay and the inland Ariake Sea', in R. Hartung and B.D. Dinman (eds), *Environmental Mercury Contamination*, Ann Arbor Science Publishers, Ann Arbor, Michigan, pp. 79–81.

Tarr, J.A. and Ayres, R.V. (1990), 'The Hudson-Raritan Basin', in B.L. Turner, W.C. Clark, R.W. Yates, J.F. Richards, J.T. Mathews and W.B. Meyer (eds), *The Earth as Transformed by Human Action*, Cambridge University Press with Clark University, Cambridge, pp. 623–41.

Thornton, I. (1977), 'Geochemical aspects of heavy metal pollution in England and Wales', in *Inorganic Pollution and Agriculture*, MAFF Reference Book 326, pp. 105–25.

Thornton, I. (1990), 'Soil contamination in urban areas', *Palaeogeography, Palaeoclimatology, Palaeoecology (Global and Planetary Change Section)*, 82, 121–40.

Timbrell, J.A. (1989), *Introduction to Toxicology*, Taylor and Francis, London, 155 pp.

Trefry, J., Metz, S., Trocine, R.P. and Nelsen, T.A. (1985), 'A decline in lead transport by the Mississippi River', *Science*, 230, 439-41.

Turekian, K.K. and Wedepohl, K.H. (1961), 'Distribution of the elements in some major units of the earth's crust', *Bulletin Geological Society of America*, 72, 175-92.

Wehr, J.D., Say, P.J. and Whitton, B.A. (1981), 'Heavy metals in an industrially polluted river, the Team', in P.J. Say and B.A. Whitten (eds), *Heavy Metals in Northern England: Environmental and Biological Aspects*, University of Durham, Department of Botany, pp. 99-107.

Welch, W.R. and Dick, D.L. (1975), 'Lead concentration in tissues of roadside mice', *Environmental Pollution* 8, 15-21.

Wittmann, G.T.W. (1983), 'Toxic metals' in V. Forstner and G.T.W. Wittmann, *Metal Pollution in the Aquatic Environment*, Springer-Verlag, Berlin, pp. 3-68.

Wolfenden, P.J. and Lewin, J. (1978), 'Distribution of metal pollutants in active stream sediments', *Catena*, 5, 67-78.

Wolff, E.W. and Peel, D.A. (1985), 'The record of global pollution in polar snow and ice', *Nature*, 313, 535-40.

Wood, J.M. (1974), 'Biological cycles for elements in the environment', *Science*, 183, 1049-52.

10

Emergency planning and pollution

Bill Myers and Paul Read

Since the Industrial Revolution the use of noxious chemicals in industrial processes has risen steadily as economies have expanded and developed. Now, in the latter part of the twentieth century when usage has become even more pronounced, the benefit to society which this is perceived to bring has been increasingly weighed against the damage, or potential damage to the environment which would be caused should there be a catastrophic chemical release caused by a process failure.

10.1 Chemical emergencies and their environmental effect

In this context, and with the evidence provided by those incidents which have occurred during the last half-century in a number of different countries, it has become increasingly obvious that prevention, or containment, must be a planning priority. Failure in either or both of these areas has, as this chapter will show, led to demands for industry to plan for the safety of people and the environment in the vicinity of plants deemed to present the greatest risk. The European Community, Australia and the United States in particular have all recognized these risks and have responded through legislation to the pressures exerted by their respective populations. Catastrophic chemical releases are due in almost every instance to the major hazards of fire, explosion or accidental release, and it is the aim of institutions responsible for emergency planning to minimize cause and to mitigate effects.

Naturally, chemical engineers will plan to *avoid* such releases and safety engineers will plan to *contain* any incident within the safest possible limits. However, in the event that an incident is so large that its effects are felt outside the immediate plant area, then it is essential that a suitable emergency plan should be in place and capable of dealing with the event. It is this off-site incident which concerns the Emergency Planning Officer and it is his planning which must aim to moderate the effects of any incident on the environment or on people.

10.1.1 Elements of 'safety first'

Lees (1980) characterizes such a safety-first approach as:

1. a concern with the depth of technology and associated major hazards;
2. an emphasis on management;
3. a system rather than a trial and error approach;
4. a concern to avoid loss of containment resulting in major fire, explosion or toxic release.

Table 10.1 The long-term health effect of certain categories of chemical incidents

Category	Example	Incident
Carcinogenic	Primary liver cancer (suspected)	Polychlorinated biphenyls (Japan)
Teratogenic	Cerebral palsy syndrome	Organic mercury (Japan)
Immunological	Abnormal lymphocyte function	Polybrominated biphenyls (Michigan)
Neurological	Distal motor neuropathy	Tri-o-cresyl phosphate
Pulmonary	Parenchymal damage	Methyl isocyanate (India)
Hepatic	Porphyria cutanea tarda	Hexachlorobenzene (Turkey)
Dermatological	Sicca syndrome	Toxic oil syndrome (Spain)

Source: Baxter (1990)

The emergency planner will assume that his colleagues in safety engineering will have given particular emphasis to point 4; his concern must be primarily with the first three elements. The increased probability of an accident occurring in the chemical and industrial processing industries is undisputed and is of course inevitable given the greater and more widespread use of chemicals and other noxious (or potentially noxious) substances such as oil and natural gas. How such disasters might occur is an element of the emergency planner's approach to contingency planning; however, the fact that they *will occur* is recognized and, as Keller (1990) has shown, can be classified into possibly three primary groups:

(a) natural disasters, which are generally beyond the ability of society to produce influence or prevent;
(b) man-made disasters in which the loss of life rarely exceeds several hundred; examples are train crashes, air crashes mining and marine disasters, fires or explosions;
(c) hybrid disasters which are a combination of the other two where the activity of man has significantly aggravated the effects of a natural disaster, e.g. the spread of disease from a community within which the disease is endemic to a community with no natural immunity, such as the introduction of European influenza to the Eskimos.

Although there has been a recognition by industrialists of their responsibilities to society, it is probably fair to say that this has to some degree been forced upon them by consumer pressure and growing public awareness of the probable long-term damage to the environment which even the threat of an accident poses. Although most incidents which occur in the chemical process and storage industries are minor and are dealt with on-site with minimal environmental damage, there is always the chance that a major, off-site incident will occur, no matter how infrequent this may be.

10.1.2 Chemical incidents

As Keller demonstrates there is a continuing history of chemical incidents occurring in recent times and originating in several countries, although it is true that even in these cases the size of the incident has varied. Certainly in terms of a major incident there may very well be significant problems affecting people and the environment which, as in the case of Bhopal, India in 1984, culminated in the deaths of about 2500 people and the long-term illness of many thousands more. Baxter (1990) has demonstrated the long-term health effects caused by some of these incidents (Table 10.1).

In practical terms, then, society has reached a crossroads in terms of applied science, protection of the environment, progress (however defined) and collective responsibility. The recognition of these pressures, together with the experience of recent industrial disasters, has prompted many countries to examine much more closely the way in which their industrial processes are permitted to operate. Indeed, the threat to the health of the population posed by major chemical incidents is of such large and insidious proportions that it can only be mitigated and controlled by legislation.

10.2 Major incidents: chemical emergencies and the planning response

A 'major incident' can be defined as: a serious unforeseen occurrence causing disruption to the normal life of a community which happens with little or no warning and which:

(a) causes or threatens death or injury to members of the public;
(b) causes damage to, or destruction of property;
(c) causes contamination of the environment;
(d) renders persons homeless in numbers in excess of that which could normally be dealt with by the public services operating under normal conditions, and therefore necessitating special mobilization and organization.

Some aspects of major incidents which need to be considered when plans are being developed include the following:

1. Initial chaos. This can be expected as the incident occurs and planners must take it into account when drafting instructions designed to minimize loss of life, injury and damage to property and the environment.
2. Need for information. Future trends in a particular industry and the off-site effects of particular substances are obviously of primary importance when considering how the effects upon people and places can be mitigated.
3. Plans must point to the means of enhancing the process of controlling and containing the incident.
4. There must be due consideration of the means by which external authorities can be co-ordinated.

10.3 Major incidents: some case-studies

Table 10.2 is a selection from Lees (1980) of some major incidents which have occurred over the last 75 years. The list is not a comprehensive chronicle of man-made environmental disasters, nor is it meant to be, but it illustrates very well the mounting scale of the risk of damage by industry on the environment. The table refers to several case-studies which are considered by Lees in his book. However, for the purposes of this chapter incidents in the table are taken as best illustrating the wide-spread effect which accidental releases can have upon the environment.

10.3.1 Seveso

The incident at Seveso occurred on the morning of Saturday, 10 July 1976 when a reactor used to produce a herbicide and a bactericide went out of control. The chemical process used to produce the above products also resulted in an unwanted by-product, 2,3,7,8,-tetrachlorodibenzoparadioxin, commonly referred to as TCDD or dioxin. Dioxin can be taken into the body by ingestion, inhalation or skin absorption and it is carcinogenic and highly poisonous.

The incident caused immediate contamination of the land and vegetation. An area of about 2 square miles was declared contaminated and people were warned not to eat anything from this area. Within a month, the area deemed to be contaminated was increased fivefold. The local population soon began to show symptoms of dioxin poisoning and the Italian authorities were compelled to

Table 10.2 Some major chemical emergencies since 1917

Year	Location	Emergency
1917	Wynandotte, USA	Chlorine release from storage tank 1 death
1928	Hamburg, Germany	Phosgene release from storage tank 10 deaths
1939	Zarnesti, Romania	Chlorine release from storage tank c. 60 deaths
1947	Texas City, USA	Ammonium nitrate explosion 552 deaths, 3000 injuries
1962	Cornwall, Canada	Chlorine release from rail car 89 injuries
1966	Feyzin, France	Propane fire and explosion 13 deaths, 31 injuries
1967	Santos, Brazil	Coal gas explosion 300 injuries
1969	Crete, USA	Ammonia release from rail car 3 deaths, 20 injuries
1974	Flixborough, UK	Cyclohexane explosion 28 deaths, >53 injuries, 3000 evacuated (see below)
1976	Seveso, Italy	TCDD release from factory £10 million of damage (see below)
1979	Bantry Bay, Ireland	Explosion on oil tanker at terminal 50 deaths

Source: After Lees (1980)

evacuate more than 600 people from the area. In addition as many as 2000 people were treated for dioxin poisoning. The effects on the immediate environment were devastating.

10.3.2 Flixborough

At about 4.53pm on Saturday 1st June 1974 the Flixborough Works of Nypro (UK) Ltd were virtually demolished by an explosion of war-like dimensions. Of those working on the site at the time, 28 were killed and 36 others suffered injuries. If the explosion had occurred on an ordinary working day, many more people would have been on site, and the number of casualties would have been much greater. Outside the Works, injuries and damage were widespread but no-one was killed. Fifty-three people were recorded as casualties by the casualty bureau which was set up by the police; hundreds more suffered relatively minor injuries which were not recorded. Property damage extended over a wide area and a preliminary survey showed that 1821 houses and 167 shops and factories had suffered to a greater or lesser degree. (Department of Employment, 1976)

The disaster at Flixborough, of which the quotation above is a concise description, was a turning-point in the UK approach to dealing with the threat of major hazards. It led directly to the legislative control of such sites by Parliament through the setting up in 1974 of the Advisory Committee on Major Hazards.

The Flixborough plant was situated just south of the Humber estuary in Lincolnshire and produced caprolactam for use in the production of nylon. The raw materials for the process were phenol and cyclohexane. The explosion, when it occurred, involved the cyclohexane process and was equivalent to the detonation of 15–45 tonnes of TNT. The environmental damage was confined entirely to blast damage; there were no long-term contamination problems. However 3000 people were evacuated from the surrounding area.

10.3.3 Chernobyl

The incident at Chernobyl occurred on Saturday, 26 April 1986 but it was not until Monday, 28 April 1986 that anyone outside a small coterie of highly placed Soviet scientists and Party officials knew that a major disaster had taken place. The incident was the result of a controlled engineering experiment in the reactor which went badly wrong, and which led to an uncontrolled release of radioactive materials into the atmosphere. Over the following days an airborne plume of radioactive material spread across Europe passing over the United Kingdom during the weekend of 2–4 May 1986.

The UK government had no contingency plans to deal with such an emergency and official reaction was at first sporadic and *ad hoc*. The Ministry of Agriculture, Fisheries and Food (MAFF) began to sample milk; the Department of the Environment (DOE) and National Radiological Protection Board (NRPB) joined in collating and interpreting the various monitored data, while the Meteorological Office prepared forecasts of the plume's path. The House of Commons Select Committee on Agriculture (1988) in its second report, *Chernobyl: The Government's Reaction*, comments that 'Some lack of cohesion was apparent'.

The forecasts prepared by the Meteorological Office indicated a reasonable probability that the weather pattern prevailing at the time would carry the plume into Britain on Friday, 2 May 1986. A historical analysis of the weather situation at the time is presented by Smith and Clark (1989), who show that the weather patterns during the ten days of the release were very complex with variable winds that carried the radioactivity from Chernobyl over almost all parts of Europe (Figure 10.1).

The plume posed both short-term and long-term threats to the environment. The levels of iodine 131 peaked in a matter of days but levels of Caesium 134 (HL 2.1 yrs) and Caesium 137 (HL 30 yrs) have persisted in some hilly areas where precipitation was high during the passage of the plume, and where sheep-farming is the only economic agricultural pursuit. In any case, by June 1986 radioactivity level in sheep from the worst-affected areas began to exceed the action levels of 1000 Bq/kg and restrictions had to be imposed by the government on their sale, movement and slaughter. Some six years later, restrictions remain in force in parts of Cumbria in England, North Wales, South West Scotland and Northern Ireland. Nevertheless, the Report of the Agriculture Committee states quite clearly that lamb carcasses with levels of contamination above the arbitrary limit of 1000 Bq/kg have entered the food chain in the United Kingdom.

10.4 Emergency planning legislation in the United Kingdom

The history of legislation covering the manufacturing, storage and transport of chemicals was until fairly recently concerned solely with the means of avoiding the occasion of accidental release. The response to such releases by emergency services and civil authorities has been recognized in the last three decades as crucial to dealing with the result of an accident. In the last decade especially, there has been growing acceptance that major incidents resulting from catastrophic releases require concerted efforts to mitigate the effects and that the prime means of achieving this is by well-written and comprehensive plans.

The impetus to write these plans has come from legislation passed by governments in response to public pressure consequent upon the effects of the increasing frequency of major incidents; in this sense it can be argued that all emergency planning legislation in the United Kingdom is disaster-derived. In Europe and the United Kingdom, the relevant legislation stems mainly from Directive 82/501/EEC of the European Community (EC 82/501/1982).

However, in the United Kingdom, after the accident at Flixborough, Lincolnshire in 1974, the Health and Safety Commission

Emergency planning and pollution

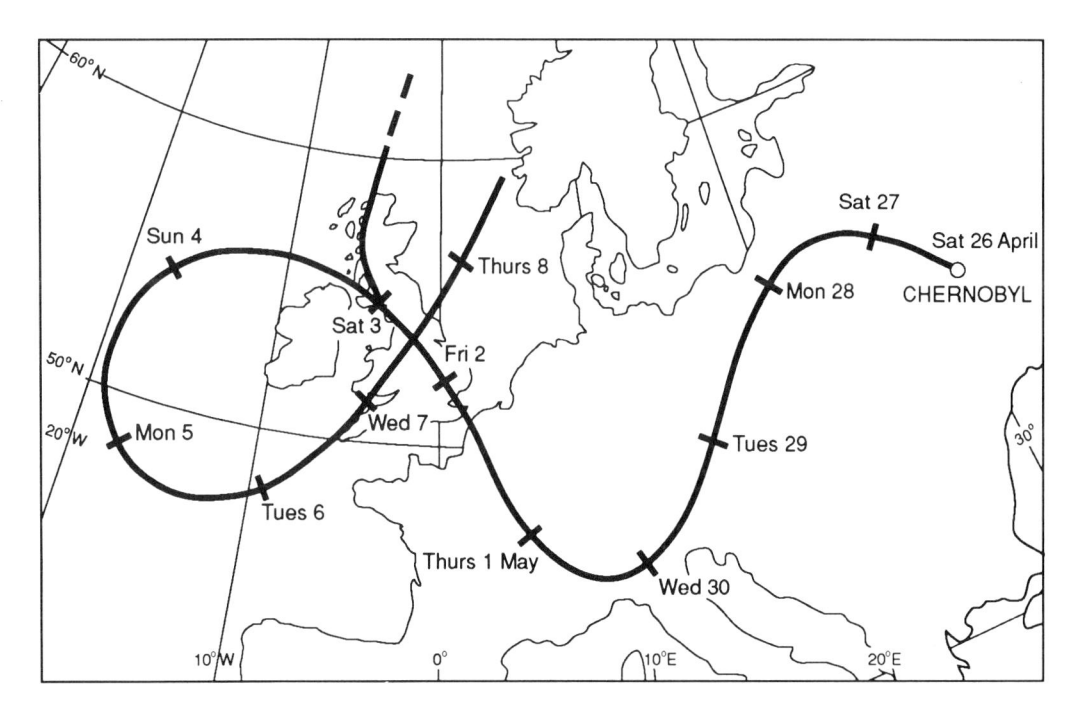

Figure 10.1 The spread of the Chernobyl 'cloud' from the reactor site across western Europe. (Smith and Clark, 1989)

(HSC) had already recognized the need for official action and had appointed a committee of experts — the Advisory Committee on Major Hazards (ACMH) — which had recommended a notification scheme for installations where dangerous substances were stored (ACMH 1972, 1976). This recommendation was accepted by the HSC and, in 1981, the Notification of Installations Handling Hazardous Substances (NIHHS) Regulations was published (Statutory Instrument, 1357, 1982).

During the same period, other accidents, at Beek in Holland in 1975 and Manfredonia in Italy in 1976, led to consideration by the European Commission of the need to control major accident hazards. As a result of these deliberations, the Commission issued the Seveso Directive, article 3 of which was accepted by the UK government, and in which Member states agree to 'adopt the provisions necessary to ensure that in the case of any industrial activities specified in Article 1, the manufacturer is obliged to take all necessary measures to prevent major accidents and to limit their consequences for man and environment' (82/501/EEC). This Directive led directly to the implementation of the Control of Industrial Major Accident Hazards (CIMAH) Regulations 1984.

10.4.1 Control of Industrial Major Accident Hazard (CIMAH) Regulations and the emergency planning response

The CIMAH Regulations (Statutory Instrument 1902, 1984), made under the Health and Safety at Work Act 1974, define such terms as 'industrial activity'; 'major accident'; 'manufacture'; 'site'; etc., and describe the requirements which operators are bound to follow in producing on-site emergency plans. The aim of the Regulations is to prevent major chemical industrial accidents and to limit the effects on people and on the environment of any accidents which may occur. They apply to the process and storage of specified dangerous substances at fixed

201

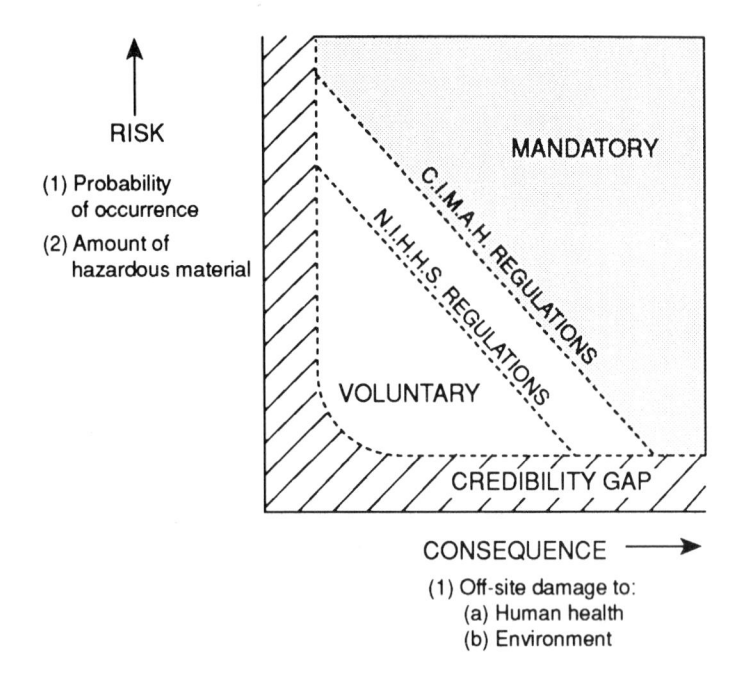

Figure 10.2 Levels of consideration for implementing chemical emergency planning related to the risk and consequences of an emergency

sites and involve many of the activities in much of the petrochemical, chemical and allied industries (including mixed-goods warehousing). Users of large quantities of fuel gases are also within the scope of the Regulations.

The Regulations also make it a local-authority duty to produce off-site emergency plans based upon information supplied by the site operator, and by the HSE. A set of guidance Notes (HSE, 1984) issued to explain the various responsibilities and duties of all three parties, gives administrative form to the SI and is used to produce the outline plans for all CIMAH sites.

Nevertheless, planning to mitigate the effects of major incidents or disasters is an imprecise science and, indeed, in many respects is more akin to the gypsy with the crystal ball rather than the laboratory scientist. Figure 10.2 expresses the criteria behind the planning process in graphical form, although the mathematician or geographer will look askance at a diagram with no scale and more than one variable along each of the axes. However, the use of the figure is convenient in that the intersection of the axes represents zero (0,0) and each axis indicates an increase in the variable with distance from the origin.

Following the graphical notation, it will be seen that no off-site emergency plan is needed if either axis is zero. On the one hand, the amount of hazardous material on-site or the high probability there may be of an accident involving such material is irrelevant if the worst accident that can happen will not result in any discernible effect beyond the site boundary. Conversely, if there is no possibility of an accident under any circumstances, then it matters not how large-scale the effects would be from the release of hazardous materials present on the site. There will, of course, always be a requirement for the site operator whose inventory includes hazardous materials to compile an on-site plan for the protection of his work-force.

The major complexities of the planning criteria occur in the area which we have termed the credibility gap (Figure 10.2). For convenience and simplicity this area between the written plan and the risk/consequence axes is shown as being constant. In reality, this is not the case; variation will occur because of the actual hazardous material and the site location with respect both to topography and micro-climate and also to adjacent population levels.

Pragmatism dictates that not in all cases will the off-site emergency plan be written for the worst-possible accident that might occur, but rather for the worst *credible* accident. This concept, of the worst-credible incident, has been introduced as an expedient when it became obvious that to implement the CIMAH Regulations in terms of off-site planning would cause, in some cases, insoluble problems in terms of resources necessary to implement the plans.

The explanation follows the general lines that a catastrophic accident involving the simultaneous release of the total on-site inventory is so remote as to be not worth consideration. This planning assumption is made on the basis of the safety measures taken by the site operator or industry in general: the safety ethos of the work-force and the quality assurance procedures in place. The consideration of terrorist attack/sabotage or the risk of an aircraft crash initiating a catastrophic release of the total inventory at an industrial site are officially to be considered too remote or not credible. Credibility is thus somewhat loosely related to probability of occurrence, as it is subjective and therefore an anathema to most emergency planning authorities; its validity will doubtless eventually be tested in the courts, when the role of the HSE as adviser to the planning function (and then prosecutor if the plan is deemed inadequate) will be brought into sharp focus.

Below the level of inventory of hazardous material which requires a site to have a mandatory emergency plan under the CIMAH Regulations, off-site emergency plans may be formulated on a voluntary basis. However, in the United Kingdom since the Cardiff judgment of 1988 it can be argued that the author of such a plan carries legal responsibility for its efficacy.

10.4.2 Duties of the site operator, manufacturer or local authority

In general, manufacturers are required to identify the major hazard associated with their sites and to take adequate steps to prevent major accidents, to report those that do arise, and to take steps to limit the consequences to persons and the environment. These duties are defined in Sections 4 and 5 of the Regulations. Manufacturers must also prepare a written report known as the 'Safety Case' on the hazard associated with the substances used in their processes and on the way in which the industrial activity is being carried on safely. The safety case is essentially an abstract of relevant information about the major hazard aspects of manufacturers' activities that should contain sufficient information to enable the HSE to judge whether the significant risks have been identified and are being properly managed. So as far as the local authority is concerned, the manufacturers must provide enough information to enable a detailed off-site emergency plan to be written based upon the worst-possible credible event.

The local authority will take into account when writing the plan the information supplied by both the manufacturer and the HSE. Contemporary methods of assessing the probability of a major incident having off-site effects, and the size and duration of such effects, can now be modelled using computer programs (e.g. 'WAZAN' — an acronym for the 'World Bank Hazard Analysis Program' developed by Technica, and 'PHAST' — 'Process Hazards Assessments Screening Tool', also developed by Technica). The more detailed information which these programs produce has allowed off-site plans to be devised which are more likely to equate with probable consequences than it has been possible to do hitherto. Such models make no allowance for local topography and use only the most

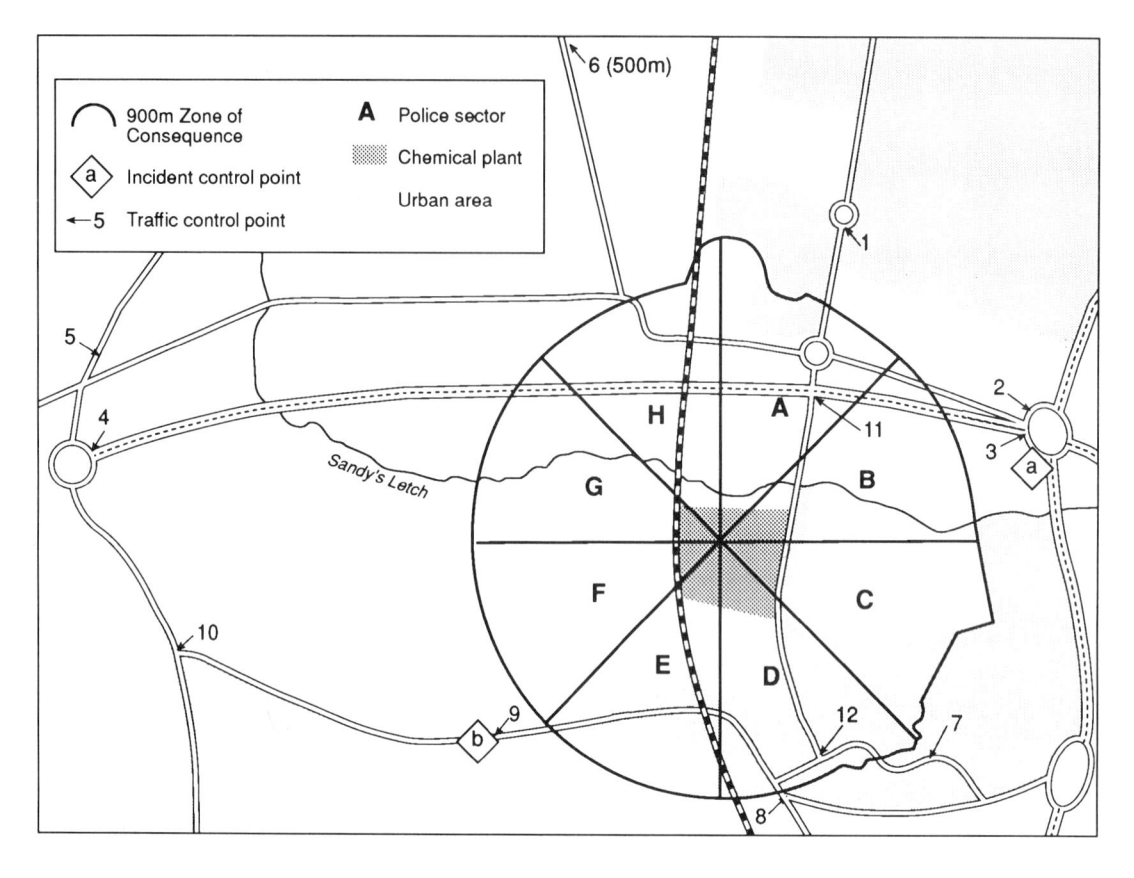

Figure 10.3 The zone of consequence around a hazardous manufacturing site in a built-up area

generalized meteorological data, necessitating input at the analysis stage by an emergency planner knowledgeable in geographical interpretation. Such vital areas of assessment as the Zone of Consequence, within which emergency methods and procedures would be required, can thus be more accurately determined within the parameters set not only by the forecast periodicity on the incident itself, but also as a consequence of varying meteorological conditions and environmental effects. A typical representation of the Zone of Consequence is shown in Figure 10.3 and from this it will be seen that the perimeter has been modified to take into account the influences of topography and population patterns.

Figure 10.4 is an example of a chemical site layout which would be included in any off-site plan for the information of the emergency and rescue services. With the advent of the CIMAH Amendment Regulations 1990 (SI 2325, 1990), a much greater emphasis has been placed upon the mitigation of environmental damage as an aspect of off-site planning. Table 10.3 is an off-site plan schema which takes this into account and which is the basis for all new plans and revisions in the Tyne and Wear conurbation in England.

10.5 Real time environmental information for chemical emergencies: the CHEMET scheme

In the event of an accident in the United Kingdom which involves the release of toxic

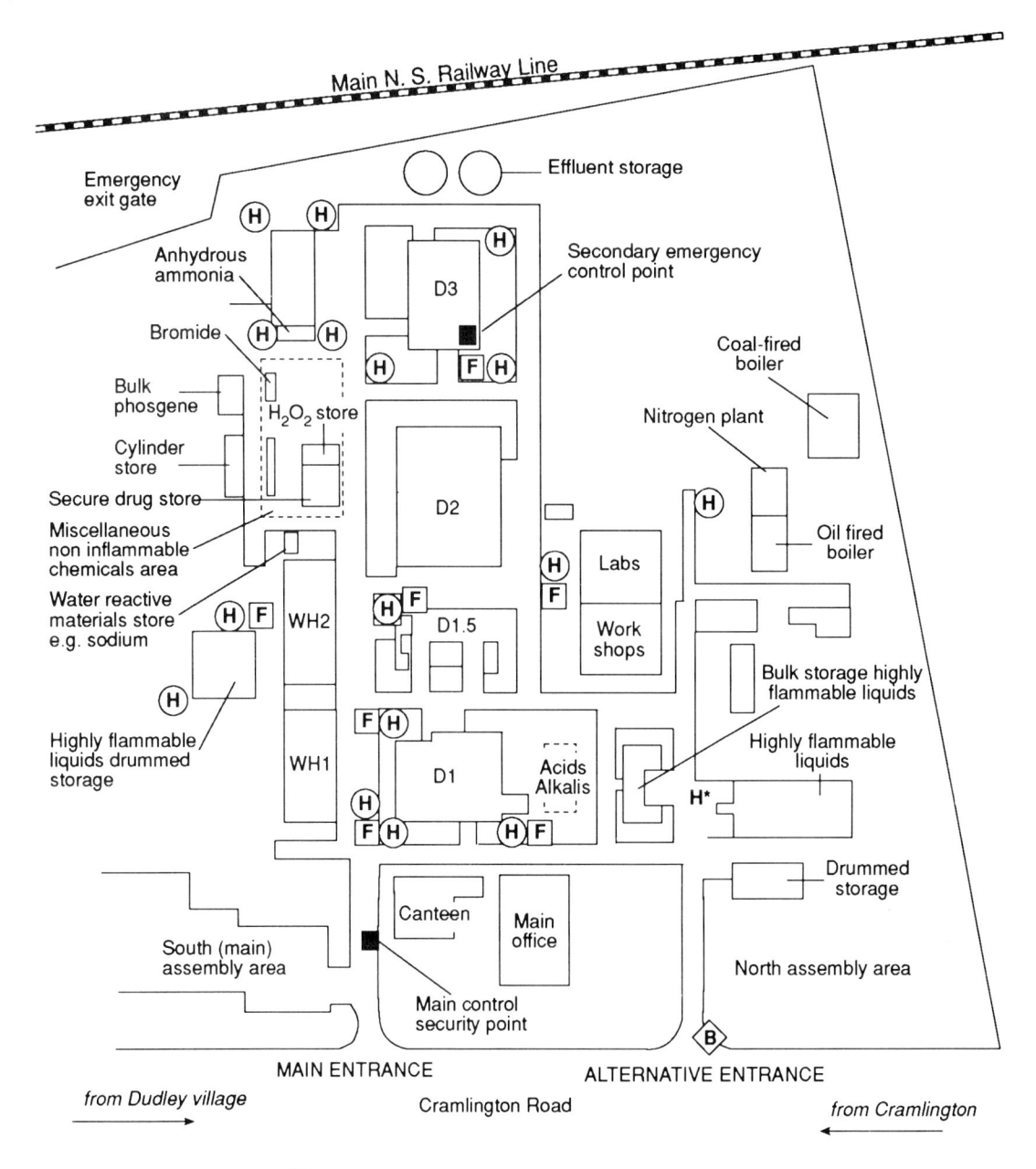

Emergency planning and pollution

Figure 10.4 Example of an emergency site plan

Table 10.3 CIMAH plan outline

Title page/Plan/Copy number	SECTION 5
Amendments	EMERGENCY SERVICES CONTROL
Distribution/Contents page	POINTS AND COMMUNICATIONS
SECTION 1	RVP (Emergency Services Rendezvous Point)
INTRODUCTION	ICP (Emergency Services Incident Control Post)
General introduction	FCP (Emergency Services Forward Control Post)
Aim	Site operations centre (on-site)
Objectives	Incident communications
Consultation list	GTPS/ACCOLC
SECTION 2	Site radio facilities, etc.
INSTALLATION DETAILS	SECTION 6
Name/Address/Grid ref.	SUPPORT SERVICES
Site location and topography	(amend list as appropriate)
Nature of activity (brief)	District council
Warning siren	British Rail
Key staff/roles	Water company
Shift capacity/persons on site	National Rivers Authority
Key facilities and location	Emergency planning unit
SECTION 3	Voluntary organizations
HAZARD DATA	DOE, MAFF, N. Gas, N. Elec., NCC, etc.
Notifiable substance(s); storage area	SECTION 7
Chemical and physical properties	MEDICAL RESPONSE
Chemdata printout	Designated receiving hospital(s)
Hazard analysis	Hospital decontamination room
Release events	Effects on health
Respond times	First aid treatment/decontamination
Emergency chemical advice	Regional poisons unit
Monitoring/detection equipment	Temporary mortuaries
Meteorological information	SECTION 8
SECTION 4	PUBLIC PROTECTION
PLAN ACTIVATION AND EMERGENCY ACTION	Area at risk (i.e. Zone of Consequence)
Alerting flow chart	Public warning siren
Alerting statement and procedures	Public self-protection measures
Plan activation	Evacuation
Action by (name of company)	Transport/evacuation assembly areas/routes
Action by fire brigade	Designated 'Rest Centres'
Action by police	SECTION 9
Action by ambulance service	PRESS/MEDIA
Stand-down procedures	Press and media arrangements
	SECTION 10
	ENVIRONMENTAL PROTECTION
	Enviromental survey

chemicals into the atmosphere there may be a requirement for meteorological advice, particularly information on the wind speed and direction, and the rate at which the cloud of material will disperse. In the United Kingdom the Meteorological Office is able to provide such advice through the CHEMET (CHEmical METeorology) scheme which has been designed to support the emergency procedures of police forces and fire brigades. The meteorological advice in CHEMET is also available to other organizations involved in chemical emergencies.

During the course of a chemical emergency many different organizations may require meteorological advice (Figure 10.5) and it is anticipated that this advice will vary as the emergency evolves:

Emergency planning and pollution

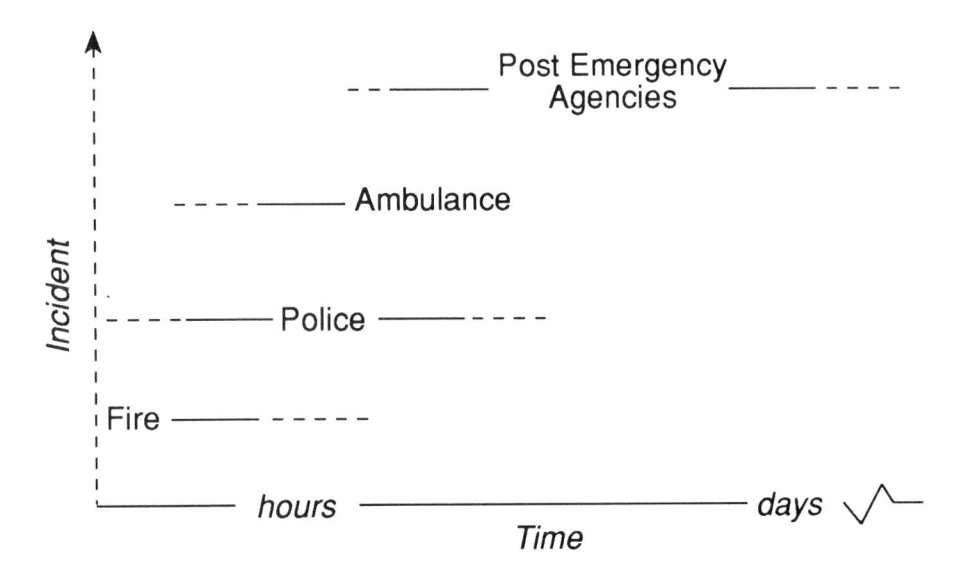

Figure 10.5 The involvement of the emergency services over the period of an emergency

(i) At an early stage the police force and fire brigade will require an immediate response giving details of surface wind speed and direction.

(ii) During a prolonged emergency there is a need for warnings of impending changes in local weather conditions.

(iii) On a longer, ecological time scale, specialist meteorological advice may be required to help such organizations as Chief Emergency Planning Officers and MAFF to determine the areas that have been affected, and where the clean-up operations may be required.

Advice under the CHEMET scheme is therefore divided into two parts, Part A and Part B.

Part A. CHEMET Part A is designed for rapid response to give the best estimate of the wind speed and direction at the site together with a brief description of the expected behaviour of any released material. It also gives details of any impending changes in the weather. It is available within two or three minutes from the initial contact and the information can be passed by telephone. An example of the form used for Part A is given in the box following Part B.

The form used for Part A requests meteorological information from the site of the accident. Although it is unlikely that this will be available at the time of the initial call, it should be passed to the Meteorological Office as soon as possible, especially if the incident is likely to be a lengthy one. It will be readily appreciated that meteorological conditions vary from site to site and that the quality of the advice from the Meteorological Office will improve if local observations are available.

Part B. CHEMET Part B provides much more detailed information and can be made available within 20 to 30 minutes. An estimated 'area-at-risk' is presented as a map. There is also a list of information which describes the meteorological conditions by facsimile which is the preferred means of distribution. In addition, local experts are available to provide continuing advice on local conditions.

1. Notification made by ....

2. Constabulary/Brigade ....

3. Date/Time of making call .... (clock time)

4. Call back numbers: * Telephone ....
   *Underline   * Telex ....
   preferred method   * DOCFAX ....

5. Time of chemical release .... (Date/Clock time)

6. Site of release .... (6-fig. OS ref)
   .... (Location)

7. Site topography ....

8. Name of chemical ....

9. Buoyancy of chemical ....

10. Nature of release: Continuous ....
    Instantaneous .... Fire at site ....

11. Site weather details: Wind (from) ....
    Cloud amount .... Weather ....

Enter weather forecast given by Meteorological Office:

WIND (from) .... degrees   Speed .... km/h

Plume behaviour: The plume is liable to disperse ....
Slowly .... Rapidly ....
....
....

10.6 Moving hazards and other considerations

The CIMAH Regulations in the United Kingdom are of course paralleled by similar legislation in Europe, North America and Australia. The latest amendment (1990) of the Regulations recognized the strong case which exists for ensuring that sufficient countenance of the possibility of environmental damage should be a feature of all off-site emergency planning. Although the initial pressures for action following Flixborough and Seveso resulted mainly from the perceived threat to human life, it has become apparent from the research carried out into the causes and effects of major incidents which have occurred in the last 50 or so years that this, although the most important aspect of accident prevention and containment, has in fact been the least frequent eventuality of the possible effects.

Although the threat to the environment from a 'loss of containment'-type incident is normally as a result of a failure in a fixed installation/storage unit, countenance must also be taken of large loads of toxic and flammable liquids transported by road and rail which also present very real dangers to human life and the environment. These loads are in effect mobile CIMAH sites, but they are not covered by the CIMAH Regulations.

There are of course various codes of conduct as well as specific legislation (HM Government, 1982, SI 1985, 1987) covering the transportation and identification of such loads (e.g. the UK HAZCHEM code) but, apart from the Major Incident Plans which the emergency services and many local authorities possess, no other planning to deal with such an incident is either a national statutory requirement or an international obligation.

Other aspects of pollution control legislation which recognize the threat to the environment are extant in the statutory controls governing the discharge at sea or in rivers and estuaries of a range of toxic or noxious substances but principally oil. The policing of this is undertaken by the UK National Rivers Authority and HM Coastguard. The response to a pollution incident arising from an oil or chemical spillage could be undertaken by a variety of agencies, but in fact there is no clear statutory duty placed on any specific agency to deal with such pollution until it comes ashore, either on the river bank or the foreshore. In some cases port authorities might deal with slicks in harbours, estuaries or rivers if, in their opinion, the slick constitutes a hazard to navigation, but it is unlikely that damage to the environment would weigh in their decisions.

The transportation of dangerous goods by sea and air is also a matter of increasing anxiety to those concerned with the care of the environment. Pressure upon national governments has helped to widen public awareness of such dangers and has encouraged governments to pay heed to the need for legislation designed to protect people and the environment from the worst effects of any incident.

The problems presented by hazardous waste dumped at sea, or a deliberate washing of oil tanks by unscrupulous ships' captains, have also led governments to recognize the damage which such actions cause to the environment and to give greater precedence to laws and regulations designed to obviate such unnecessary pollution. The International Maritime Organizations has recognized the need for action and in the United Kingdom local government has responded by setting up a small working party to examine the effects in the first instance of dangerous goods washed ashore.

The carriage by air of hazardous substances is now well documented and has long been a part of the air carriage trade. Nevertheless, the growing pressure for action designed to mitigate or even obviate the threat to life and the environment which such cargoes present has led now to the issue of guidance notes (ICAO, 1989) on the dangers which such activities pose.

This chapter has attempted to show how the United Kingdom in particular has responded to the growing awareness of the dangers posed to life and the environment by modern methods of manufacture, trade and commerce. It may be necessary in future for such pressure to be even greater as the list of substances dangerous to man expands. In the final analysis, though, only industry itself can determine the difference between a clean and healthy environment and one which is polluted.

References

Advisory Committee on Major Hazards (1976), *First Report*, HMSO; see also *Second Report*, HMSO (1972).
Baxter, P.J. (1990), 'Review of major chemical incidents and their medical management', in V. Murray (ed.), *Major Chemical Disasters: 'Medical Aspects of Management'*, Royal Society of Medicine Services, London, p. 8.
Department of Employment (1976), *The Flixborough Disaster*, Report of the Court of Inquiry, Judge Parker, HMSO, London.
Health and Safety Executive (1985), *A Guide to the Control of Industrial Major Accident Hazard Regulations 1984: Health and Safety Booklet HS(R) 21*, HMSO, London, ISBN 0 11 883767 2.
House of Commons Select Committee on Agriculture Second Report (1988), *Chernobyl: The Government's Reaction*, HMSO, London.
ICAO (1989), *Technical Instructions for the Safe Transport of Dangerous Goods by Air*.
Keller, A.Z. (1990), 'The Bradford disaster scale', in A.Z. Keller and H.C. Wilson (eds), *Disaster Prevention, Planning and Limitation*, The British Library, London, p. 5.
Lees, F.P. (1980), *Loss Prevention in the Process Industries*, Vol. I, Butterworth and Co., London, p. 7.
Official Journal of the European Communities (1982), L230, vol. 25, 5 August.
Statutory Instrument (1982), no. 1357, *Notification of Installations Handling Hazardous Substances*.
Statutory Instrument (1984), no. 1902, *The Control of Industrial Major Accident Hazards Regulations*.
Statutory Instrument (1987), no. 37, The Dangerous Substances in Harbour Areas Regulations 1987; The Civil Aviation Act 1982; Air Navigation (Dangerous Goods) Regulations 1985, SI 1985/1939.
Statutory Instrument (1990), no. 2325, Control of Major Industrial Accident Hazards (Amendment) Regulations.
Smith, F.B. and Clark, M.J. (1989), 'Airborne debris from the Chernobyl incident', Meteorological Scientific Paper no. 2, HMSO, London.
Tyne and Wear Fire and Civil Defence Authority (1990), *The Tyne and Wear County Major Incident Plan*, Newcastle upon Tyne.

Part 3 Futures

11
Radiation and the environment: types, sources, impacts and management

Stan Openshaw

Technocentric futures imply risk, none better perceived nor worse understood than that associated with artificial sources of radiation. The term 'radiation' tends to strike immediate fear into people. It is assumed to be associated with radioactivity and thence a short step to nuclear power, Chernobyl, nuclear weapons, and Japanese Second World War bomb victims, etc. However, the term is much broader in scope than this narrow connotation with atomic radioactivity. The world in which we live is bathed in many different types of natural and man-generated radiation. Concern might be expressed that the man-made sources may be adding significant extra, and in some cases rapidly increasing, amounts of artificial radiation that may be either harmless or harmful; the problem is that currently we simply do not know. Much of the increase in the total radiation background is recent, so recent that were there to be major health impacts it might well be a few years before anyone started to notice, assuming of course that anyone is looking! Indeed, herein lies a major dilemma. It seems that no one looks until there is a good reason to do so; ideally a more cautious exploratory mentality would be more relevant when no one knows the medium- and long-term impacts of man's activities. When dealing with the unknown, or the little known, it is important to avoid 'shocks' by the precautionary principle of surveillance. The reassurance factor is well worth the effort; negative findings are even more useful than positive ones. Environmental science should not forget the Yorkshire Television Sellafield lesson. A news-hunting TV journalist working on something different was able to 'spot' an apparent cancer cluster in a very circumstantially obvious location (near the Sellafield nuclear reprocessing plant) yet it was a place where previously no one else had thought to look! It would be easy now to dismiss such activity as little more than doorstep epidemiology but it provides a very useful lesson; it would be extremely worrying if TV journalists were the only ones looking for problems. Equivalents to Sellafield may well exist, but where and how to find them are questions that need to be faced.

11.1 A geographical view of radiation and the environment

The term 'radiation' covers a broad spectrum of energy emissions, most of which at present occasion little or no public or medical or scientific concern. Indeed it is by no means improbable that at least some other types of radiation will one day be viewed as even more dangerous than nuclear radiation. The problem at present is that we are effectively blinkered by fear of nuclear radioactivity (which is in fact fairly easily measured, extensively monitored and heavily regulated) and almost totally blind to many other sorts of non-nuclear radiation which are often far more difficult to measure and far less well regulated, if at all. Our obsession with nuclear radiation scare stories may well be making us blind to other forms of radiation, some of which may possess the potential to be even more dangerous.

The purpose of this chapter is to try and restore a degree of balance and at least pose the question: 'How safe are the many different sources of radiation?' Related questions might be: 'Does it matter?' 'What health effects do they have?' 'What legacies might we be leaving to future generations?' 'What, if anything, needs to be done?' and 'What can we, as geographers, suggest should be done?' If black spots exist, 'How do we find them?' 'Where do we look?' 'What do we look for?' and 'When should we be looking?'

11.1.1 Types of radiation environment

In considering the possible effects of the different types of radiation, it is useful to identify different forms of environment. Medical science increasingly talks about microenvironments, which are essentially those in which people live and work. A person's microenvironment is only partially related to the external physical or natural environment in which it is located because there are various barriers and filters that serve to reduce, or amplify, and channel the direct effects. For instance, complex food production and distribution networks tend to mix contaminated with less contaminated produce and serve to reduce, to hide, or to remove the immediate and geographic impact of a particularly polluted local environment. Air pollution is of less direct consequence to workers in air-conditioned offices than to those working on top of a coke oven. On the other hand, different types of microenvironment can amplify pollution effects: for instance, the smoke pollution from cigarettes on children living in the confined space of a house or the possible consequences of indoor cooking in poorly ventilated kitchens. The point to bear in mind here is that the effects of exposure to all the different pollution sources are integrated over time and summed over the many different environments in which they occur during a person's life history. Life-cycle effects, residential mobility, changes in lifestyle, all alter and change these exposures. Additionally, various other complications, some genetic and some physiological, critically influence any immediate, short-term, long-term or genetic effects. Clearly, nothing in this area is simple and it will be a very long time, probably hundreds of years, before the causal mechanisms are understood to any adequate degree (see simple toxicology in Chapter 2).

11.1.2 Monitoring, measuring, mapping, modelling and geographical analysis

Meanwhile, life has to continue and the available tools for analysis and management are broadly limited to:

(1) monitoring health databases and then performing epidemiological studies that attempt to relate unusual mortality and morbidity patterns to possibly linked sets of common events and backgrounds; and
(2) broad spectrum environmental monitoring and associated information systems which are designed to measure and identify temporal changes in pollution or contamination levels.

There are important roles for geography in both these management tasks, e.g. to aid the small area analysis of geographical patterns in data. The information technology revolution together with geographic information systems are creating a wealth of geographically referenced health and environment databases. However, new tools are urgently needed to analyse the data to identify any key linkages and find pattern or spatial anomalies in the geographic patterns of health, death and disease. Methods are needed which can explore data without being told in advance either precisely what to look for or where to look for it or when to look. Medical science seldom knows enough about the various diseases to create good a priori hypotheses. Maybe there is an opportunity here for geographers to develop the necessary spatial data exploratory technologies to create new insights that may help answer some of the basic questions and narrow down the areas of search.

In addition to the need for new analytical tools, there is an urgent need for a major comprehensive environmental radiation pollution monitoring and measuring programme. Civilization is causing massive change to the environments in which we live. It is increasingly important that we can identify and measure significant changes in key indicators relevant to human health and wellbeing. At the very least such a programme should seek to meet the following criteria:

(a) A fine mesh of long-term monitoring points is needed at a sufficient level of spatial resolution so as to identify local abnormalities and hot spots. Chernobyl demonstrated the need for accurate local knowledge about radiation levels.
(b) The development and application of measurement sensors that can cover the whole spectrum of the chemical, biological and electromagnetic activity, and not just that handful of toxic substances of current interest. What is often regarded as a minor contaminant today may well be viewed as a major pollution problem in a few years' time. The technology exists to measure and quantify the presence of thousands of substances and quantities to an accuracy measured in parts per billion.
(c) It is immensely important to seek to build up space–time data series with a view to the next hundred years and not just the next five. Many pollutants accumulate in the environment, radiation exposure is also cumulative, and both may be dynamic in terms of nature and location. The effects of the Roman lead miner can still be found, as indeed can the fallout from 1950s nuclear weapons. So it is essential for future control and management purposes to build up high-resolution space–time data series.
(d) Finally, it is also necessary to develop in parallel to the data collection and measurement activities some degree of understanding of the spatio-temporal processes that are responsible for the patterns that are observable. Modelling of biological response and health impacts are important parts of this activity.

It is useful therefore to view the subject of radiation in the broader context of environmental monitoring and disease data analysis. This is important because we are still at the early stage of trying to demonstrate health impacts. The entire leukaemia debate of the 1980s was based on an excess of four cases over a 25-year period in a rural area near to a major nuclear installation. The statistics suggested that this was extremely unusual; immediately a major research programme was set up and more stringent constraints placed on radioactivity emission levels — 'just to be safe', because there was in fact no hard scientific basis to most of it. Nevertheless, this 'just to be safe' paradigm is an important response. It is not always sensible to wait for proof when the available evidence is sufficient for action, even if the intellect is not satisfied. The same paradigm could be viewed in relation to this chapter, although the problems are harder. Non-ionizing radiations tend to be more ubiquitous with perhaps fewer obvious geographically concentrated sources of generation; although this is probably a reflection of our current state of ignorance. On the other hand, before the Black Report, nuclear radiation was also considered to be fairly widespread and evenly distributed because of fallout from the atmospheric testing of nuclear weapons in the 1950s and 1960s. In-depth, localized measurement of environmental radiation simply did not exist. Once radiation measurements start to be made, then the spatial unevenness of radiation exposure starts to become apparent. The same 'need to make a case' exists in this chapter. The risks are twofold. There is a Type I Error in needlessly engendering public fears and concerns; but there is also the Type II Error of failing to find major problems because no one looked or, if they did, neglected to interpret the evidence correctly.

11.1.3 Types of radiation

In seeking to develop a better understanding of radiation, a useful starting point is to recognize that there are, broadly speaking, three major types: mechanical energy in the form of ultrasound as well as the more usual non-ionizing and ionizing sources. When various types of radiation pass into a material, they interact at the atomic scale. If the radiation particles or photons have enough energy then they can remove electrons from these atoms. This is a definition of ionizing radiation normally associated with nuclear substances. However, in order to ionize, the particles need to possess energy greater than that of ultraviolet (UV) radiation; anything else is non-ionizing radiation. Examples of ionizing radiation are: X-rays, gamma rays, neutrons, alpha particles, beta, etc. In reality the distinctions are somewhat artificial as they are all part of the same electromagnetic spectrum. Nevertheless, it might be helpful to treat each major type of radiation individually and then discuss some of the implications for the environment.

Table 11.1 presents a conventional classification of the electromagnetic radiation spectrum. Differences in wavelengths are important even within a particular radiation type, because it affects the ability of the radiation to damage the human body. The areas affected may well depend also on wavelength and type. Unfortunately, there is very little knowledge as yet of the nature of possible health impacts, and any ideas about dose–effect relationships and latency periods are almost completely absent outside the ionizing radiation area.

Table 11.1 Principal types of electromagnetic radiation

Type of radiation	Approximate wavelengths
Ionizing	<100 nanometres
Ultraviolet (UV)	1–400 nanometres
Infrared	0.7–1000 micrometres
Conventional lasers	0.2–20 micrometres
Microwaves (MW)	1–1000 mm
Radar	5–1300 mm
Radiofrequency (RF)	1–3000 m

11.2 Non-ionizing radiation

Non-ionizing radiation (NIR) is less energetic than ionizing types and it interacts with human tissues primarily by heat energy. Like ionizing radiation, it pervades the entire environment and, apart from the visible part of the electromagnetic spectrum, it is invisible and unperceived by human senses unless felt as heat. So NIR is mainly invisible and it has the capability of causing all manner of biological damage. There has been a rapid growth in NIR emissions in the latter half of the twentieth century and despite its ubiquity and the rapid increase in exposure levels, it has been almost entirely neglected as a pollution source capable of harm. Indeed, current knowledge of the effects of NIR is broadly similar to what was known about ionizing radiation towards the end of the nineteenth century.

11.2.1 Ultraviolet radiation

Exposure to ultraviolet (UV) radiation is widespread. The principal natural source is the sun, whilst lights (gas discharge and incandescent) are important artificial sources. There are different types of UV radiation with different characteristic wavelengths and different effects. The principal concerns at present involve:

(1) the depletion of the ozone layer which increases UV exposure in middle to high

lattitudes in both Northern and Southern Hemispheres; and
(2) life-style effects that interact with natural sources to change exposure levels; for example, outdoor activities and sunbathing.

It is to be noted that UV affects plants, animals, and man. It damages the ability of crops to grow and in man is thought to be a major cause of skin cancer. However, in common with many other complex pollution processes, the health effects are stochastic in that other factors are also responsible so there is no simple or direct relationship with exposure. It is possible that should the ozone layer be depleted in a major way then additional screening against harmful UV radiation will be needed by many life-forms and many forms of outdoor agriculture may also need shielding. The patterns may well be geographically structured on a regional rather than a local scale.

11.2.2 Optical radiation including lasers

Lasers are now widely used both commercially and domestically. The eyes are particularly vulnerable as is the surface of the skin. It is thought that most low-powered lasers are fairly safe and no doubt this type of thinking will increase the laser radiations we receive because of other people's activities over which we have no control: for example, range-finding, bar-chart readers in retail establishments, and laser displays. The use of lasers in satellite platforms and by aircraft might be other examples. It is presumed to be 'safe' but until there is a vast increase in laser-related diseases (namely, eye cataracts) during the next 50 years no protective action will be taken. Note also that laser radiation may well interact with other factors before there is any noticeable effect, so that even if there is a recognizable effect the linkage to laser radiation may by then be extremely well disguised and difficult to find.

11.2.3 Infrared radiation (IR)

Here the energy levels are sufficient to produce heat when they are absorbed by matter. Humans can readily detect sources of heat. Currently there are many applications of IR in domestic control devices. Maybe this is not a problem but it might become one if there are interaction effects with other types of radiation or chemicals. Again we do not know, and if we do not know then why assume it is safe until proved otherwise?

11.2.4 Microwaves (MW) and Radiofrequency (RF) radiation

When a biological system is exposed to either MW or RF energies, electric and magnetic fields are induced which can or may disrupt their normal functioning. The amount of energy absorbed depends on wavelength. There is currently very little medical evidence about actual effects other than extreme cases, although there is a wide range of possible effects of low exposures on biological systems. Indeed, there have already been a number of epidemiological scare stories, for example, the observation that there seems to be an excess cancer incidence near to overhead domestic power lines in the United States. In Britain, a cancer cluster in Gateshead was seen to be what can only be described as circumstantially located near to substations and overhead power lines. The problem here is that proximity to substations and overhead wires may well be a surrogate for other more directly relevant but missing variables; for instance, maybe the only people living close to overhead power lines are those living in poor council estates. Additionally, the range of any effects will probably be small, say 20–50 metres; whereas the positional accuracy of current geo-referencing systems used to locate postal addresses in cancer databases is probably an order of magnitude worse. Additionally, in Britain the precise location of all substations, their internal electrical characteristics, overhead wires, and the location of major underground cables are not public knowledge. On the other hand, the electrical utilities have all been investing heavily in conversion of their paper map records into digital format, so that by the late 1990s it should be possible to start linking up accurate electromagnetic field information with health databases.

Another problem is that MW and RF radiations are now ubiquitous. Our world is full of RF emissions to the extent that there are few available frequencies left to be allocated. At the same time, there are many sources of MW; in the home (TV and ovens) and elsewhere from radar in planes and satellites, from intelligent traffic lights that use MW to detect approaching cars, police speed traps, secret military systems, etc. The widespread use of electricity in the home results in universal exposure to electricity fields and other electromagnetic phenomena which continually vary in intensity, as well as in space and time. Indeed, anyone who sleeps with an electric blanket switched on is lying inside a huge electricity field. However, there are also areas of concentrated exposure, for instance, near to overhead cables or electricity substations, or underneath MW corridors, etc. If it were possible to measure such things, then integrated personal dosimetry would probably vary tremendously depending on location, behaviour and age. How can anyone tell whether or not it was an unfortunate combination of a sudden (but unnoticed) surge of electromagnetic radiation that caused an individual's DNA repair mechanism to miss a single malignant cell that caused his cancer? The point is that whilst these radiations are ubiquitous personal exposure rates will vary tremendously, probably with a strong geographic component.

Of all the types of radiation that exist, it is RF and MW that are least regulated and yet more is known about their potential biological dangers than of all other non-ionizing types. The challenge is now either to demonstrate by empirical observation and measurement that highly irradiated areas are no different in their health and disease profiles, or to identify by geographical analysis potential ill-health-causing effects. Until recently this task was impossible because computers were not fast enough, the health data were not geo-referenced, and GISs did not exist. Since the late 1980s this is no longer the case. An important challenge now is to demonstrate in some way that there is really no cause for concern here. Until then, is it really sensible to continue pretending that unscreened substations are as innocuous as a garage and that the possible health impact of power lines (both under- and over-ground), MW towers, radar installations, etc. should continue to be ignored? Here is a potentially major source of pollution. Do we have to wait for the TV journalists to highlight the risks? What would happen if the key affected group turn out to be pensioners rather than children?

11.3 Ionizing radiation

Our world is naturally radioactive and has always been so. The contribution from man-made sources is fairly recent. It is hardly surprising then that ionizing radiation is the best known type of radiation. What is worrying is that almost all sufficiently energetic ionizing radiation is harmful to humans, and the proved link with cancer makes this type of radiation extremely fearful. It is perhaps an indication of the strength of this nuclear neurosis that other similarly carcinogenic chemicals have not yet engendered equivalent levels of fear. Maybe the nuclear weapons link is one answer; another is probably ignorance of the dangers of many other non-ionizing pollutants.

With ionizing radiation, the amount of damage caused depends on: the amount of energy deposited, the method, the rate, and perhaps, especially if the dose rate is small, on the characteristics of the victim. The human body possesses repair mechanisms that constantly try to undo the damage caused by radiation. Additionally, human DNA is fault-tolerant to a degree with code redundancy and error-checking systems but it is not totally foolproof. As a result, there are probably no safe dose levels that can be specified which are meaningful. Currently, safety limits continue to fall as science discovers more about the effects of even very low levels of radiation (Caufield, 1989).

11.3.1 Damage by ionizing radiation: measurements

There are two characteristics of ionizing radioactivity that cause concern: first, it is easy to measure in parts per billion (or less) and this is not true for many other substances; second, the apparent complexity of the subject requiring a mix of physics, chemistry, biology, mathematics and engineering means that there are very few experts in more than a small part of it.

It is at first alarming to discover that sensitive radiation-measuring devices can count the disintegration of individual atoms; indeed, the standard unit of measurement is the Becquerel (Bq) which is the number of disintegrating atoms per second! In the mid-1980s a major crisis in public confidence was occasioned by a switch from the previous measurement unit of the Curie. The problem was that 1 Curie does not seem much, but this is equivalent to 37 billion Bqs and that sounds bad. The innocuous nature of Bqs receives another major blow when it is necessary to convert Bqs into a measure of the amount of radiation absorbed by a body. It is not necessarily the total number of Bqs you receive but its nature. The absorbed dose is an attempt to take into account the quality and biological significance of different types of ionizing radiation. The unit of measurement is the Gray (Gy), equivalent in energy to 1 joule per kilogram. This is then translated into a dose equivalent which reflects its supposed biological significance. For example, a dose of 1 Gy of fast neutrons is thought to be ten times more damaging biologically than 1 Gy of X-rays. The units of radiation that will produce the same biological effect are called Sieverts (Sv) and usually, units of 1 millisievert (mSv) are used. One problem is immediately apparent in that the precision of the physical measurements of Bqs is translated in an approximate and subjective way into some average statement of biological significance that everyone then believes. It might be questioned whether the conversion is either always correct or that simple.

Nevertheless, it is clear that we live in a radioactive world. The 'average' UK resident is calculated to receive 87 per cent of radiation from the natural background, 11.5 per cent from medical, 0.5 per cent from weapons fallout, and a miniscule 0.1 per cent from the nuclear industry. However, beware of the meaningless average. This does not imply that there are not major regional and within-region differences, although we do not know much about them because of the absence of a fine-resolution monitoring network. Most of the natural radiation comes from the sun and cosmic sources (14 per cent), 19 per cent from the earth itself, 17 per cent is received from within ourselves, 32 per cent from radon, and 5 per cent from thoron (see Hughes and Roberts, 1984). Of the artificial sources, the medical profession is mainly to blame with 11.5 out of a total of 13 per cent. Such statements say nothing about the geography of radiation and, indeed, we know very little about the detailed local variations. There is need for a major research project here.

11.3.2 Nuclear regulation

The 13 per cent of artificial radiation is heavily regulated. The nuclear industry (responsible for 0.1 per cent) is beset by all manner of draconian regulations; by contrast, the medical profession escapes much more lightly. Yet the potential for harm might be considered similar and both could cause major public disasters — for example, the medical profession by performing too many diagnostic X-rays or by using old, leaky and heavy-dose equipment, and the nuclear industry because of the potential for major accidents. The latter is well known in the post-Chernobyl era, but the risk of excess cancers as a result of faulty X-ray equipment or too liberal usage is just as real and might also result in geographically localized effects. Does anyone look? The answer at present is no.

In general the international regulations that all civilized countries subscribe to suggest that there should be no exposure to radiation unless there is a benefit. In the medical context, the crisis atmosphere of an emergency will justify almost anything. In the nuclear power context, the need for electricity and increasingly the need to avoid burning fossil fuels will justify almost any level of nuclear power expansion. Indeed, it can be argued that the world is on the verge of a major move, by international concensus, towards a nuclear-powered future (see Chapter 12).

The other principle in radiological protection is that any doses should either be as low as reasonably achievable (ALARA) or increasingly as low as reasonably possible (ALARP). The problem here is one of cost–benefit analysis (see Chapter 5). It costs money to reduce public exposure to radiation. There comes a point, the experts argue, when the cost (in health terms) of a low dose of radiation is far less than the financial cost of reducing it. Nuclear safety is a gigantic cost–benefit analysis gamble. Since nothing in this world is completely safe, there comes a point when it costs too much money to make something already incredibly safe, even safer. The nuclear industry used to use the analogy with building hospitals. The cost of reducing the 'safe' emissions from Sellafield in order to save maybe a few cancers over the next hundred years, it is argued, is equivalent in money terms to building ten new hospitals which could presumably save many thousands of lives over the same time-scale. This argument has some merit but only for illustrative purposes because the money used for reducing the output of radionuclides into the Irish Sea would not otherwise be available for building hospitals. Likewise, the cost of giving cancer to people who may not otherwise have had it is seldom attributable to the polluter.

It is reassuring to know that the maximum public dose allowed in the United Kingdom from nuclear power sources is 5 mSv per year, and the current total dose at present averages at about 1.9 mSv per year. Nevertheless, there is considerable uncertainty about the health impact of very low-level radiation doses. The effects are dominated by stochastic factors and rely heavily on calculations concerning the dose–response relationships associated with the two Japanese atomic bombs. However, at the same time, radionuclides (like many other pollutants) accumulate in the environment so that in theory at least the scope for future releases of radiation will be affected by a combination of what is already there and historically reducing 'safe' limits. Indeed, the half-lives of some radionuclides are extremely long (see Table 11.2). The longer-lived ones tend to be characteristic of high-level nuclear waste. For comparison, it should be noted that the estimated age of the earth is about 4,500 million years, about the same as the half-life of uranium-238. On the other hand, caesium-137 only has a half-life of 30 years but this is really quite long enough to present considerable future management problems to humans.

The tendency of radionuclides to accumulate in the environment, in people, and to be transported by food chains is a very important feature. This is characteristic also of other toxic substances, such as mercury and lead. It leads to the concept of pathways and critical groups. It really does not matter much if radionuclides are deposited in the local area around a nuclear power station if there are no people living there, or no crops or animals that might be eaten. The doses received by the public will reflect a number of alternative distribution systems:

(1) inhaling radioactive particles;
(2) exposure due to proximity from either a passing cloud containing radionuclides or by walking on ground or living in a house that is contaminated with radioactive particles; and
(3) ingesting food that is contaminated.

Table 11.2 Half-lives for selected radionuclides

Radioisotope	Half-life
Krypton-85	10 years
Strontium-90	29 years
Iodine-131	8 days
Caesium-137	30 years
Carbon-14	5,730 years
Zinc-65	245 days
Cobalt-60	5 years
Plutonium-239	24,400 years
Tritium	12 years
Technetium-99	210,000 years
Neptunium-99	2,100,000 years
Iodine-129	17,000,000 years
Potassium-40	1,300,000,000 years
Uranium-238	4,500,000,000 years
Thorium-232	14,000,000,000 years

The latter food pathways can be major sources of dose because the radiation can often be (a) concentrated and (b) spread far beyond the area of origin. It also seems that longer-lived radionuclides can be stored in soils and biomass with a continuing release into food chains over long periods of time.

The critical role of nuclear pollution pathways has been known about for a long time. The nuclear industry sets its authorized discharge limits at levels which would not expose the most sensitive critical group to excessive doses. For example, fishermen eating radioactive fish on the Cumbrian coast should not receive an excess dose. In theory, it is possible to work backwards to estimate a 'safe' discharge limit. The problem here is one of uncertainty. Do the models work that well? Are the diffusion mechanisms well understood? Have all the pathways to the critical group been identified? Have all the critical groups been identified? The answer is no! Indeed, herein lies one of the major problems with the nuclear industry. The physics and chemistry of what they are doing are fairly well understood and capable of being accurately modelled. By contrast, the biological, environmental and social science aspects are actually far more complex. Although they are not well understood, they have traditionally been dealt with as if they were susceptible to a similarly rigorous scientific approach. Chernobyl provides a good example here. The available computer models predicted a much more intense localized impact than that which was recorded. The problem with the nuclear industry is that it mixes systems which are well understood with those which are partially understood with some that are not understood at all. The pathways concept is a nice theoretical idea but virtually impossible accurately to operationalize in a space–time dynamic manner that is needed if it is to be of any real use.

11.4 Nuclear power

In general nuclear power is very safe. The industry is extremely safety-conscious, much more so than many others. It is also heavily regulated and run by professional experts dedicated to scientific principles (see O'Riordan, 1987). The problem is that safety cannot be guaranteed. Indeed, it is only because the industry is so advanced that it can actually quantify the risks associated with it; this is not true for nearly all other industries, some of which are also potentially capable of great harm. As a result, it is currently estimated that the death risk is less than one in a million for the latest reactors (Kelly et al., 1983). This compares extremely favourably with most other risks associated with modern life; for example, the probability of death per person per year in a car accident is currently 0.00014 and death by murder at 0.000012 (see Table 11.3). So is this not good enough? The problem is also how to explain what these risks mean to the public and the nuclear industry has so far failed to discover how best to achieve this goal (see Openshaw, 1988a).

The counter-arguments are as follows:

(1) These risk estimates are based on assumptions and models that are incorrect and rely on accident scenarios that may be too optimistic; see, for example, Marshall et al. (1983).

Table 11.3 Risks of death in Britain

Cause of death	Number per million per year
Average: all causes	11,900
All cancers	2,800
Violent causes	396
Road accidents	100
Gas	1.8
Lightning	0.1
Nuclear power	0
ICRP maximum radiation dose	1–10

(2) The probabilities are also not historical facts, since the nuclear industry has a history of less than 50 years.
(3) Reactor accidents severe enough to kill one person will kill many more and this is quite different from a car accident which can by definition can only kill a few at a time thereby invalidating the simple comparison of risk probabilities.
(4) Extremely improbable accidents have already happened at Three Mile Island and Chernobyl which must cast doubts on the meaningfulness of the probabilities.

The Chernobyl reactor accident very nearly signalled the political demise of nuclear power for good (Openshaw, 1988b). It was rescued by a series of fortuitous factors: few media-visible deaths, a cause due to gross operator error, extreme bravery which reduced the consequences, good luck in that the effects were distributed over a much wider area than might otherwise have been expected, and the fact that Chernobyl was a very remote site by European standards with the nearest major town, Kiev, being over 100 miles away. Growing concern about global climatic change is also another important factor because it leads to public doubt over conventional power generation.

11.4.1 Siting nuclear power plants: the geography of radiation sources

If the argument is accepted that nuclear power is unavoidable mainly for environmental reasons, then it would seem sensible to start planning for the long term (see Openshaw, 1988c). Current nuclear power sites in the United Kingdom were largely selected in secret in the 1950s, with little or no public debate, to meet the demand for bulk electricity supplies in the 1960s (Openshaw, 1986). The siting criteria used would today be regarded as very inadequate for reactors which possessed no secondary self-containment. With the first generation of so-called remote sites, remoteness was defined as being about ten miles from a major population area, and all but two of the existing sites date from this era. The subsequent era of relaxed siting, from 1968 onwards, allowed sites to be developed much closer to major population centres (Figure 11.1), in the case of Hartlepool on the edge of a major industrial area. It all seemed quite reasonable because the worst-case accident for the AGR design was thought to have no off-site consequences whatsoever.

Unfortunately, the need to be near to major load centres and thereby minimize transmission costs overruled the safety need to avoid being too near to too many people. Table 11.4 shows the population living at various distances from current UK nuclear sites. It is obvious that some of these sites are poorly located should major accidents ever happen in the United Kingdom. The problem is, therefore whether or not the worst-case planning accident is in fact the worst-case accident that could occur under plausible assumptions. As Openshaw (1986, 1992) points out, it is the big accident scenarios, currently ignored in safety studies, that dominate the long-term risks of nuclear power. It is also the risks of sabotage or terrorism which are quite high-probability events and yet are theoretically capable of matching Chernobyl in terms of radiation releases, but these scenarios are currently

Radiation and the environment

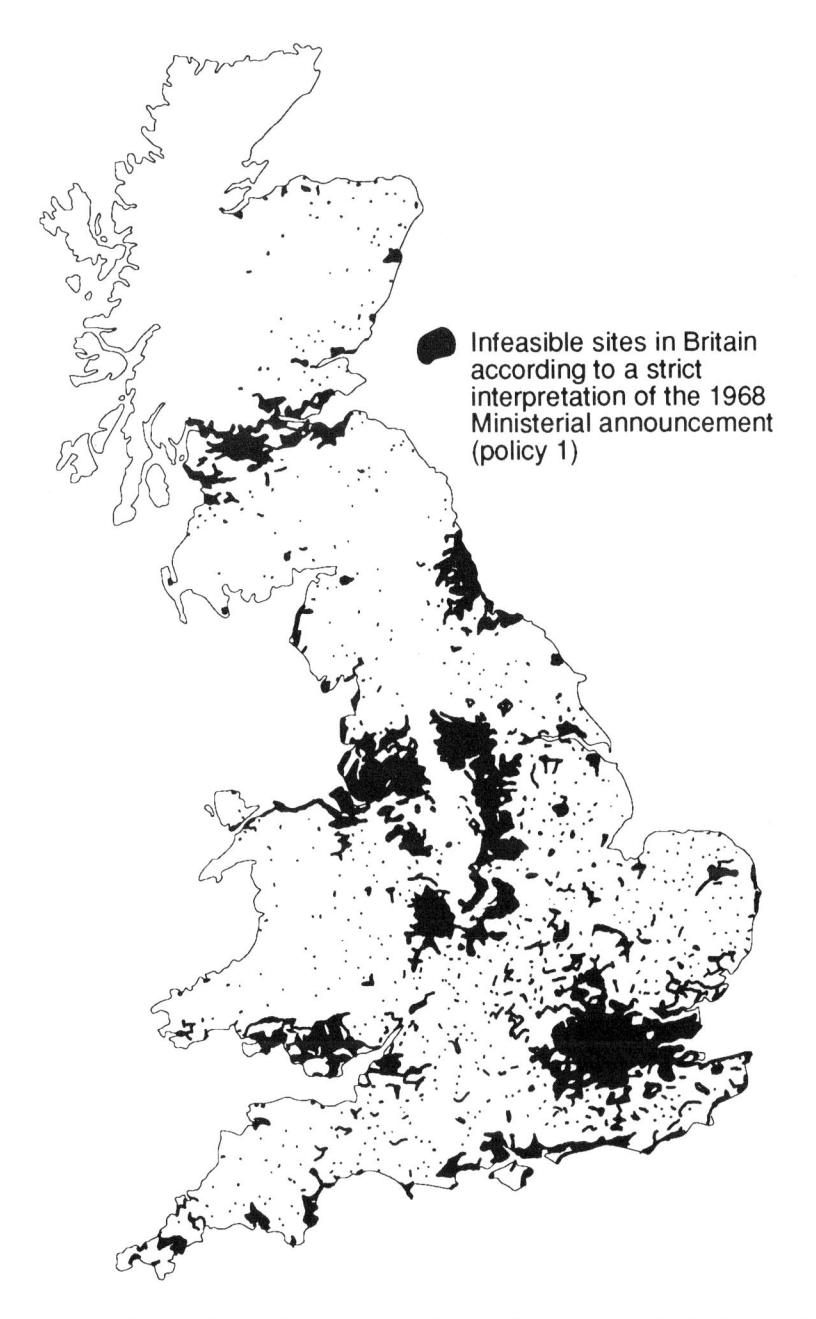

Figure 11.1 Infeasible sites for nuclear power stations under the relaxed criteria post-1968. (Openshaw, 1986)

Table 11.4 Populations near to UK reactor sites, 1981

Power station	Population within: 16 km	30 km
Berkeley	125,316	876,304
Bradwell	128,637	731,143
Calder Hall	54,258	123,670
Chapelcross	26,678	185,441
Dounreay	10,975	15,644
Dungeness	18,256	240,510
Hartlepool	430,941	852,272
Heysham	146,258	524,538
Hinkley Point	78,937	407,205
Hunterston	87,555	409,367
Oldbury	168,017	969,989
Sizewell	29,524	130,204
Torness	9,611	44,759
Trawsfynydd	18,201	64,032
Winfrith	75,660	392,047
Wylfa	23,952	64,040

excluded from safety studies and risk assessment statements (see Openshaw, 1992). The Gulf War of 1991 demonstrated the obvious, that nuclear plant are major military targets and in the event of regional conflicts they would be legitimate targets (Krass, 1990; Ramberg, 1980). The one-in-a-million death risk is essentially only based on random component failure and ignores everything else.

Another important aspect to consider is the extremely long-term nature of the nuclear power business. It is likely that full decommissioning is currently infeasible and that the simplest strategy is to leave the old reactor hulks for 150–200 years prior to final dismantling. This is not what the nuclear industry originally intended. Meanwhile, the problem is hidden by building new reactors next to the old ones, thereby initiating a process of perpetual site succession which will create a nuclear archaeology that subsequent generations will have to manage (see Openshaw, 1990a). The problem is whether it is sensible to permit site succession at locations which are now marginal in relation to power needs and potentially very dangerous should a major accident ever occur.

The risk is that big accidents will not be one in a million per year events but probably once a decade. As the number of reactors increases in the world as a whole, so the overall probability of accidents somewhere will rapidly grow. Let us assume that there are 1000 reactors each with a major accident risk of one in a million per year; this is a very conservative statement because most old designs will not be that safe. Unfortunately, with 1000 reactors, the risk of a major accident happening somewhere is now 1 in 10,000 per year (because there are 1000 reactors in operation). If the average risk is 1 in 100,000 then the annual worldwide reactor accident rate would be 1 in 1000; and these estimates relate only to what might be termed 'natural causes'.

Table 11.5 Feasible nuclear power sites in Britain in terms of demographic criteria

Siting criteria	Feasible Coastal area	Feasible Total land area
Original remote siting, 1955	61.3%	67.6%
Later remote siting, 1962	51.9%	45.5%
Relaxed siting, 1968	97.3%	94.3%

With the benefit of hindsight, it is now obvious that even in this densely populated country reasonably remote sites could have been found if the nuclear industry had been prepared to trade off short-term convenience against long-term public and political acceptability issues. Table 11.5 suggests that, contrary to the past and prevailing wisdom, there was never any shortage of feasible nuclear sites and that it would have been, and would still be, possible deliberately to seek the 'safest' possible sites should the impossible big accident ever happen. The nuclear industry should have sought to 'future proof' their nuclear investments by selecting sites that were likely to be acceptable to both public and government both now and in 200 years' time. If the nuclear imperative is dominant then the sites selected should be as remote as possible, housing the safest possible reactors.

The task of gaining and retaining long-term public acceptability is a most crucial factor. It is also least amenable to rational debate and most sensitive to external events elsewhere in the world. A key criterion would be remoteness as measured by the number of people living within the short-term death radius, approximately 30–50 miles. This assumes that early deaths rather than delayed cancers are the critical factor. The result of failing to appreciate these non-engineering arguments is a situation where about 36 million people in the United Kingdom live within 50 miles of at least one nuclear reactor; yet it could have been a few hundred thousand. When the next big accident happens in the world, a large number of people are going to feel threatened in a very real way. This is not a sensible locational strategy for a future essential industry.

The real danger is that another big accident anywhere will threaten nuclear power everywhere. It is no use, therefore, planting future essential reactors in locations which would risk threatening harm to millions should a major accident occur there. A long-term strategy with a 200-year perspective instead of a 20-year one would seek to locate new reactors only in accident-tolerable, extremely remote locations. Openshaw (1986) makes some suggestions as to where these locations might be. Some will not be acceptable to local communities yet any other strategy could be regarded as putting in jeopardy the very technology that our future may well have to depend upon. So whether we like it or not, there may soon be no option whether or not to have many more nuclear power plants but there will always be a choice as to where to put them. The argument is therefore that remote siting is probably the only strategy likely to maximize public acceptability both now and in the future, given the nuclear industry's inability to ensure that there is never again any accident anywhere in the world.

11.5 Nuclear waste

The need to find suitable radwaste sites is in many ways similar to the nuclear power siting problem. The radwaste problem too cannot now be avoided. Whether we like it or not, Britain has a sizeable amount of radwaste that needs to be isolated from the environment by careful management for varying lengths of time, from a few years to many millions (see Table 11.6). Radwaste is usually categorized into low-, intermediate- and high-level wastes. The low-level wastes (LLW) are only slightly radioactive but are high in bulk. Typically they are dumped in landfill sites or sent to Drigg (Sellafield, Cumbria — a waste facility); they were once discharged into the sea. An LLW dump may well require careful management and land-use restrictions for 100 years after dumping ceases. The main danger is either fire or escape of radionuclides into groundwater. Intermediate-level wastes (ILW) are much more radioactive and some will require isolation for geological time-scales. Once many of these wastes were dumped in the Atlantic but since 1983 this has been prohibited. The high-level wastes (HLW) are both highly radioactive and generate heat due to fission. It is estimated that about 97 per cent of all the radioactivity present in spent fuel is concentrated in HLW. They are currently stored in surface tanks at Sellafield.

It is possible to ask why is there so much waste from such a small nuclear power programme; for instance, the current total UK nuclear capacity is equivalent to about six modern PWRs. The answer is threefold: partly the historic need to develop nuclear weapons (weapons waste is not distinguished from civilian wastes), mainly the dream of the nuclear power pioneers in the 1950s of the fast breeder reactor, and (very slightly) the use of nuclear isotopes in research and medicine. The fast breeder dream resulted in both the MAGNOX and AGR reactor fuel cycles being designed to generate plutonium and resulted in a spent fuel handling system orientated towards reprocessing rather than storage. The MAGNOX spent fuel, once

Table 11.6 Sources of total projected nuclear wastes

Type of waste	AD2000 total	AD2000 decomm.	AD2030 total	AD2030 decomm.
Low level	423,560	—	1,411,000	250,000
Intermediate	99,770	—	259,200	65,000
High level	1,193	—	3,030	—

Source: RWMACC (1988), pp. 53, 55.

stored under water, has to be promptly reprocessed. The AGR spent fuel is currently stored pending reprocessing in the new THORP facility at Sellafield. Again, once it is stored under water, it will probably have to be reprocessed albeit on a more leisurely scale. The reprocessing recovers the plutonium that was once needed for nuclear weapons and also to load the fast breeders of the future. The fast breeder programme is still some time off. Unfortunately, the combination of military necessity and fast breeder dreams has created a considerable amount of radwaste to dispose of. It is interesting that in the United States, with a much larger nuclear power programme, none of the civilian spent fuel has yet been reprocessed.

Openshaw et al. (1989) provide a detailed account of the search for a radwaste site in Britain. They argue that with modern GIS technology (see Figure 11.2) it should have been possible to use a more publicly visible and politically accountable siting process than that used by NIREX, the nuclear industry's waste executive. In fact NIREX tried and failed twice with the same sort of siting strategy that was characteristic of the old CEGB when seeking nuclear power sites in the 1950s and 1960s. Basically, the engineers would spot a suitable location, usually one of several hundred alternatives. They would claim that this was the only suitable site. The subsequent Public Inquiry would then focus on the site-specific aspects and exclude the choice of site and the possible existence of alternatives from active consideration. This strategy failed when NIREX picked an old salt mine under Teeside as the ideal geological repository for medium-level waste. It failed again by picking politically expensive sites and then mistakenly restricted their usage to low-level and short-lived wastes, making them financially infeasible. Currently, they are investigating both Sellafield and Dounreay locations, an obvious rationalization. Again, the best long-term locations from a public acceptability point of view will be sites which are as far away from as many people as possible. This would seem to preclude Sellafield.

Again there is no shortage of potentially suitable sites. Table 11.7 and Figure 11.2 give an illustration of the areas of the country that would appear to satisfy a preliminary site screening using GIS. Of course many of these areas would be unsuitable on other grounds, but not all of them; or at least not to the extent of there only ever being one or two left, as no doubt a future public inquiry will be expected to believe. There is no benefit in being a neighbour to a radwaste dump containing trillions of Bqs of long-lived radioactivity which by definition will be slowly leaked into the environment, with luck on a geological time-scale. The sooner the nuclear industry realizes that it should be playing safe rather than playing short-term commercial games with mass propaganda campaigns to change public opinion, then the sooner it will end up with future-proof and change-of-government-proof siting strategies.

11.6 Nuclear disease?

Apart from Chernobyl, the other cause of a near nemesis of the nuclear industry was the

Radiation and the environment

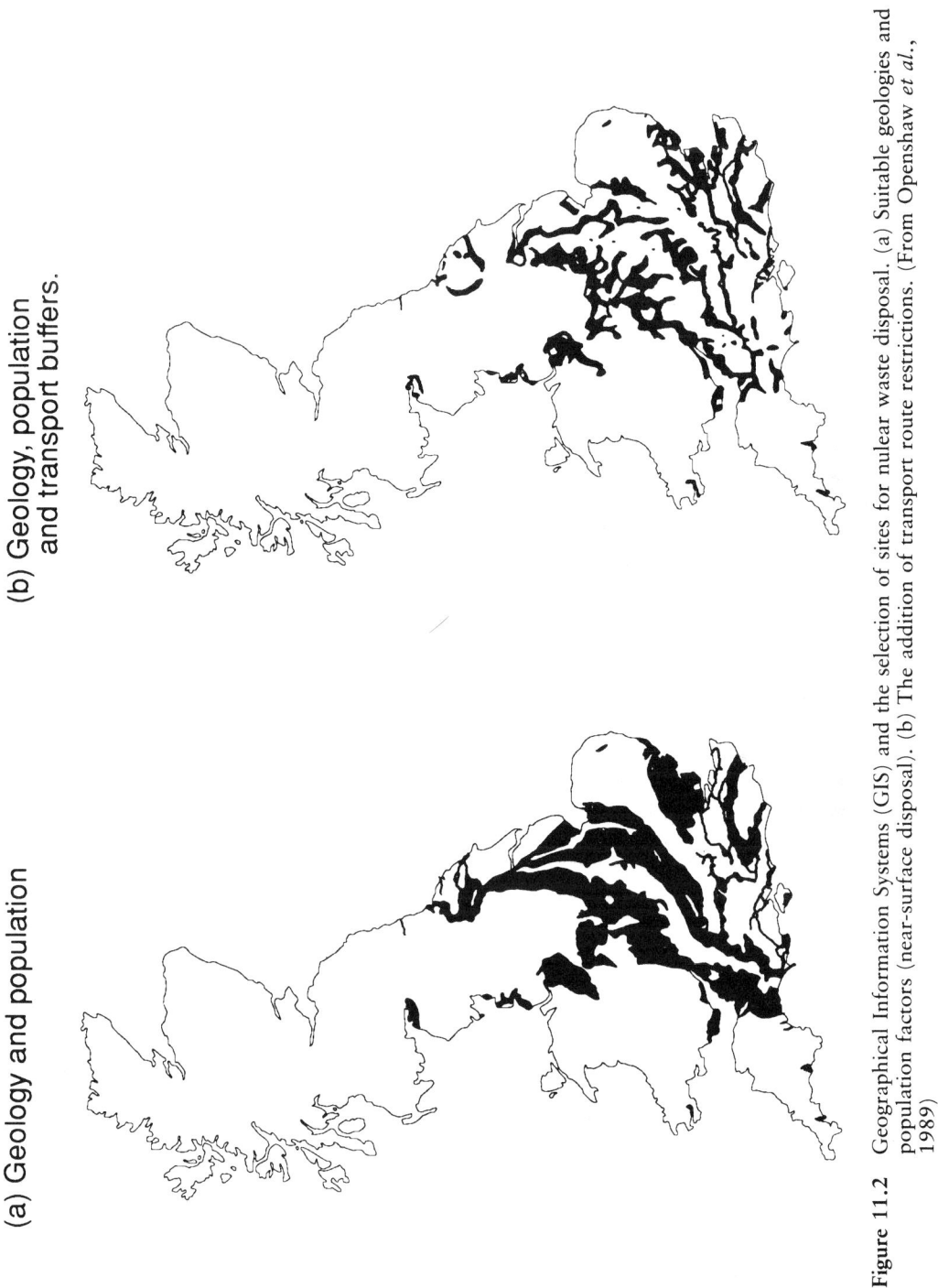

(a) Geology and population

(b) Geology, population and transport buffers.

Figure 11.2 Geographical Information Systems (GIS) and the selection of sites for nulear waste disposal. (a) Suitable geologies and population factors (near-surface disposal). (b) The addition of transport route restrictions. (From Openshaw *et al.*, 1989)

Table 11.7 Feasible deep repository radwaste sites in Britain

Geology	25
Suitable demographic	24
Near railway	8
Conservation-free	7

Source: Openshaw, Fernie and Carver (1989), p. 165

discovery of an apparent cluster of childhood leukaemias near Sellafield in 1983, and another a little later near Dounreay and a few other nuclear installations. The Sellafield claims caused a major panic (see Craft and Openshaw, 1987). What could be more emotive than innocent children being harmed by nuclear emissions that were supposed to be safe, and may well be. The problem is proof. If the cancer clusters can be proved there would be grounds for pausing nuclear expansion both here and elsewhere, because clearly the current international safety standards would be shown to be suspect. The entire subject is immensely complex. The original clustering report has been verified but also put into a broader context by the discovery of another, bigger cluster, in an area remote from nuclear facilities (see Figure 11.3 and Openshaw et al., 1987, 1988, 1990). Again this merely demonstrates that there is likely to be more than one cause of cancer clusters, if indeed such clusters are not merely a natural characteristic of a rare disease.

Alternatively, there could be a common cause other than radioactivity; for instance, there were incinerators at both Sellafield and Gateshead. At present we simply do not know what is happening. There is unlikely to be much more progress until there is a major improvement in cancer databases and particularly in the provision of demographic data for small areas that are accurate and similar in temporal resolution to the already available cancer data. New analytical tools are also needed that can search for relationships between disease patterns and environment. A start has been made; see for example, the Geographical Analysis Machines (Openshaw, 1990b; Openshaw and Craft, 1991); the Geographical Correlates Exploration Machine (GCEM) of Openshaw et al. (1990); and the Space Time Attribute Analysis Machine (STAM) (Openshaw et al., 1991). At the very least it would appear that we do now know what questions to ask! Furthermore, as Draper (1991) illustrates, the necessary data are starting to become available at the national level. On the other hand, the current surge of interest in large-scale incineration in urban locations suggests that a new generation of polluters may be appearing. Should any of these plant be the source of a major cancer cluster, then it could well take between 30 and 50 years with current database and analysis technology to spot it. Yet if we locate the same activities in a rural location, we should now be able to spot health effects in less than ten years, assuming of course that the latency periods are short rather than long. This raises the interesting question as to whether it would be possible to find locations that would offer optimal properties in that they would 'hide' any likely health effects.

11.7 Ultrasound

This survey of radiation would not be complete without a brief mention of noise pollution and particularly ultrasound. This is not electromagnetic radiation but a form of mechanical energy with the mechanical vibrations being at a frequency beyond the hearing range of humans. Ultrasound is invisible and unsensed yet it is also everywhere, ranging from the effects of water running through a pipe or being pumped to its uses in industry, in the home, and in medicine. Ultrasound may damage DNA but there is little real evidence as yet. Nevertheless, there may well be a number, maybe even a large number, of neurophysical and neurobehavioural effects of low-frequency noise. It may also exist as a complement to other types of radiation. The questions to be posed are: 'Is ultrasound associated

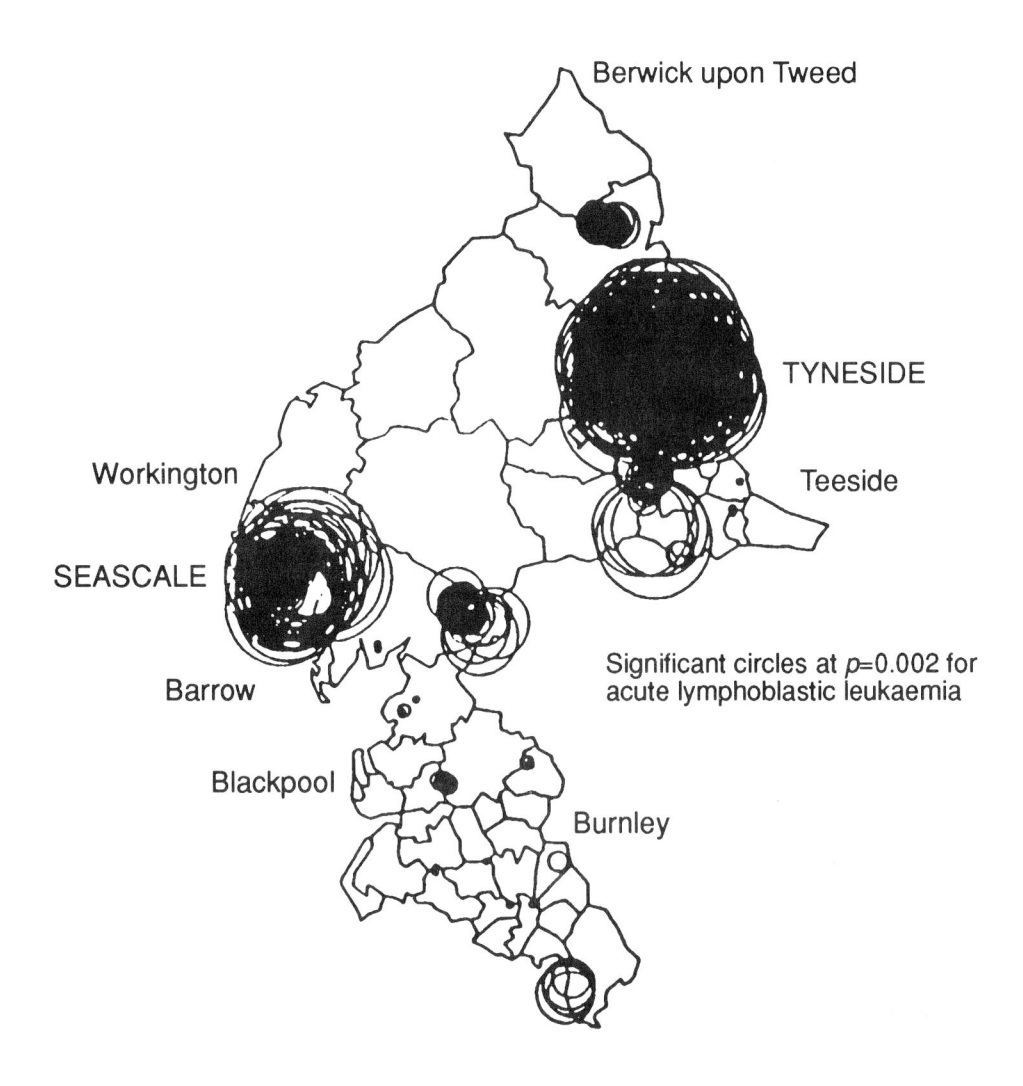

Figure 11.3 Statisticallly significant clusters of cases for acute lymphoblastic leukemia (Openshaw, S., Charlton, W., Wymer, C. and Craft, A., 1987, 'A Mark 1 Geographical Analysis Machine for the automated analysis of point data', *Int. Journal of GIS*, 1 (4), 335–8

geographically with any ill effects?' 'Does it vary spatially in intensity?' and 'How do you start making a case for taking it seriously?' Maybe this is something best left for the next century and a new generation of geographers.

11.8 Conclusions

This chapter is designed to scare! It covers types of radiation that are mainly invisible and beyond what humans can sense. It raises the spectre of fear of the unknown. In theory the same fears that are attached to ionizing radiation could be transferred to non-ionizing types, although caution is advised in assuming that non-ionizing radiation is safe until proved otherwise. The problem at present is that probably not many people are actually looking at possible effects. It is important, therefore, to ask the obvious question: 'Are there any major sources of pollution that may well be a significant cause of ill health but are currently largely ignored?' The suggestion is made that certain types of non-ionizing radiation are, a priori, potential candidates. Additionally, it might well be argued that the same philosophy can be applied to several other areas of environmental pollution; for instance, inorganic and organic chemicals, viruses of various forms, and harmful bacteria. Fortunately, mankind is at least gradually developing the measurement technologies to observe what is there. The problem here is that as more types of pollution can be accurately and continuously measured at more and more sites so more and more is revealed of the health-threatening environment in which we live. There are many major problems that are currently 'awaiting discovery' which may well have various morbidity and mortality consequences. What are we waiting for? There are undoubtedly good 'political' reasons for not looking, but even better health and public reasons for looking even more carefully and closely. It is one legitimate role of the geographer to point out the importance of this total (not just broad) spectrum monitoring and measurement task as being very important if major surprises are to be avoided. Many complex diseases have long latency periods. The extremely rapid rate of change in many technologies that are altering our environments is greatly reducing the time available for scientific investigation and action. It is most important that the necessary databases and analysis systems are developed so that it is possible continually and intelligently to monitor and watch over our polluted world, forever alert to the risks of new problems caused by our polluting ways. At the same time, sensible locational strategies need to be determined for major pollution-causing industries. Industrial location should no longer be purely a matter of maximizing short-term economics. When health risks are involved, and the prospect of litigation starts to appear, then conventional transportation cost minimizing strategies will need to be replaced by more future-proof planting.

Eventually, mankind will learn to become more environmentally friendly; we shall have to. Meanwhile it is important to discover how best to manage and survive in an increasingly polluted and contaminated world. We probably know at present very little about the long-term damage we may be causing. Survival is only going to be possible by eternal vigilance. GISs, environmental databases, comprehensive pollution monitoring networks, access to the totality of available mortality and morbidity data, and clever analysis and modelling technologies are all essential ingredients for survival management.

References

Craft, A.W. and Openshaw, S. (1987), 'Children, radiation, cancer, and the Sellafield Nuclear Reprocessing Plant', in A. Blowers and D. Pepper (eds), *Nuclear Power in Crisis*, Croom Helm, London, pp. 244–71.

Caufield, C. (1989) *Multiple Exposures. Chronicles of the Radiation Age*, Penguin, 304 pp.

Draper, G. (1991), 'The geographical epidemiology of childhood leukaemia and non-Hodgkin lymphomas in Great Britain, 1966–83', *Studies in Medical and Population Subjects*, OPCS, HMSO, London.

Kelly, G.N., Charles, D., Broomfield, M. and Hemming, C.R. (1983), *The Radiological Impact on the Greater London Population of Postulated Accidental Releases from the Sizewell PWR*, National Radiological Protection Board Report 146, HMSO, London.

Krass, A.S. (1990), 'The release in war of dangerous forces from nuclear facilities', in A.H. Westing (ed.), *Environmental Hazards of War*, Sage, London, pp. 10–29.

Marshall, W., Billington, D.E., Cameron, R.F. and Curl, S.J. (1983), *Big Nuclear Accidents*, UK Atomic Energy Research Establishment Report 10532, HMSO, London.

Openshaw, S. (1986), *Nuclear Power Siting and Safety*, Routledge, London.

Openshaw, S. (1988a), 'Post Chernobyl prospects for nuclear power in the UK', *Environment and Planning C: Government and Policy*, 6, 251–68.

Openshaw, S. (1988b), 'Making nuclear power more publicly acceptable', *Journal of the British Nuclear Energy Society*, 27: 131-6.

Openshaw, S. (1988c), 'Planning Britain's long term nuclear power expansion programme', *Land Use Policy*, 5, 7-18.

Openshaw, S. (1990a), 'Nuclear archaeology: the influence of decommissioning on future reactor siting in the UK', in M.J. Pasqualetti (ed.), *Nuclear Decommissioning and Society*, Routledge, London, pp. 143-58.

Openshaw, S. (1990b), 'Automating the search for cancer clusters: a review of problems, progress, and opportunities', in R.W. Thomas (ed.), *Spatial Epidemiology*, Pion, London, pp. 48-78.

Openshaw, S. (1992), 'The safety and siting of nuclear power plant when faced with terrorism and sabotage', in E. Millar (ed.), *Technological Disasters*, Pennsylvanian Academy of Science, Pennsylvania.

Openshaw, S. and Craft, A.W. (1991), 'Using Geographical Analysis Machines to search for evidence of clusters and clustering in childhood leukaemia and non-Hodgkin lymphomas in Britain', in G. Draper (ed.), *The Geographical Epidemiology of Childhood Leukaemia and Non-Hodgkin Lymphomas in Great Britain, 1966-83*, Studies in Medical and Population Subjects No. 53, OPCS, HMSO, London.

Openshaw, S., Cross, A. and Charlton, M. (1990), 'Building a prototype Geographical Correlates Exploration Machine', *International Journal of GIS*, 3, 297-312.

Openshaw, S., Fernie, J. and Carver, S. (1989), *Britain's Nuclear Waste: Siting and Safety*, Belhaven Press, London.

Openshaw, S., Wymer, C. and Cross, A. (1991), 'Using neural nets to solve some hard problems in GIS', *Proceedings of EGIS 91*, EGIS Foundation, Utrecht, Vol. 2, pp. 788-96.

Openshaw, S., Charlton, M., Craft, A.W. and Birch, J.M. (1988), 'Investigation of leukaemia clusters by the use of a Geographical Analysis Machine', *The Lancet*, i, 272-3.

Openshaw, S., Charlton, M., Wymer, C. and Craft, A.W. (1987), 'A Mark I geographical analysis machine for the automated analysis of point data sets', *International Journal of GIS*, 1(4), 335-58.

O'Riordan, T. (1987), 'Assessing and managing nuclear risk in the United Kingdom', in R.E. Kasperson and J.X. Kasperson (eds), *Nuclear Risk Analysis in Comparative Perspective*, Allen and Unwin, Boston, pp. 197-218.

Ramberg, B. (1980), *Nuclear Power Plants as Weapons for the Enemy: An Unrecognized Military Peril*, University of California, Berkeley.

RWMACC (1988), *Ninth Annual Report*, HMSO, London.

12

Global environmental implications for future energy supply and use

Ian Fells

The last three years have seen a rapidly growing awareness of the damaging effect the expanding use of fossil fuels is having on the global environment. Acid rain and the even more insidious 'greenhouse' effect ignore national boundaries, causing damage to agriculture, fisheries and buildings running into many billions of dollars and destabilizing the weather machine with unknown long-term consequences. At the same time rapidly expanding populations, with high expectations of industrial growth and improved lifestyles, accelerate energy demand growth. The two biggest polluters are road transport and electricity generation, with inefficient use of fuel in buildings running a close third. Matters are exacerbated by the destruction of the rain forests with short-term agricultural gain in mind.

Despite heightened awareness of the problem, concerted international action to curb fossil fuels use is proving difficult to achieve. The obvious strategy is to use all energy forms more efficiently, by switching to renewable energy such as wind, tidal and, of course, hydroelectric; more contentiously, to expand nuclear power, develop efficient public transport systems, eliminate the motor car and, even more difficult, try to curb population growth.

It is clear that the fashionable market forces energy strategy will do nothing to help the environment and ironic that, as the centrally planned economies crumble and switch to privatization, a global energy strategy becomes imperative.

12.1 History and geography of energy development and energy-related pollution

Civilization began when man discovered how to make and control fire. The first fuel was probably wood and even today fuel wood provides just over 10 per cent of man's energy requirements. Nearly 80 per cent is produced by burning fossil fuels; oil, coal, gas and associated hydrocarbons. They are a non-renewable resource and took 200 million years to form. Man started to burn them around 1000 years ago when simple coal-mining techniques were developed, although seepages of oil and tar had provided a primitive fuel many hundreds of years earlier. Once the Industrial Revolution got under way towards the end of the eighteenth century, the rate of consumption increased. We now know that in an economy undergoing industrialization an increase of 1 per cent in gross domestic product (GDP) requires a 1.5 per cent increase in energy use (known as the energy coeft.).

The earliest fuel consumption figures available (Fells, 1980) relate to the United Kingdom when in 1800 the UK energy coeft.

was just about 1.5. In the latter years of the twentieth century this figure has fallen to 0.5 in the United Kingdom and the rest of Western Europe and is typical of mature industrialized economies. In rapidly expanding and industrializing economies such as those around the Pacific Rim the energy coeft. is still 1.5 and with some economies growing at between 7 and 8 per cent the increasing demand for fossil fuels in these developing countries could well dominate world energy demand early next century. Matters are exacerbated by massive increases in world population. Numbers have risen from 1.7 billion in 1900 to 5 billion today, and will rise to 10 billion in 2060. Each additional person requires his or her energy ration and anticipates a steadily improving lifestyle. All these pressures conspire to increase the world's energy demand and use.

The implications of the effect on the environment of burning fossil fuels were realized in England early in the fourteenth century when coal, brought from Newcastle to London, the result of a charter to dig coals in Newcastle granted by Henry III in 1239, led to serious air pollution in central London. Burning coal was seen as 'a public nuisance which corrupted the air by its smoke and noxious vapours'. Fines were levied and John Evelyn the diarist proposed to rescue the city from that 'hellish, dismaille cloude of sea coal' by banishing all brewers, dyers, soap and salt boilers and lime burners to below Greenwich. It took another 750 years before the intolerable London 'smogs' of the 1940s and 1950s caused the Clean Air Act to be passed in 1956 (see Chapter 6).

So began an era of local environmental protection which affected the way we generate and use energy; subsequent developments have been on a much larger geographical scale. In 1985 Joe Farman and his colleagues collected evidence at Halley Base in Antarctica to show that chlorinated fluorocarbon gases (CFCs), used in refrigerators and aerosols, were destroying the ozone layer which protects the earth from damaging ultraviolet rays. A conference held in Canada in 1987 led to the Toronto Protocol which limits the release of CFCs. The CFC gases are also remarkably potent 'greenhouse' gases and along with methane and in particular carbon dioxide, these gases pose the most serious environmental threat to the planet, 'global warming'.

This introduces the alarming prospect of destabilizing the weather machine and consequential dramatic changes in weather patterns as well as sea level rises, and the implications of ever rising rates of fossil fuel combustion to provide the energy requirements of both developed and developing national economies (see Chapter 13).

Appreciation of the 'global' nature of some forms of environmental pollution was brought sharply into focus in April 1986 when an RBMK nuclear reactor at Chernobyl, near Kiev, exploded throwing radioactive debris 2 km into the atmosphere and spreading a radioactive cloud of pollution over Scandinavia, large parts of Europe and even further afield. Russia's accident rapidly became the world's accident (see also Chapters 10 and 11).

The damage to the ozone layer caused by CFCs, acid rain and global warming are now perceived as problems requiring international co-operation, but the difficulties of organizing worldwide action to ensure 'damage limitation' by controlling emissions to the atmosphere are extreme. The cost of environmental damage is difficult to assess, although research into the problem is being pursued vigorously; until it is possible to count the cost it is difficult to apply the 'polluter must pay' principle.

12.2 Taking stock: energy reserves and use

There are two major problems. First, what is the extent of the world's reserves of fossil fuels? How long will they last? Second, all fossil fuels burn to form carbon dioxide; each tonne of oil burnt produces 3.3 tonnes of carbon dioxide, for example. Carbon dioxide has been steadily building up its concentration in the atmosphere since the

Table 12.1 World energy reserves and consumption, 1987 (expressed as tonnes of oil equivalent)

	Industrial countries	Developing countries	World
Economic reserves (btoe)			
Coal	380	160	540
Oil	16	100	116
Gas	50	46	96
Total fossil fuel	446	306	752
Uranium, non-communist world:			
Thermal reactors			33
Fast reactors			>2000
Consumption (mtoe)			
Coal	1200	1190	2390
Oil	2060	880	2940
Gas	1250	300	1550
Total fossil fuel	4510	2370	6880
Nuclear	370	40	410
Hydro/geo	330	190	520
Wood	100	1100	1200
Total (btoe)	5.3	3.7	9.0
toe/cap	4.8	0.9	1.8

start of the Industrial Revolution and has risen from 270 parts per million (ppm) in 1700 to 350 ppm today. Together with methane, to a lesser extent, and other gases this increase in concentration gives rise to the enhanced 'greenhouse effect' which is perceived as a serious threat to the environment, causing global warming and destabilization of the weather machine. Sulphur and nitrogen oxide emissions from coal and oil combustion cause acid rain which seriously damages forests, crops, buildings and health.

The finite nature of the world's fossil fuel resources, taken in conjunction with the destructive effect on the environment caused by their combustion, makes an unassailable case for conserving and constraining their use. On the other hand, the expectations of the rapidly rising world population require ever increasing supplies of energy to 'fuel' their aspirations. This presents the governments of the world with a political as well as a resource dilemma which will steadily get worse. The extent of world fossil fuel reserves is always a matter of judgement and figures, particularly in the oil industry, have sometimes been distorted for commercial reasons. The best and most objective figures are those provided by the World Energy Council and published in 1989. Reserves and consumption figures are set out above in terms of tonnes of oil equivalent (toe). One toe equals 42 giga joules (GJ). Figures are divided between industrialized countries (IC) and developing countries (DC) and given in Table 12.1. The figures given here refer to proved recoverable reserves. Estimates of proved and probable reserves are sometimes quoted but should be accepted with caution.

A simplistic lifetime for the different fossil fuels can be obtained by dividing the reserves by current annual consumption. The results are given in Table 12.2. Figures for uranium are included as it is a fossil fuel, although not

Table 12.2 Reserves/current consumption (yr)

	World
Coal	225
Oil	40
Gas	60
Thermal reactors	80

Table 12.3 Fuel scenarios for 2020

	IIASA (1981) btoe	%	WEC (1986) btoe	%	WRI (1988) btoe	%
Coal: high	4.6	28	6.0	31	1.4	13
low			4.6	30	1.4	17
Oil	4.4	27	4.4	23	3.1	29
			3.2	21	2.3	29
Gas	2.7	16	3.4	17	3.1	29
			2.6	17	2.3	29
Nuclear	3.5	21	2.4	12	0.5	5
			1.7	12	0.5	6
Renewables	1.3	8	3.4	17	2.6	24
			3.0	20	1.5	19
Total	16.6	100	19.6	100	10.7	100
			15.1	100	8.0	100

always thought of as such. There is a striking difference in uranium reserve potential compared with other fuels if its energy release is calculated, assuming it is used in the highly efficient fast-breeder reactor where the abundant uranium 238 isotope is converted to plutonium for use as a fissionable fuel rather than merely burning up the uranium 235 isotope in the currently commercial thermal reactors (see also Chapter 11). The multiplying factor is 60.

12.3 Energy demands and development

Of course these 'lifetimes' shorten dramatically if, as seems likely, annual consumption rates increase. The attempts to predict future energy demand made by three well-known and broadly based foundations, the International Institute for Applied Systems Analysis (IIASA), the World Energy Council (WEC), and the World Resources Institute (WRI), have been compiled by AEA Technology and are shown in Table 12.3.

There are noticeable differences in the predictions for 2020. IIASA is pessimistic about the contribution of renewables (8 per cent) and optimistic about the future nuclear component (21 per cent); both WEC and WRI give renewables a 20 per cent share, whereas nuclear is 12 per cent for WEC and a low 5 per cent for WRI. In any event the brunt of demand will be taken by the fossil fuels coal, oil and gas; all the scenarios, whether high or low, give a figure of between 68 and 71 per cent. WRI is more enthusiastic about gas and less enthusiastic about coal, probably for environmental reasons. All predict an increase in energy demand over the 1987 figure of 9 btoe, ranging from 46 per cent IIASA, 11 per cent or 67 per cent WEC to 18 per cent WRI, except for the WRI 'low' scenario which is not a projection at all but a 'target' involving enormous improvement in the efficiency of energy use.

These predictions were made over the period 1981 (IIASA) to 1987 (WEC) and 1988 (WRI). More recently (1989) the WEC has published *Global Energy Perspectives 2000–2020*. In it the figure for world energy consumption between 1985 and 2020 has been revised to grow between 50 and 70 per cent. The rise in demand will be uneven with spectacular rises in developing countries and particularly Centrally Planned Asia.

Again there will be wide differences in per capita consumption, although the world average will stay at around 1.6 toe. In 1985 average per capita consumption in the south, including non-commercial energy sources such as fuel wood stood at 0.65 toe as against 4.25 toe in the north. By 2020 the corresponding figures will be 0.8/0.9 toe in the south and 4.45/5.15 toe in the north, so there is hardly any improvement in the ratio for the south in terms of consumption. The situation is particularly bad in countries such as Sub-Saharan Africa and South Asia where per capita consumption will rise from 0.36 toe in 1985 to only 0.39/0.46 toe in 2020.

Of this, non-commercial energy would still constitute between 30 and 45 per cent of demand.

It is clear that chronic poverty in the energy field will continue to exist in a region whose population is expected to double from 1.4 billion to 2.8 billion by 2020, that is, one-third of the world population.

12.4 Energy supply

On the fuel supply side the 1989 WEC predictions are that coal demand will rise after 2000 to between 3 and 4 btoe but that before 2000 there will be strong competition from hydrocarbons, particularly gas. Nevertheless coal will only have around 30 per cent of the market in 2020. Outlets are restricted and environmental constraints constitute a check on development.

Natural gas, because of its environmental advantages in not containing sulphur and producing less carbon dioxide per unit of heat output than coal or oil, is better placed than oil to maintain its share of world demand at current levels.

The nuclear power sector, despite its environmental advantage in producing neither carbon dioxide nor acid rain, suffers from public acceptability problems since Chernobyl and also increasing financial difficulties, so the World Energy Conference (1989) report has scaled down the anticipated nuclear contribution from 12 per cent to between 7 and 8 per cent compared with 4 per cent in 1985.

Non-commercial energy sources will increase in the south and will still provide between 15 and 25 per cent of Third World energy needs in 2020 as compared with 33 per cent in 1985. The inevitable consequence is increased pressure on agriculture, society and the environment.

Overall, renewable energy sources, that is hydropower and non-commercial sources taken together, are unlikely to provide more than 20 per cent of energy demand by 2020, despite the enthusiasm of their supporters. The position of oil will be maintained to a higher degree than foreseen in earlier reports and will still be meeting 26–28 per cent of demand by 2020 compared with 32 per cent in 1985. In particular, Third World demand will rise from less than 0.7 btoe in 1985 to 1.4/1.6 btoe in 2020 and the south's share of world consumption will increase from 26 per cent to 43–44 per cent in 2020, whilst industrialized countries will have stabilized their demand at 1985 levels.

12.5 World trade in energy

Oil is the largest single commodity traded in the world today. In 1989 trade rose by 6 per cent over the 1986 figure to 28.5 million barrels/day (b/d) (7.2 bls of oil = 1 tonne). OPEC alone obtained net export revenues of £117 billion in 1989. The United States is the single largest importer of crude oil and in 1989 imported 7.8 million b/d, an increase of 19 per cent on the 1988 figure. Figure 12.1 shows the distribution of oil reserves in different countries. The role of OPEC is crucial to the continuous supply of oil in the world, particularly to the industrialized countries. It is ironic that OPEC was formed in 1960 because of the intransigence and greed of the large international oil companies, particularly Exxon. The fact that the major reserves of oil lie in a politically highly unstable area and that oil looks set to provide the major fraction of world energy well into the next century bodes ill for price and supply stability.

The rising fortunes of natural gas with its environmental advantages has led to a broad geographic spread of gas supplies. Trade in natural gas is to some extent constrained by the development of gas pipelines although a network of pipelines in the North Sea and from the former Soviet Union into Europe are rapidly extending trade across national boundaries. Worldwide, ten countries export to 25 others and eight nations export liquefied natural gas. Once again world natural gas reserves are concentrated in particular locations; the former Soviet Union and the Middle East between them have 67

Global environmental implications

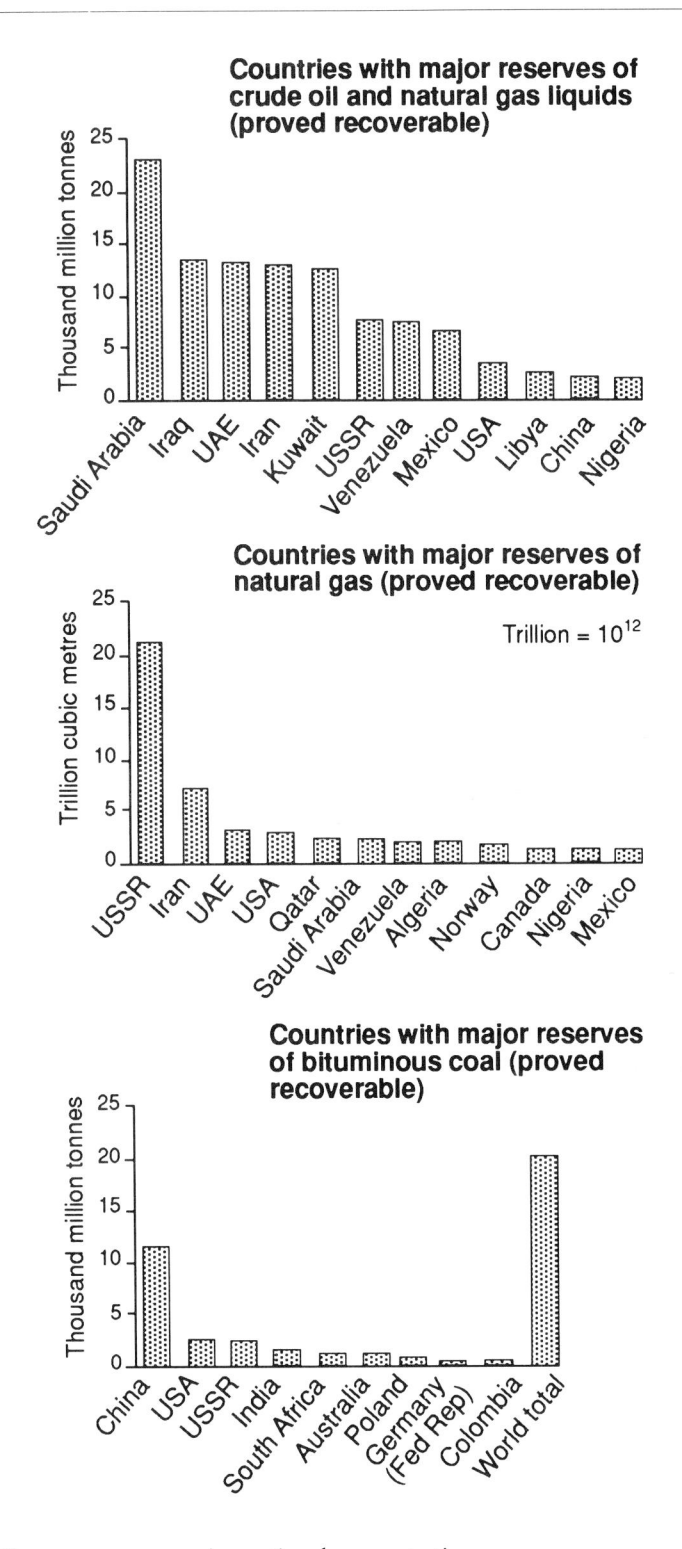

Figure 12.1 Energy reserves: major national concentrations

per cent of world reserves, Western Europe an important 5 per cent. Of the world's ten largest gas fields, five are in the former Soviet Union and two in Western Europe. As far as Europe is concerned, a gas market controlled by Norway, Russia and possibly Holland is a strong possibility. A better arrangement would be the growth of a 'spot' market in gas which could constrain prices, although they will inevitably rise, possibly quite quickly in the late 1990s.

About 360 million tonnes of coal is traded worldwide; that is, about 11 per cent of production. Australia is the largest exporter with the United States not far behind, followed by South Africa. China has great but unrealized coal export potential. Coal's major competitor in the world market is now natural gas and this will constrain growth in the coal trade to around 2 per cent for a time until the price of natural gas gives coal an advantage when burnt using clean coal technology.

12.6 Fuel technology

Perhaps the most important chemical reaction that exists is the combustion of carbon-based materials in air to form water and carbon dioxide with the generation of heat. Wood, coal, oil and gas can all be easily burned and have provided first heat and later electricity via heat engines. Combustion provides a reliable and continuous source of energy which makes our complex civilization possible. Without heat and electricity we spiral down into chaos within as little as 24 hours, leaving people stranded in lifts, hospital life-support systems paralysed, and intolerable conditions of cold or overheat as heating and air conditioning systems fail.

As is often the case, people used fire and later boilers and furnaces before the chemistry of combustion was understood. After Priestley and Lavoisier, the work of Humphrey Davy at the Royal Institution on the combustion of methane following the Felling Colliery disaster in 1812 (ironically a disaster involving the extraction of another fuel, coal) started a train of combustion research by chemists that continues to this day. A great international industry is built around research into combustion in gas turbines, fluidized bed combustors, furnaces, internal combustion engines and combined heat and power schemes, all with the intention of improving our understanding of the complex combustion process with its chain reactions and free radical mechanisms, and at the same time of improving efficiency of fuel usage. Indeed, the whole development of chemistry has been strongly influenced by fuel and combustion.

The treatment of fuels once they have been extracted from the earth involves a wide range of chemical techniques. Originally the primary fuel was coal; in the United Kingdom alone 287 million tonnes were mined in 1913 and 100 million tonnes exported. The Navy was only just in the process of switching to oil. The chemistry of coal utilization included the analysis of coal, both proximate and ultimate, as well as many years of work in trying to establish the structure of the coal molecule. This information embodying the measurement of calorific value, volatile content, propensity to cake during heating and so on was included in several methods for the classification of coal. This enables different coals to be directed to particular uses such as steam-raising coals, coal suitable for conversion into strong metallurgical coke or, more appropriate, for gasification and the production of smokeless fuels. The techniques of carbonization and gasification are central to the development of coal technology. Coal is carbonized with the main object of converting it into products which can be utilized with improved efficiency and recovering valuable by-products which are generally lost when raw coal is consumed. An important product is coal gas and in the late eighteenth century coal gas was produced for illumination and this role of gas continued until well into the twentieth century. Gas gradually came to be used as a clean and convenient fuel in the latter part of the nineteenth century. The by-products of coal gas manufacture were active coke which

Sources of man-made carbon dioxide emissions 1980-85

Carbon emissions from fossil fuel use by region 1987

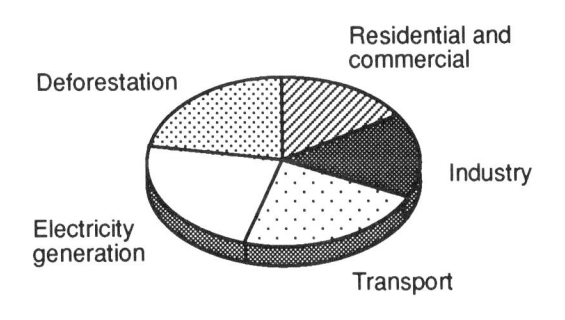

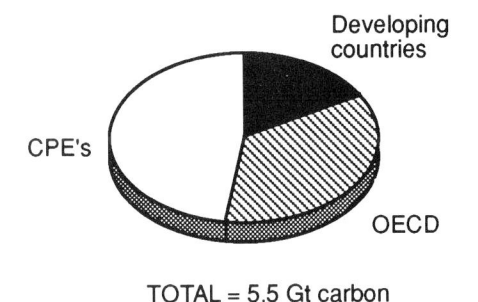

TOTAL = 5.5 Gt carbon

Figure 12.2 Fossil fuels and carbon dioxide emissions
Source: UK Department of the Environment.

Cutting the UK's carbon dioxide emissions
contribution towards target

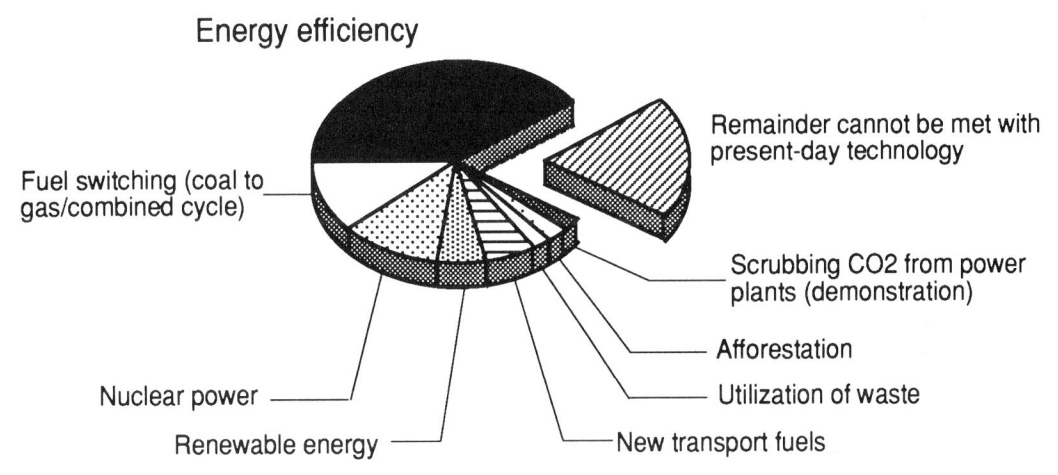

Target 50% reduction in 30 years

Figure 12.3 Potential means for reducing UK carbon dioxide emissions
Source: Energy Technology Support Unit UKAEA (1989).

could itself be used as a clean, solid fuel and a host of organic chemicals, such as phenol, and oils which could themselves be converted into fuels, including petrol, and spawned an important chemical industry. This role was only taken over by the petrochemical industry, based on oil, in the late 1930s despite the discovery of oil early in the nineteenth century, when Col. Drake drilled the first oil well in Texas (although he was looking for water at the time).

The chemistry and chemical engineering required to process oil had originally been developed in Scotland in the early nineteenth century by Young who retorted oil shale (Kerogen rock) available in the Bathgate area and then went on to distil and chemically treat the products. As the oil industry grew, particularly in the United States, a powerful group of techniques, including catalytic and thermal cracking, reforming, acid treatment and so on, brought chemistry to bear and the petrochemical industry was born, leading to polymers, fibres, paints and a host of essential chemical products. Rather later, around the mid-1950s, steam reforming of naphtha led to a process to produce town gas which replaced coal gas. Natural gas brings its own chemical problems: the formation of clathrate compounds with water and the removal of sulphur compounds from sour gas.

12.7 Energy generation: pollution and environment

The externalities of energy production and use have now become a major control on future supply, demand, trade and technology, the subjects of the four preceding sections. The development of the oil industry and, more particularly, the coal and coking industries, was carried out initially with little thought for the depredation and dilapidation they might cause to the environment. In the United Kingdom the Clean Air Act of 1956 triggered the national conscience into a belated interest in cleaning up the environment, but industry was reluctant to embark on an expensive clean-up campaign without legislation. Even now it has required the European Community (EC) to adopt the Large Combustion Plants Directive (1988) and other legislation concerning dumping of waste and so on to force industry and others to clean up our polluted environment. Problems outside the EC are highlighted by countries like Poland, Czechoslovakia and the former East Germany where the environment has been seriously, perhaps irrevocably, damaged. In Los Angeles the damage to the local atmosphere caused by the pollution from burning gasoline in cars is all too apparent as photochemical smog chokes the city. The fuel industries must shoulder a good deal of the blame for global as well as national pollution. The production of acid rain by burning coal and oil on a large scale for electricity generation, leading to the destruction of forests and agriculture, as well as sterilizing hundreds of thousands of lakes, is only dwarfed by the damage caused by the production of carbon dioxide which enhances the 'greenhouse effect'. Most of this carbon dioxide is the result of burning one fossil fuel or another, as Figure 12.2 shows.

Whilst technology can do something to alleviate the formation of acid rain by scrubbing out the sulphur and nitrogen oxides from the flue gases, the prospect of removing carbon dioxide and then disposing of it in some acceptable way has proved an insuperable problem. Figure 12.3 shows some of the techniques which could be used to reduce carbon dioxide emissions in the United Kingdom, but the cost would be considerable.

12.8 The environment and future energy supply patterns

Fossil fuels have sustained the growth of civilization for several thousand years but the accelerating demand for more and more energy, encouraged by fast-improving lifestyles and exacerbated by rapidly rising populations, particularly in the south, are beginning to emphasize the finite nature of fossil fuel reserves. Nevertheless, forward predictions suggest that fossil fuels will still supply some 70 per cent of our ever increasing energy demand well into the next century. More efficient use of energy is a paramount priority but even then the increasing combustion of fossil fuels leading to environmental damage by acid rain and, more insidiously, by carbon dioxide via the greenhouse effect points to an uneasy and uncomfortable compromise between environmental protection and economic growth. In the medium to long term, without population control and downgrading of lifestyle expectations, nuclear power will remorselessly take over from fossil fuels as reserves become depleted and world energy demand continues to accelerate.

In the short term the most noticeable development over the next decade will be a rapid increase in the use of natural gas, particularly in the Pacific Rim countries and Europe. Rapid extension of gas pipeline systems will take place with the development of 'spot markets' in natural gas prices as transnational boundary gas movements are made easier. The anticipated European Energy Charter will have a profound effect on the ease of movement of both natural gas and electricity throughout the area and beyond. A drive towards increased energy efficiency to limit the use of fossil fuels will accelerate the ordering of combined cycle, gas fired, power plant and combined heat and power generation. This will inevitably increase gas demand and consequently price. Formation of gas cartels are easier than oil cartels. Coal use, seen as environmentally dirty, will be depressed for a time but clean coal technology will re-establish growth in coal consumption. It is unlikely that renewable energy will be able to provide more than 20 per cent of world energy demand; fortunately the use of biomass fuel is 'greenhouse-neutral'.

It is becoming increasingly clear that international regulations and pressure will be required to underpin national energy strategies aimed at minimizing environmental damage. Techniques for costing environmental damage caused by different energy sources are being urgently researched. Once the true costs are known fiscal techniques, financial incentives and taxes will be used more extensively to control energy use. It is becoming increasingly clear that a simplistic, market-led energy policy does almost nothing to protect the environment.

Reference

Fells, I. (1980), 'Energy, implications for economic growth', *Chemistry & Industry*, 6, 655–8.

Data sources

World Energy Council Publications, *1989 Survey of Energy Resources*.
BP Statistical Review of World Energy.
Shell Briefing Service Publications, *Global Climate Change* (1990), *Prospects for Natural Gas* (1990).
Data compiled by AEA Technology from projections by Institute for Applied Systems Analysis (1981), World Energy Council (1986), World Resources Institute (1988).

13

Natural environments of the future: adapting, conserving, restoring

Anthony Stevenson and Malcolm Newson

13.1 How we got to here: humans and their environment

Many of today's landscapes which people value are the direct or indirect consequences of attempts by humans to alter or control their environment, processes which have progressively become stronger and more successful. Initially, human populations were controlled by the environment either directly or indirectly via their food supply. The earliest human societies adopted a hunting-fishing-gathering (often known as hunter-gatherer) strategy and this remained the dominant mode of production from the evolution of the genus *Homo*, some two million years ago, until the advent of agricultural modes of existence 5000–6000 years ago.

13.1.1 Hunter-gatherers

The hunter-gatherers' ability to control and alter their environment involved the development of sophisticated tools with which to catch and process food and the use of fire for warmth, cooking and as an agent of habitat modification. Hunter-gatherers typically require around 26 km^2 per head to garner enough energy for survival (Simmons, 1989). The resultant low population densities combined with a semi-nomadic–nomadic form of food collection ensured that human impact on the environment during this period was minimal. The first evidence of early prehistoric human impacts on the environment is tantalizing. The earliest known use of fire by humans (*Homo erectus*) dates back 1.5 million years in Kenya (Gowlett *et al.*, 1981), from 500,000 years onwards frequent evidence of fire is found. Pollen diagrams prepared from sediments at Hoxne and Marks Tey in Essex (Turner 1970, West and McBurney, 1954) and laid down during the Hoxnian interglacial (about 200,000 years ago) show a reduction in forest taxa and increases in grassland taxa associated with fire and at Hoxne the presence of stratigraphically coincident Achuelian hand-axes. Although no hominid bones have been found at the sites this, together with the length of time of the clearance, about 350 years, has led some to suggest (Simmons, 1989: 37) that this is the first evidence of early human impact on the environment. However, it is entirely possible that an opening in the forest could have been maintained by the grazing and browsing activities of elephants and hippos.

More recent evidence for human impact has been postulated for North America, where a significant decline in large herbivore and carnivore genera (about 60) occurs at the end of the last Ice Age (*c.* 12,000 BP). These extinctions are coincident with the invasion

Natural environments of the future

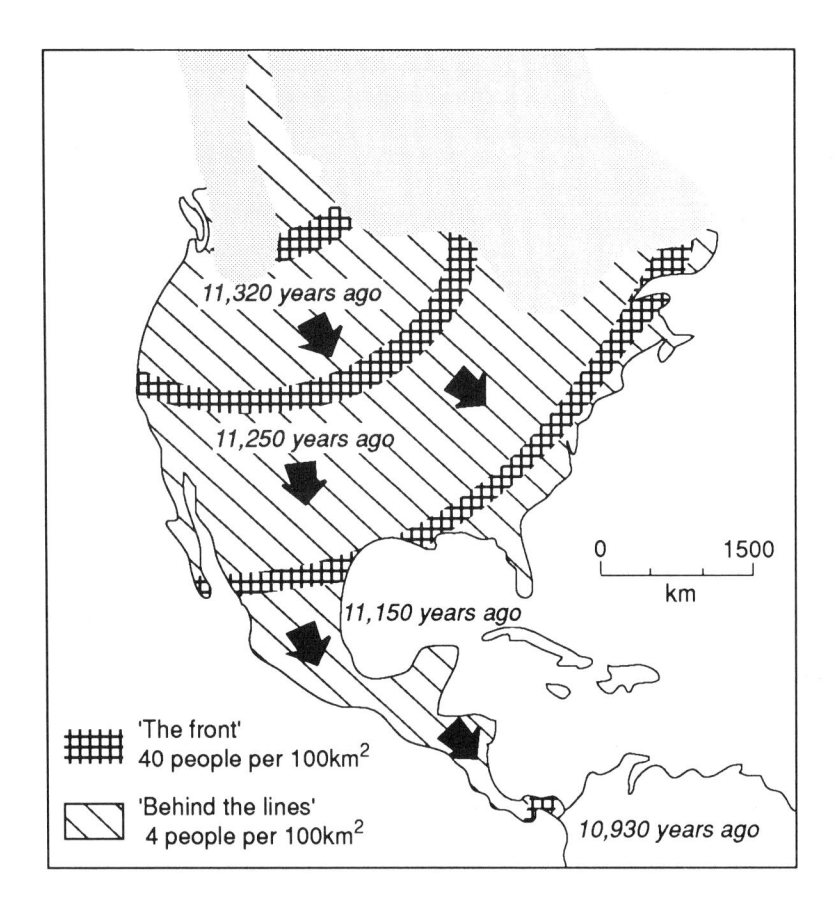

Figure 13.1 A hypothesized 'Pleistocene Overkill' as mankind colonized North America. (P.S. Martin, 'The discovery of America', *Science*, 179, 1973, 969–74)

of the North American continent by prehistoric hunter-gatherers across the Bering Straits (Figure 13.1) and an analogy drawn with the extirpation of the American Buffalo in the 1800s leads some to suggest that a similar process occurred (Martin and Klein, 1984). Although comparable patterns of megafaunal extinction have been reported from other continents (Martin and Klein, 1984), other natural changes, e.g. climate, have still not been excluded.

The first convincing examples of pre-agricultural interference in the landscape is seen in Europe where mesolithic activities (*c.* 6500 BP) were shown to be responsible for the early clearance of forest and consequent formation of the present heathland of the North York Moors (Simmons and Innes, 1981) and for the formation of blanket peats in the southern Pennines (cf. Tallis, 1991).

13.1.2 The rise of agriculture

The switch from a predominantly food collection mode of existence to a food producing system took place relatively recently in the evolution of the hominids, *c.* 5000–10,000 BP. The causes of this profound switch in strategy are not clear, although increasing population pressures together with climate changes may have been the major stimuli. An agricultural way of life allows the generation of surplus food and together with the long-term occupation of sites, allowing the development of villages and towns, leads to significant alterations of landscapes as a result of forest clearance for both agricultural and pastoral production. Geographically, the major timing and effects of this pastoral shift vary. In New Guinea horticultural practices are known to have occurred from *c.* 9000 BP and together with the regular use of fire are thought to be primarily responsible for the formation of the majority of grasslands in the region. In north-west Europe this transition did not occur until the Neolithic revolution some 3500 years later.

With this phase of agricultural activities the woodland vegetation of the northern temperate region underwent profound changes as forest was cleared for agriculture and for grazing by domesticated animals, e.g. cattle and sheep. Initially, the forest was exploited by a method of shifting cultivation known as 'slash and burn'. Here a small area of forest is cleared and burnt and crops grown for three to four years until the soil fertility is exhausted. The patch is then abandoned and the vegetation allowed to regrow and rebuild the nutrient store while another patch of forest is cleared. This mode of exploitation of the forest ecosystem gradually became more intense as population pressures increased and with the advent of better ploughs and metal implements the harder to cultivate, but inherently more fertile, heavier soils of the valley floors and sides were utilized. Apart from periods of disease or warfare the forest was prevented from re-invading by a combination of intensive cultivation and grazing by domesticated animals.

One of the best examples that can demonstrate the impact of humans on their environment is seen in a phenomenon known as the 'elm decline' in north-west Europe which was a widespread event *c.* 5500 BP resulting in a massive decline in elm populations. Many theories have been advanced as an explanation including climate change. Until recently, the most favoured one is that of selective collection of elm leaves for fodder by neolithic peoples (Troels-Smith, 1960). More recent interpretations of the evidence implicate Dutch elm disease as the major factor, similar to the fungal strain of the disease that caused widespread decimation of European elm populations in the 1970s, the spread of which was aided by prehistoric people trading diseased wood, and causing structural alterations of the forest enabling the easier spread of the beetle vector (Girling, 1988; Perry and Moore, 1987; Rackham, 1980).

In the Mediterranean region, progressive firing of the landscape for agriculture and game has encouraged the spread of pyrophytic plants like *Quercus ilex* (holm oak), *Erica spp.* (tree heathers), *Cistaceae* (rock-roses) and has caused significant changes in the vegetation of the Mediterranean basin (Ben Tiba and Reille, 1982; Reille, 1984). While on the Iberian peninsula, human manipulations of the progenitor oak and pine woodlands on the agriculturally poor, sandy soils of the south west has led to the formation of cork oak parklands (known as dehesas) (Stevenson and Harrison, in press). Here, not only can the understorey be grazed by cattle, pigs, sheep and goats, but in addition the acorns, bark and associated wood products of the cork oak trees are an important source of human and animal food and other secondary products.

In the Near East early environmental mismanagement led to the collapse of the great, hydraulically based civilizations of the Mesopotamian basin. While changing climate was a contributory factor in their downfall, archaeological evidence suggests that mismanagement of the irrigation systems led to progressive soil salinization and erosion (Rzoska, 1980). Similar problems have been implicated in the downfall of the Harrapan cultures of the Indus valley (Allchin *et al.*, 1978).

These processes of environmental change, initiated with the advent of agricultural modes of existence, were progressively exacerbated as humans became more and more successful at controlling their environment as technological improvements occurred. By AD 1000, the basic pattern of the present-day European landscape had been established.

13.1.3 Environmental change through 'natural' causes

Whilst the rise of agriculture clearly changed the land, the rise of industry changed the air and waters. This grossly oversimplified statement conveys a marked change in human impacts once the fossil fuels of the planet were harnessed. They connected to the drives provided by mechanistic science, its translation into technology and to a capitalist system orientated to innovation and development.

However, in parallel to our precision in separating contamination and pollution (see Chapter 2) we should be very careful when assuming that the only causes of environmental change since the evolution of humans has been at the hands of our own species. Lovelock's Gaia hypothesis (Lovelock, 1979) has stressed that the many profound changes of natural environments recorded in geological history have a connection to life on earth in a cybernetic (control) system of mutual interaction. We may at once, therefore, conclude both that *Homo sapiens* is not the only organism capable of influencing 'natural' environmental change and that the product of the interaction is a largely beneficial equilibrium state. We always ignore the many times bigger biomass and numbers of microscopic organisms on earth which have, in the Gaia hypothesis, the potential to effect environmental change far bigger than our species — even continental drift may be biologically facilitated. Lovelock, however, also warns that the tolerance of this equilibrium is finite and that human beings appear capable of overwhelming its key components such as tropical shelf seas and the atmosphere.

To chart broadly the environmental changes which have accompanied the rise of *Homo sapiens* on earth we can clearly label them 'natural' whether they are extrinsic (wrought by astronomical/solar/atmospheric changes) or intrinsic (activated or modulated by the organic life of the planet itself). Table 13.1 lists the climatic and vegetation sequences which we have identified by environmental reconstruction for the Holocene period and their accompanying cultural phases in the British Isles. Clearly from the 'elm decline' onwards vegetation change is never again wholly natural in this region of the world but there is little to suggest that there was a reflex effect on climate, either from temperate deforestation or semi-arid prehistoric agriculture; these are much less active zones for the global heat engine than the tropics and they appear to have occurred too gradually to affect atmospheric circulation.

Table 13.1 Holocene environmental changes in north-west Europe

Years before present	Major environmental changes
700	Little Ice Age; Greenland colonists perish
1500	Exploration of north Atlantic
2500	Alpine glaciers advance
3500	Alternations of cool and warm climate
5000	Elm decline
6500	Continuation of climatic optimum
8000	Beginning of climatic optimum — warmest period of the post-glacial

Source: Goudie (1983).

However, whilst forest clearance in the temperate zone shows no obvious global climatic response, the industry which followed clearance, settlement, agricultural surplus and technology into this zone of the Northern Hemisphere has begun to evoke one — the enhanced 'greenhouse effect' (see below). Similarly the depletion of stratospheric ozone is now established as a global environmental change brought about solely by our species. We are still unable to prove that a Gaian equilibrium will not be restored in some way — there appears to be more confidence about the 'greenhouse effect' being mitigated in this way than about ozone 'healing'.

An important point arising from this argument is that if atmospheric pollution has begun to threaten the Gaian equilibrium, in which the microscopic organisms of the oceans play a significant part, we cannot afford to continue to pollute water without extending the threat. We conduct no surveys of the numbers and vitality of these myriads of bacteria and other microscopic life-forms, but this merely means that we should interpret our impact on those which are surveyed at the very least as a 'miner's canary' to indicate the level of risk and the need for precautionary change.

13.2 The living planet: species diversity and future environments

While the initial disturbances of the world's ecosystems probably led to an initial increase in species diversity, progressively greater human impact on the landscape has led to an increased rate of species loss. This process was greatly accelerated with the advent of the industrial revolution and the subsequent growth of cities and associated pollution. In temperate regions this loss was exacerbated by the advent of intensive agriculture with heavy pesticide use to rid crops of weeds and economically harmful insects, together with a move to larger production units and consequent removal of valuable species reservoirs like wetlands and hedgerows.

Little quantifiable information can be gained about the early losses of species richness, with the best evidence coming from the fossil record of now extinct mammals (cf. the loss of taxa in the 'Pleistocene overkill', the loss of the North American Plains camel (Camelops) c. 5000 BP). From AD 1600 more robust information is available and over the last 500 years there have been 724 recorded instances of species extinction (Table 13.2). However, these estimates are for taxonomically well-known groups. For others, the number of recorded extinctions is likely to be a significant under-estimate given that we still do not actually know the total number of species alive on the earth. Estimates of the number of species yet to be described vary, but given that 80–90 per cent of 950 beetle genera found on 19 trees of the same species in a tropical forest in Panama were unknown to science (Erwin and Scott, 1982; Reid and Miller, 1989), some estimates predict that another 3–4 million species, chiefly invertebrates, remain to be described.

The high rate of deforestation and habitat fragmentation in the tropical region are the primary cause of the present increased rate of reduction in regional biodiversity. Estimates of losses of the major tropical vegetation life zones vary; in the Afrotropical realm over 50 per cent of the original dry forest and 60 per cent of the moist tropical forest has been lost (Table 13.3) and a similar situation is reported for the Indomalayan and Amazonian realms. More worrying is the fact that only 7–15 per cent of the remaining forests are in designated, legally protected areas (Table 13.3). Of the attempts that have been made to estimate the loss of species richness if present rates of tropical deforestation continue (Table 13.4, McNeely et al., 1990), the best approximations come from applying standard island biogeographic theory (Chapter 3, MacArthur and Wilson, 1967) to estimate the likely reduction in species richness as the area of available habitat declines. These generally suggest, for the worst-case scenario where only the legally protected areas remain, a 35–72 per cent reduction in species richness. The seriousness of this finding is greatly increased under Gaian interpretations because of the domination of the global climate 'heat engine' by tropical energy and moisture exchanges. The knock-on of greenhouse warming to sea-level rise will further change (though not necessarily reduce) the cybernetic potentials of tropical estuaries and shelf seas — rates of change will be critical.

13.2.1 CO_2 and the greenhouse world: global pollution

While tropical deforestation poses the greatest threat to present-day species richness the greatest threat to terrestrial, coastal and marine ecosystems in the future will be the result of the enhanced greenhouse effect.

Table 13.2 Recorded extinctions, 1600 to present

Taxa	Mainland (a)	Island (b)	Ocean	Total	% of total species extinct since 1600
Mammals	30	51	2	83	2.1
Birds	21	92	0	113	1.3
Reptiles	1	20	0	21	0.3
Amphibians	2	0	0	2	0.0
Fish	22	1	0	23	0.1
Invertebrates	49	48	1	98	0.0
Vascular plants	245	139	0	384	0.2
Total	370	351	3	724	

(a) Landmasses > 1 million km^2
(b) Landmasses < 1 million km^2

Table 13.3 Wildlife habitats on the Afrotropical and Indomalayan Realms in 1986 (%)

Vegetation	Afrotropical realm remaining	Afrotropical realm in protected areas	Indomalayan realm in reserves	Indomalayan realm in protected areas
Dry forests	41.6	15.0	27.5	10.6
Moist forests	39.7	7.1	36.5	7.7
Savanna grass	40.8	10.5	36.0	0.0
Scrub/desert	97.8	10.1	14.5	21.2
Wetland/marsh	70.9	5.4	38.8	10.3
Mangroves	44.6	2.9	42.4	8.2

Table 13.4 Predicted per cent loss of tropical forest species due to extinction

Region	Historical–1990	1990–2000	1990–2010	1990–2020	1990–2020 (2 × current rate of deforestation)	Only reserves
Africa and Madagascar	10–22	1–2	2–4	3–7	6–14	35–63
Asia and Pacific	12–26	2–5	4–10	7–17	22–44	29–56
Latin America/Caribbean	4–10	1–3	3–6	4–9	9–21	42–72

Table 13.5 O'Riordan's environmental ideologies

ECOCENTRISM		TECHNOCENTRISM	
Lack of faith in technology, central state authority and élitist expertise. Materialism for its own sake wrong		Technology provides a means of providing for human need with or without controls	
DEEP ENVIRONMENTAL-ISTS	SELF-RELIANCE SOFT TECHNOLOGISTS	ACCOMMODATORS	CORNUCOPIANS
Intrinsic importance of nature; ecology dictates morality. 'Biorights'	Small-scale community base for production and human fulfilment	Legal and economic controls on development. Environmental assessment and empowered management agencies	Political, scientific and technological ways out of all difficulties resulting from pro-growth goals

Source: Modified from O'Riordan (1981).

Since the available GCMs (General Circulation Models) have coarse spatial and temporal resolutions it is difficult, and perhaps will remain so, to predict the future pattern of rainfall and temperature in specific locations of the world. Nevertheless, the models do allow general patterns to be established of future trends in temperature and rainfall at a coarse geographical scale and these show that significant changes are likely to occur in many ecosystems. The linkage of regional estimates of predicted climate to vegetation comes from the use of various models that describe vegetation in terms of climate (e.g. Henderson-Sellars, 1991; Leemans, 1989; BIOME — Box, 1981). Simulations from these models, for future climate scenarios of a greenhouse world, illustrate some of the major changes in life-zone classes that are likely to occur (Table 13.5). Solomons and Leemans (1990) show on the basis of model simulations that a climate scenario involving a doubling of atmospheric CO_2 concentrations would result in the loss of about 40 per cent of today's 15.54 mkm^2 of boreal forest, most of which would go to steppe (12 per cent) and deciduous forest (28 per cent). Only 84 per cent of the boreal forest lost to the south would be replaced by new forests at higher latitudes and elevations. Large losses in particular are predicted for Russia, just west of the Urals, and North America, east of the Rocky Mountains.

A major problem with these existing models is the coarse geographical resolution at which they work, combined with the lack of any data on present-day land-use. While many GCMs predict an increase in the area suitable for tropical forest growth, only a small fraction of the land available for tropical forest growth is currently occupied by it as a result of centuries of logging and forest clearance for agriculture (Table 13.3). Studies that have taken into account current patterns of land-use show that large reductions in biodiversity are inevitable. A GIS-based study on the high diversity wet and seasonally dry tropical rainforest and tropical montane communities of Costa Rica show that greenhouse warming will lead to a migration of the vegetation belts upwards

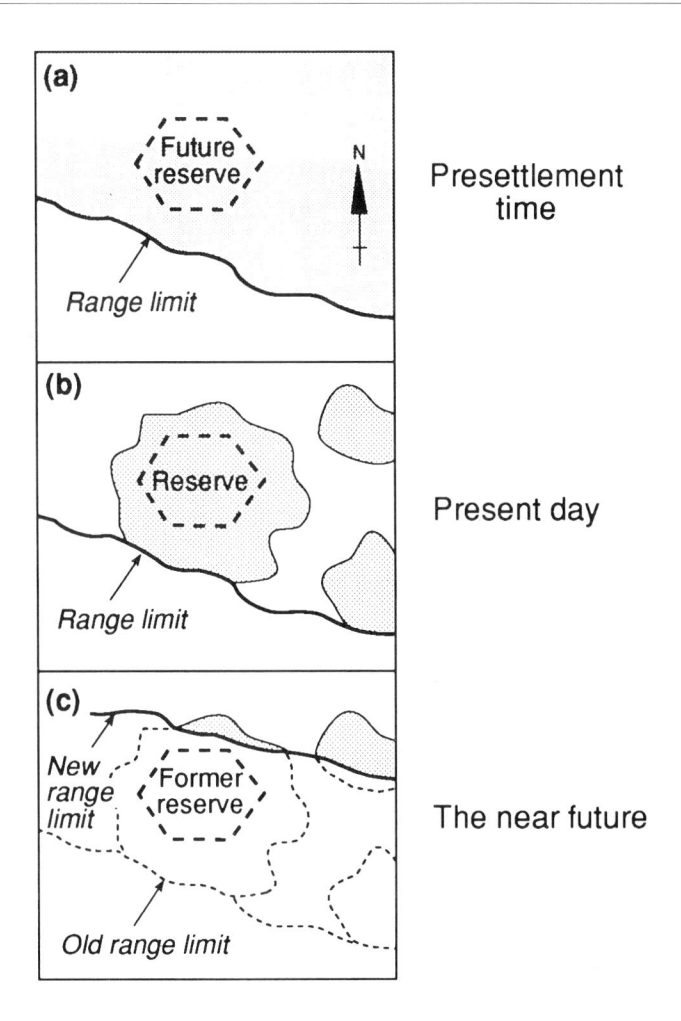

Figure 13.2 Changing climate and displacement of species range limits (RL) beyond the boundaries of a biological reserve. (Peters, R.L. and Darling, J.D.S., 'The greenhouse effect and nature reserves', *Bioscience*, 35, 1985, 707–17)

and result in the extirpation of many of these montane communities (Halpin, personal communication). These predicted changes pose enormous problems for the management of our protected areas, especially those lying on ecotonal boundaries which are at serious risk from greenhouse warming, and the resultant rapid changes in species ranges (Figure 13.2).

Moreover, most of these studies assume that species are able to migrate at the same rate at which climate change is taking place. The rate of climate warming (0.3°C per decade) for North America during the next century is predicted to result in a 3°C rise and to lead to a 300–75-km northward isothermic displacement (Houghton *et al.*, 1990). Normally, species will respond to environmental change by migration, i.e. they will expand and/or retreat along their range boundaries, maintaining their position in environmental space. Palaeoecological studies of an analogous situation at the end of the last glaciation demonstrated that the maximum migration rates of the major forest trees were around 25–40 km per century

(Davis, 1981; Huntley and Birks, 1983) with the fastest rate ca. 200 km per century (Ritchie and MacDonald, 1986). Clearly, species are well adapted to this kind of change; however, the rate of climate change in a greenhouse world is an order of magnitude higher than has been previously experienced by these tree species and it is clear that the response of the major forest trees will not be fast enough (Davis, 1988). Furthermore, since all species have an individualistic response to the environment we should not expect present-day communities to migrate as a unit (Huntley, 1990).

Palaeoecological data conclusively demonstrate that present-day plant communities are impermanent assemblages of taxa whose environmental ranges overlap in space and time (Huntley, 1990). Indeed, as was seen during the interglacial and early Holocene periods, we should expect new communities to be formed with no modern-day analogues. Those species favoured by these changes are likely to be those possessing an 'r' selective reproductive strategies (MacArthur and Wilson, 1967). Inevitably, the very species that we wish to protect tend to be those that possess a 'K' selective reproductive strategy and are not inherently adaptable to short-term environmental change.

13.2.2 Evolution or genetic manipulation?

Argument continues about the relevance of the Gaian hypothesis to organic evolution (Lovelock, 1989). Lovelock's warnings about serious disturbance to planetary equilibria are highly philosophical in that his mechanism of mutual regulation will mean simply that our species is displaced for another which fits the 'disturbed' (our perception) conditions better. However, whether Darwinian or Gaian mechanisms control organic evolution, can it keep up with the rate and/or extent of current and predicted environmental change?

Although evolutionary adaptation can occur on a decadal to centuries time frame (cf. the peppered moth (*Biston betularia*) and industrial melanism; Common Bent grass (*Agrostis tenuis*) and mine waste tolerance (cf. Bradshaw and McNeilly, 1981), evolutionary processes normally operate over far longer time scales. Over shorter time frames, species will respond to environmental change by adaptation, migration or invasion. If the rate of environmental change is beyond the rates of any of these responses then extinction will result. Clearly, it is extremely unlikely that evolutionary change can keep pace with current rates of deforestation or climate change and ways by which the world's biodiversity can be protected must be developed (cf. Reid and Miller, 1989; McNeely et al., 1990).

Although there has been a great deal of success with *ex situ* conservation and notable achievements have occurred, e.g. the successful reintroduction into the wild of the Arabian oryx and the Californian condor, and the derivation of high-yielding varieties of commercially important crops like rice, wheat and maize. But *ex situ* conservation and associated genetic engineering programmes in zoos and botanic gardens can, and will, only hold a small fraction of the world's total biodiversity. Indeed, most of these captive breeding programmes are dependent on a very small gene pool because of the small number of individuals involved and are therefore subject to the normal vagaries of stochastic genetic drift and all the inherent problems associated with it (e.g. inbreeding depression).

At present some 0.2 per cent of all the world's plant species are in long-term storage in seed banks or germplasm vaults. However, most of these are the wild relatives of important crop plants that perhaps hold important genetic traits conferring high yields and/or resistance to common fungal pathogens and other such desirable traits. The utilization of similar techniques to conserve the remainder of the world's plant species are unlikely to be affordable in the future. Even with many of these wild relatives, of economically important crops, long-term cold storage is not feasible because the seeds of many of these species are recalcitrant with poor viability after even quite short periods of storage, e.g. the potato. The cheapest and most cost-effective way of preserving this variability into and for the future must be to preserve their current habitats (cf. McNeely et al., 1990). Policy initiatives in the future must seek to explore the best ways of doing this and thereby maximize the preservation of the world's fauna and flora.

13.3 Global capacities: wilderness, commons and indigenous peoples

The theme of environmentalism in the early 1970s became one which is now often labelled 'doom-laden scenarios'; a threshold increase in data gathering and processing in the field of global resource use and pollution produced such sets of predictions as those in *The Limits to Growth* (Meadows et al., 1972). The model used by the authors of the report *World 3* predicted the following world dynamics:

(a) a sudden and uncontrollable decline in population and industrial capacity would occur within a hundred years if existing trends in world population, pollution, industrialization, food production and resource depletion continued unchanged.
(b) an equilibrium condition was, however, possible which would both guarantee the needs of each individual on earth and give them equal opportunity to realize their human potential.

The model employed exponential increases in growth and its impacts; in 1992 we can be sure that this was the wrong model and our preference has apparently changed from resource limits to growth to pollution threats to major global systems — possibly Lovelock's influence. The 'Limits to Growth' argument has become reissued as the sustainable development argument, with its neo-Malthusian (population-driven) 'doom' replaced by a measure of regulatory optimism.

For that reason we here take the view that, given a continued rapidity of technological innovation (and the accompanying high risks — see below), the resource most likely to be limiting is one of environmental capacities to absorb pollution and land-use change; the wildernesses and common resources of the planet come under sharp focus in Gaian approaches to environmental management. The modelling which followed that by the Meadows team optimistically offered options (e.g. Gribbin, 1979) or stressed the need for international data collection, analysis and diplomacy (e.g. Barney, 1982), some setting out sectoral policies (e.g. United Nations, 1990).

In the 1990s the 'doom-laden' thesis has come to us in the new form of Bill McKibben's *The End of Nature* (McKibben, 1990). In McKibben's argument it is not merely the habitat loss which gives concern, it is the fact that the technologies needed to survive will de-naturize our relationship with habitat. 'Environmentally sound is not the same as natural,' he protests.

13.3.1 Wilderness

Under the United States Wilderness Act wilderness is defined as land that 'generally appears to have been affected primarily by the forces of nature, with the imprint of man's work substantially unnoticeable'. Therefore it need not be entirely untouched or pristine. In 1989, a major conservation group in the United States, the Sierra Club, published their findings from a global survey of wilderness blocks in excess of 4000 km^2 (McCloskey and Spalding, 1989). They identified 1039 tracts of such dimensions, totalling an area of a third of the planet's land surface. They were attempting to answer the question 'How much of the land surface of this planet is predominately affected by the forces of nature?'

However, 41 per cent of the wilderness identified is in the Arctic and Antarctic;

Antarctica is 100 per cent wilderness under the working definition, Canada 64.6, Egypt 42.5, Ethiopia 19.3 and India 0.4 per cent (to quote at random). McCloskey and Spalding quote as the values of wilderness:

(a) biological value as a gene pool and benchmark against which to measure change;
(b) geophysical value to buffer climatic change and to conserve the functions of river basins;
(c) recreational value as places of adventure;
(d) moral value as the homeland of indigenous peoples and of wild creatures unhindered by humans.

Most of the settled continents are between one-quarter and one-third wild, with the exception of Asia (13.6 per cent) and Europe (2.8 per cent). Tundra communities are the dominant biome (41.7 per cent), with warm deserts and semi-deserts next (19.4 per cent). Perhaps because of the impressively large area of wilderness left, very little (13–20 per cent of the major areas) is protected. The authors' view of the vulnerability of unprotected wilderness is that 'it can slip away easily with little notice of encroachment as billions more are added to the human population' (p. 227).

13.3.2 The commons

Hardin (1968) first drew our attention to the principle of commons as an allegory for our exploitation of the whole planetary resource system and to expose the dangers of conventional economic theory when applied to the environment (see Chapter 5). The Gaian perspective of environmental management selects the global commons as the core of environmental capacity needed to sustain further growth without experiencing threshold 'flips' which would result in extinctions.

Caldwell (1990) sums up the commons as follows:

All of the earth's atmosphere and the more than 70 per cent of its surface that is covered by oceans . . . are much less amenable to political control than are its dry land surfaces. The environmental conditions of the polar regions, especially of Antarctica, severely restrict normal human occupancy. Even less amenable to human influence is the magnetosphere and the gravitational field of the earth in outer space. Thus extensive parts of the planetary environment may be affected by mankind but are not under its effective political control. (p. 257)

Diplomatic lip-service has been paid to the notion of these commons as a human heritage but there are, as yet, no all-embracing legal principles allowing their regulation to meet environmental goals. Nation states can do little to influence or control the sweep of the air or the flow of the oceans over or around them; they have the excuse, therefore, of prioritizing those elements of the natural environment which can be controlled for their active interventions. The ozone layer, however, has proved that there are frameworks for managing and safeguarding the commons (Benedick, 1991). As Benedick puts it,

On September 16, 1987, a treaty was signed that was unique in the annals of international diplomacy. Knowledgeable observers had long believed that this particular agreement would be impossible to achieve because the issues were so complex and arcane and the initial positions of the negotiating parties so widely divergent. Those present at the signing shared a sense that this was not just the conclusion of another important negotiation, but a rather historic occasion. (p. 1)

The Montreal Protocol on Substances that Deplete the Ozone Layer was agreed despite a huge measure of scientific uncertainty about cause and effect, yet it wrote off billions of dollars of investment and hundreds of thousands of jobs in the chemical industries of the world.

Benedick's 'lessons for a new diplomacy', i.e. one focusing upon the global commons, include the unaccustomed but critical role of scientists in negotiations (and this is fast becoming a characteristic of the 'greenhouse effect' diplomacy). Governments may, however, need to act before the scientists are certain and in such an atmosphere public opinion is a strong indication of the outcome. Strong leadership from a major country is also necessary (in the case of ozone it was the United States). Economic and structural inequalities among countries must be recognized in any regulatory scheme. Further case studies are provided by Carroll (1988).

13.3.3 Indigenous peoples

For centuries we considered indigenous peoples as 'primitive' and saw the development process for them as reforming, accompanying an imposition of Christian or other Western codes. However, two revelations have brought about a change of perception:

(a) Indigenous peoples inhabit large areas of the planet but they are threatened by development; 200 million, or 4 per cent of world population, remain. They are, therefore, candidates for conservation.
(b) Indigenous peoples have a complex relationship with their habitat, largely uncodified in our terms but sustainable by any standards.

Beauclerk and Narby (1988) list the characteristics of indigenous peoples:

(a) sustainable use of resources;
(b) land held in common
(c) wealth shared equally
(d) societies rooted in kinship;
(e) vulnerability.

In terms of vulnerability, it is difficult to know how we may explore (and even exploit) the bases of their sustainable codes and protect them against the disruption and occupation of their territory without damaging their isolation through well-intended adoption. For this reason Beauclerk and Narby's book is subtitled 'a fieldguide for development'. Many developing nations move their indigenous peoples in 'resettlement' schemes connected to, for example, dam-building schemes; this is particularly ruinous if the basis of the sustainable environmental use has been nomadism.

If indigenous peoples are to be introduced gently to development it is the dream of some environmentalists and economists, as well as non-government charities, that the developed world can buy from them the extensive resources of their habitat. For example, tropical forest products and the related crafts are attractive in the developed world; the wildlife resources of the savannas could be saved by a world trade (stressing a proper reward for producers) in plant or animal products.

The Prescott-Allens (1982) have compiled both a range of products and their potential and actual value to the economies of the developing world. Their evidence is that preservation of wildlife is impossible but with a less squeamish attitude to, for example, zebra steak on our menus and by paying the producers for the drug products of the forests we could achieve sustainable development for the developing nations, conservation of wildlife and conservation of the indigenous peoples.

13.4 The end of nature or recovery, restoration, enhancement?

It is hard to refute McKibben's assertion that continued development, wherever it occurs on the planet and however much it is deemed 'environmentally acceptable', will produce an increasing distance of both substance and spirituality between human beings and nature. The stark facts are, however, that to succour the estimated world population in AD 2050 we need to increase global production of food and goods/services sevenfold.

13.4.1 Nature, development and attitudes

At this point we need guidance from O'Riordan (1981) in segregating human attitudes towards the future of development into manageable groups (a process which is also helpful in rationalizing the myriad political and pressure groups — see Chapter 1). Table 13.5 (p. 248), as we have seen, illustrates the widely used O'Riordan classification (further used in modified form in Chapter 5). Attitudes to the essential rise of productivity may be taken basically as optimistic and pessimistic according to whether, respectively, one believes in human inventiveness combined with the duty of care or the inevitability of 'the end of nature'. O'Riordan separates technocentrism from ecocentrism as doctrinal labels for these groups, but they need further subdivision, as the table shows.

We have already indentified the critical need for new knowledge bases and Santos (1990) presents ways in which these should concentrate upon the problems of carrying capacities, with evolution in forms of international law feeding on that knowledge under what Santos calls, quoting Potter, 'new wisdom which will provide the knowledge of how to use knowledge'. This is a technocentric accommodation but it neglects one key feature of technocentric futures: risk.

Roberts and Weale (1991), in prefacing their collection of essays *Innovation and Environmental Risk*, point out that a key element of capitalism is ceaseless innovation; the Austrian economist of the 1930s, Joseph Schumpeter, called this 'creative destruction'. New goods, new markets, new methods of production and new organizations are an essential part of this 'progress' (or what Thoreau called 'improved means to unimproved ends'). Innovation is risky and as it becomes an essential element of global carrying capacity as well as typical of the dominant economic system the risks rise, either because they are spread wider or, as in the case of nuclear power and its waste products, they are concentrated.

Petroski (1982) titles his book about the fallibility of engineers *To Engineer is Human*. We have already observed that we are poor as a species at anticipation and prediction in the environmental sciences (though in the next chapter we come to depend upon planning!). Petroski claims that engineers learn by their mistakes and that there is no safe innovation. Thus, whilst the population badly needs further education on the nature of risk, fallibility builds in an extra, non-mathematical element which may make even technocentrists sceptical!

The major implication of this revelation for Gaian interpretations is that the cybernetic properties of natural systems will be tested fairly frequently by what society will call 'disasters'. As Chapter 10 demonstrates, we are already putting in place socio-economic structures to recognize our increasingly self-imposed hazardous lifestyle. The natural system will, however, be tested: dispersion, dilution and other processes are very powerful in the pollution pathway following a disaster, and targets reproduce and recolonize.

Environmental managers will, however, often be faced with monitoring and assisting recovery of ecosystems and their habitats. As the following section shows, our increasing knowledge of processes and techniques may assist both in the recovery from 'disaster' and in enhancing natural systems as the 'down-payment' for future developments (see Chapter 14).

13.4.2 Natural and assisted recovery in ecosystems

Given enough time most ecosystems will recover naturally once the cause of disturbance is removed. In most cases assistance is required to alter conditions to favour the development of the pre-disturbance fauna and flora. This may involve: filling in drainage ditches; growing reeds to remove excess nutrient inputs; provision of water treatment stations; removal or moderation of grazing regimes; provision of pre-disturbance species because of dispersal problems. A few ecosystems will have been irretrievably transformed by the actions of humans and little prospect of natural recovery can be held unless a large amount of, often heavy engineering, restoration assistance is given, e.g. restoration of lowland raised mires after horticultural peat extraction. Some systems will recover naturally without assistance but the recovery time scale can vary enormously from tens to hundreds of years in the case of recovery of Brazil's caatinga forest from slash and burn agriculture, to thousands of years for sites that have been previously cleared by bulldozer (Uhl et al., 1982).

Since the determination of the pre-disturbance condition of an ecosystem is difficult and since ecosystems are dynamic entities, complete ecosystem restoration is unrealistic. Most difficulties arise from problems in defining what we mean by the word 'natural'. Some of the communities that we seek to conserve are themselves the result of previous anthropogenic actions (cf. the species-rich hay meadows of the Yorkshire Dales). Most conservation exercises these days are moving away from attempts at restoration of specifically targeted species to restoration of something approaching the original community. These efforts, if successful, offer us the best way of conserving biodiversity.

We have already seen some examples where ecosystem recovery is driven by a target organism approach. In the case of Loch Fleet the successful restoration of the fishery was achieved at the expense of the catchment vegetation (cf. Chapter 3) and little attempt was made to adjudge whether the lake had been restored to anything approaching its pre-disturbance condition. Similar parallels exist in the conservation of butterfly populations. One of the main reasons for the loss of the swallowtail butterfly from Wicken Fen in Cambridgeshire, in addition to a general decrease in winter flooding, was a shift in the management of the fen some 30 years ago from cutting the fen vegetation in late spring/early summer to late summer/early autumn (Rowell, 1986). This prevented the chief food plant (*Peucedanum palustre*) of the swallowtail from flowering and regenerating; a change in the cutting regime reversed this decline. However, the success or otherwise or the restoration is adjudged by the successful reintroduction of the butterfly, not whether the conservation of the whole fen was enhanced as a result of the manipulation. In addition, the lowering of the water table had also led to an increase in species diversity and the establishment of two rare calcifugous plants — *Hookeria lucens* and *Sphagnum fimbriatum* — in an otherwise calcareous fen. A proposal to increase the winter flooding of the fen will inevitably reverse some of these changes and exposes the complexities and dilemmas of practical conservation management (Rowell, 1986).

A further example can be seen in southern Spain, where the saline lake of Fuente de Piedra, in the province of Almeria, has its water level manipulated precisely to achieve maximum breeding success for flamingoes. As in the other examples, no holistic overview is taken as to whether the saline lake ecosystem is better for the manipulation and some see the manipulations as no more than the creation of a flamingo farm.

In contrast, elsewhere large scale restoration programmes are seeking to restore the original tropical rainforest. The most sophisticated and extensive of these is occurring in Costa Rica where the 800-km^2 Guanacaste National Park is being created in an area that was once covered by tropical dry forest — the most threatened of all major lowland tropical forest communities (Janzen, 1986, 1988a). Some 700 km^2 is now being restored from pasture and forest fragments. The restoration relies primarily on natural processes, however, to hasten the process, fires will be suppressed and some cattle may be allowed to graze initially since reducing grass height enhances tree seedling survival (Janzen, 1988a, 1988b).

The restoration of Guanacaste can occur because there are existing forest fragments which can provide a natural source of native plants and animals. In some cases the restoration will have to rely on the ability of *ex situ* conservation to provide the relevant fauna and flora. We should not underestimate the extent to which botanic gardens, zoos and germplasm vaults play a key role in the survival of a significant number of species.

In conclusion it is clear that the Gaia hypothesis is a substitute framework of guidance to environmental futures to that provided by the socio-economic models of the 'Limits to Growth' era. Equally clearly, however, a Gaian concentration upon carrying capacities, risks and recovery mechanisms requires considerable socio-economic and political refinement before it can become the basis of 'knowledge to use knowledge'.

References

Allchin, B., Goudie, A. and Hegde, K. (1978), *The Prehistory and Palaeogeography of the Great Indian Desert*, Academic Press, London.

Barney, G.O. (director) (1982), *The Global 2000 Report to the President*, Penguin Books, Harmondsworth, 766 pp.

Beauclerk, J. and Narby, J. (with Townsend, J.) (1988), *Indigenous Peoples. A Fieldguide for Development*, Oxfam, Oxford, 121 pp.

Benedick, R.E. (1991), *Ozone Diplomacy: New Directions in Safeguarding the Planet*, Harvard University Press, Cambridge, Mass. and London, 300 pp.

Ben Tiba, B. and Reille, M. (1982), 'Recherches pollenanalytiques dans les montagnes de Kroumirie (Tunisie septentrionale): premiers résultats', *Ecologia mediterranea*, 8, 75–86.

Box, E.O. (1981), *Macroclimate and Plant Forms: An Introduction to Predictive Modelling in Phytogeography*, Junk, The Hague, 258 pp.

Bradshaw, A.D. and McNeilly, T. (1981), *Evolution and Pollution*, Edward Arnold, London.

Caldwell, L.K. (1990), *International Environmental Policy: Emergence and Dimensions* (2nd edn), Duke University Press, Durham, NC and London, 460 pp.

Carroll, J.E. (ed.) (1988), *International Environmental Diplomacy*, Cambridge University Press, Cambridge, 291 pp.

Davis, M.B. (1981), 'Quaternary history and the stability of forest communities', in D.C. West, H.H. Shugart and D.B. Botkin (eds), *Forest Succession*, Springer-Verlag, New York, pp. 132–53.

Davis, M.B. (1988), 'Ecological systems and dynamics', in *Toward an Understanding of Global Change*, National Academy Press, Washington, DC, pp. 69–106.

Erwin, T.L. and Scott, J.C. (1982), 'Seasonal and size patterns, trophic structure and richness of Coleoptera in the tropical arboreal ecosystem: the fauna of the tree Luehea seemannii Triana and Planch in the Canal Zone of Panama', *Coleopterists Bulletin*, 34, 305–22.

Girling, M.A. (1988), 'The bark beetle (Scolytus Scolytus-Fabricius) and the possible role of elm disease in the early Neolithic', in Martin Jones (ed.), *Archaeology and the Flora of the British Isles*, Oxford University Committee for Archaeology, Oxford.

Goudie, A. (1983), *Environmental Change* (2nd edn), Oxford University Press, Oxford, 258 pp.

Gowlett, J.A.J., Harris, J.W.K., Walton, D. and Wood, B.A. (1981), 'Early archaeological sites, hominid remains and traces of fire from Chesowanja, Kenya', *Nature*, 294, 125–9.

Gribbin, J. (1979), *Future Worlds*, Abacus, London, 225 pp.

Hardin, G. (1968), 'The Tragedy of the Commons', *Science*, 162, 1243–1248.

Henderson-Sellars, A. (1991), 'Developing an interactive biosphere for global climate models', *Vegetation*, 91, 149–66.

Holdridge, L.R. (1967), *Life Zone Ecology*, Tropical Science Center, San José, 206 pp.

Houghton, J.T., Jenkins, G.J. and Ephraums, J.J. (eds) (1990), *Climate Change: The IPCC Scientific Assessment*, Cambridge University Press, Cambridge.

Huntley, B.H. (1990), 'European vegetation history: palaeovegetation maps from pollen data — 13,000 yr BC to present', *Journal of Quaternary Science*, 5, 103–22.

Huntley, B.H. and Birks, H.J.B. (1983), *An Atlas of Past and Present Pollen Maps for Europe: 0–13,000 Years Ago*, Cambridge University Press, Cambridge.

Janzen, D.H. (1986), *Guanacaste National Park: Tropical Ecological and Cultural Restoration*, Editorial Universidad Estatal a Distancia, San José, Costa Rica.

Janzen, D.H. (1988a), 'Tropical dry forests: the most endangered major tropical ecosystem', in E.O. Wilson and F.M. Peter (eds), *Biodiversity*, National Academy Press, Washington, DC, pp. 130–7.

Janzen, D.H. (1988b), 'Management of habitat fragments in a tropical dry forest: growth', *Annals of the Missouri Botanical Garden*, 75, 105–16.

Leemans, R. (1989), 'Possible changes in natural vegetation patterns due to global warming', in A. Hackl (ed.), *Der Treibauseeffekt: das Problem — Mögliche Folgen — Erforderliche Massanahmen*, Akadamie fur Umwelt und Energie, Laxenburg, Austria, pp. 105–22.

Lovelock, J.E. (1979), *Gaia: A New Look at Life*

on Earth, Oxford University Press, Oxford, 157 pp.
Lovelock, J.E. (1989), *The Ages of Gaia: A Biography of Our Living Earth*, Oxford University Press, Oxford, 252 pp.
MacArthur, R.H. and Wilson, E.O. (1967), *The Theory of Island Biogeography*, Princeton University Press, Princeton.
McCloskey, J.M. and Spalding, H. (1989), 'A reconnaissance-level inventory of the amount of wilderness remaining in the world', *Ambio*, 18(4), 221–7.
McKibben, B. (1990), *The End of Nature*, Penguin Books, Harmondsworth, 212 pp.
McNeely, J.A., Miller, K.R., Reid, W.V., Mittermeir, R.A. and Werner, T.B. (1990), *Conserving the World's Biological Diversity*, IUCN, WRI, CI, WWF-US, World Bank, Gland, Switzerland and Washington, DC.
Martin, P.S. (1973), 'The discovery of America', *Science*, 179, 969–74.
Martin, P.S. and Klein, R.G. (eds) (1984), *Quaternary Extinctions: A Prehistoric Model*, University of Arizona Press, Tucson.
Meadows, D.H., Randers, J. and Behrens, W. (1972), *The Limits to Growth*, Universe Books, New York and Earth Island, London.
O'Riordan, T. (1981), 'Environmental Issues', *Progress in human geography*, 5, 393–407.
Perry, I. and Moore, P.D. (1987), 'Dutch elm disease as an analogue of neolithic elm decline', *Nature*, 326, 73–3.
Peters, R.L. and Darling, J.D.S. (1985), 'The greenhouse effect and nature reserves', *Bioscience*, 35, 707–17.
Petroski, H. (1982), *To Engineer is Human*, St. Martin's Press, New York, 247 pp.
Prescott-Allen, R. and Prescott-Allen, C. (1982), *What's Wildlife Worth?*, Earthscan, London, 92 pp.
Rackham, O. (1980), *Ancient Woodland: Its History, Vegetation and Uses in England*, Edward Arnold, London.
Reid, W.V. and Miller, K.V. (1989), *Keeping Options Alive: The Scientific Basis for Conserving Biodiversity*, World Resources Institute, Washington, DC.
Reille, M. (1984), 'Origine de la végétation actuelle de la Corse sud-orientale: analyses polliniques de cinq marais côtiers', *Pollen et Spores*, 26, 43–60.
Ritchie, J.C. and MacDonald, G.M. (1986), 'The patterns of post-glacial spread of white spruce', *Journal of Biogeography*, 13, 527–40.

Roberts, L. and Weale, A. (1991), *Innovation and Environmental Risk*, Belhaven, London, 186 pp.
Rowell, T.A. (1986), 'The history of drainage at Wicken Fen, Cambridgeshire, England, and its relevance to conservation', *Biological Conservation*, 35, 111–42.
Rzoska, J. (1980), *Euphrates and Tigris: Mesopotamia Ecology and Destiny*, Junk, The Hague.
Santos, M.A. (1990), *Managing Planet Earth: Perspectives on Population, Ecology and the Law*, Bergin and Garvey, 172 pp.
Solomon, A.M. and Leemans, R. (1990), 'Climatic change and landscape ecological response: issues and analyses', in M.M. Boer and R.S. de Grouf (eds), *Landscape Ecological Impact and Climatic Change*, IOS Press, Amsterdam, pp. 239–316.
Simmons, I.G. (1989), *Changing the Face of the Earth: Culture, Environment, History*, Blackwell, Oxford.
Simmons, I.G. and Innes, J.B. (1981), 'Tree remains in a North York Moors peat profile', *Nature*, 294, 76–8.
Stevenson, A.C. and Harrison, R.J. (in press), 'Ancient forests in Spain: a model for land-use and dry forest management in S.W. Spain from 4000 BC to 1900 AD', *Proceedings of the Prehistoric Society*.
Tallis, J.H. (1991), 'Forest and moorland in the South Pennine uplands in the mid-Flandrian period, III. The spread of moorland — local, regional and national', *Journal of Ecology*, 79, 401–16.
Troels-Smith, J. (1960), 'Ivy, mistletoe and elm: climatic indicators-fodder plants', *Danmarks geologiske Undersolgelse*, IV(4), 1–32.
Turner, C. (1970), 'The middle Pleistocene deposits at Marks Tey, Essex', *Philosophical Transactions of the Royal Society B*, 257, 373–440.
Uhl, C., Jordan, C., Clark, K., Clark, H. and Herrera, R. (1982), 'Ecosystem recovery in Amazon caatinga forest after cutting, cutting and burning, and bulldozer clearing treatments', *Oikos*, 38, 313–20.
United Nations (1990), *Global Outlook 2000*, United Nations Publications, New York, 340 pp.
West, R.G. and McBurney, C.M.B. (1954), 'The Quaternary deposits at Hoxne, Suffolk and their archaeology', *Proceedings of the Prehistoric Society*, 20, 131–54.

14

Planning, control or management?

Malcolm Newson

Throughout this book we have raised issues of concern for the future of the global environment; working as scientists we have largely written in the hope that our work will become part of sound policies for management of the natural environment. However, as indicated by those chapters which focus upon management and policies (e.g. Chapters 4, 5, 10), society does not regard its scientists as soothsayers. Indeed, were politicians even more eager listeners than they seem, one of the current crises of development is to yield sufficient financial wealth to support the requisite research activity. Those scientific studies which are carried out on environmental systems often end in controversy because of the way they are structured and financed (Collingridge and Reeve, 1986).

In deciding, therefore, whether policy is best guided by moral pressure, economics or the law we need to enquire how each of these management perspectives is resourced with sufficient knowledge to promote action acceptable to society and along which routes such action will move. Our simplistic division here is between largely proactive planning and, for the most part, retrospective legislation. 'Management' is seen as a flexible compound of these and other, e.g. fiscal, approaches.

14.1 Knowledge bases for future environmental interventions

Consider the hitherto successful relationship between science and technology, enshrined in the popular word 'invention'. The image of this relationship, often created by TV programmes such as 'Tomorrow's World' in the United Kingdom, is of a scientist working long hours in the laboratory and discovering, sometimes by chance, a useful application of a basic principle. Trials prove that it works, a manufacturer sees that buyers will be satisfied and it is quickly in our homes. Improved versions may follow and some customers may be disappointed but few doubt the value of the scientific 'discovery' or conceive of the scientist as being 'biased' — the work is largely neutral. Environmental science is different!

14.1.1 Holist and reductionist frameworks: a brief guide

Part of the difference between traditional laboratory science and environmental science lies in the use of the results for social decision-making. Medicine is a socialized science but has largely retained its technical decisions within professional circles. Environmental science quickly 'goes public' because of the public's interest and often their concern.

This is often a difficult process since lack of the controls possible in laboratory science inflicts considerable degrees of uncertainty upon the results of field studies or monitoring; the environment and its management require 'decision-making under uncertainty'. If one adds the fact that environmental scientists may be environmentalists too, by personal choice or because they work for a protection agency, society may well feel that, alongside the uncertainty, there is also bias.

Cannot environmental science behave in a more traditional manner? In order to give a negative answer we must first dismember a small section of the history of science and its relationship with society in order to prove that environmental science cannot fit this mould. Francis Bacon (1561–1626) advocated experimental science, in the modern sense, in his *Novum Organum* (New Instrument); this was essentially applied science in support of the 'relief of man's estate' and its rapid, lasting success is testified by the Royal Society's establishment role in science ever since Bacon's writings helped found it in the 1660s (Ronan, 1983). The Baconian approach became, through its success, something of an ontological straitjacket: 'real' knowledge was accumulated about nature by experiment and to perform an experiment one needs to select elements from a complex overall picture.

The pattern became set, therefore, for what is now often scorned by historians of science, sociologists and environmentalists — the reductionist approach. At its centre is the analysis of components in order to synthesize the whole. It is, after all, one of the basic forms of human endeavour to 'split up the problem', be it house-painting or nuclear physics.

For as long as conventional science has followed reductionism, with great success and with public respect, a number of writers (particularly in the natural sciences and in geography) have rejected it; they have promoted holism. Holism takes the wide view first, e.g. of a landscape, community or ecosystem, and retains it for as long as it can be practically studied. As a methodology it perhaps more closely resembles the arts: few portrait painters begin with two marvellously detailed ears!

The holistic approach, however, produces a very heavy burden upon the knowledge of individuals and it is to individuals that holism has owed much while science has been rigidly assembled along disciplinary lines. One of the desirable outcomes of the recent pressure on financial resources for science has been a greater willingness to break down barriers within knowledge and to work in interdisciplinary teams on 'big issues'; social scientists are also involved. One of the suspicions held of holism, beside the fact that its methodology breaches disciplinary (and therefore professional) boundaries, is that it seeks only to discover more problems. However, as those who have championed a holistic framework have proved, holism manifestly helps to categorize and prioritize. By thinking holistically James Hutton (1795) was able to create the time framework for modern geology, Charles Darwin (1859) was able to combine his biological observations with this geological time frame, and latterly James Lovelock (1979) has added the importance of life to the chemical systems of the earth; by doing so he has shown how some pollution problems pale into insignificance beside those affecting the atmosphere/ocean coupling.

Both frameworks for environmental science are acceptable; battling between them is potentially negative. Reductionists find that the public will not accept remedies for cities based upon what happens in a test-tube; holists find that the critical processes of the universe require small-scale tests for validation — perhaps in that same test-tube! Meanwhile, we tend to conduct inadvertent 'ecosystem experiments' through those monitored 'crises' or 'disasters' produced by global development. Mooney *et al.* (1991) have recently catalogued examples such as tropical deforestation, acid rain, desertification and the natural 'El Niño' climate oscillation.

Table 14.1 Types of uncertainty and styles of problem solving

Type of uncertainty	Conditions for decision making	Prerequisite for authority	Source of authority	Problem-solving style
Technical	Risk	Information	Instruction	Reductionist
Methodological	Uncertainty	Respect	Experience	Pragmatic
Epistemological	Indeterminacy	Faith	Revelation	Holistic

Source: O'Riordan and Rayner (1991)

14.1.2 Knowledge and the politician

Political processes lie across the path of any environmental scientist who seeks incorporation of his or her knowledge into planning or legislation. The processes have been seen as a 'filter' between research and applications. The politicians may be those controlling international affairs, national MPs concentrating on constituency matters, or local government representatives arguing at a planning meeting. They will all combine motives of public good and personal gain (by re-election) when they make decisions. They may also do so within a framework of broad support for capital or broad support for labour. As yet these dimensions have stubbornly refused to reorientate to those of, for example exploitation and sustainable development.

Whilst politicians may operate a filter in the applications for scientific knowledge they do no block progress; they are essentially society's risk-takers and it is in this spirit that they judge the environmental science input. In an unchanging natural environment (the popular assumption until the last decade) politicians operated within a framework determined by the seriousness of the phenomenon reported by the scientists, the unanimity of their conclusions and the availability and desirability of a 'cure'. Even so, they were frequently guided by the 'issue-attention cycle' and by crises (see Chapter 1) which, when they were the culmination of sustained scientific pressure, evolved results — perhaps one reason why 'command and control' strategies have proved so popular.

In the context of environmental change we cannot wait for a disaster or crisis to prompt reaction and indeed there may not be one. The risk of political decision-making, particularly because conventional definitions of national growth and development may need to be modified, becomes markedly increased. However, as a major European politician said in 1990, 'History is speeding up', and we live in a time when rapid change is part of the life-experience of all of us (in the developed world).

14.1.3 The precautionary principle

Underlying German public policy on the environment for many years has been the 'Vorsorgeprinzip', translated as the precautionary principle (Royal Commission on Environmental Pollution, 1988). This principle prescribes: minimization of risk; definition of environmental quality; and implementation of ecological principles for policy. This helps redefine the relationship between environmental science and government and has been translated into common parlance as the 'least regrets' strategy. It has been variously modified by scientists themselves, especially those concerned with work on the 'greenhouse effect', who prefer a clause from the German guidelines to the Vorsorgeprinzip, 'the avoidance or reduction of risks to the environment before specific environmental hazards are encountered'.

O'Riordan and Rayner (1991) investigate new ways of classifying and presenting environmental risks on ways which will aid politicians. As Table 14.1 shows, the classification stems from a tripartite split of the original uncertainty. Henderson-Sellers (1991) takes the argument to the more practical limits in linking the uncertainty of scientists about identifying the greenhouse effect to a nervous traveller waiting for a bus: all the information lines up except the actual arrival of the bus; in its absence one must decide the risks of going home, walking the journey, etc.

14.2 Environmental planning

If we are able to consider the future more realistically in our economics (Chapter 5) and in governmental decision-making by listening more coherently to what a majority of good scientists predict, the process of planning holds a good many keys to sustainable environmental projection. Once again, however, scientists must not consider it a wholly rational process, whether it is policy planning by corporations or land-use planning by states. It is mainly with the latter activity that this section is concerned, since most of our major development or conservation decisions are transacted in patterns of land use.

14.2.1 Source, pathway, target: where to site?

Assuming, for the moment, a technocentric argument on development, all human society is faced with blending a necessary minimum of resource-consuming, polluting and habitat-damaging projects into an environment which can only cope if this action is planned with care and supported by the best available knowledge. Experience has shown that, to be truly sustainable, such planning must also be the will of the people affected (Section 14.4) Successive totalitarian planning experiments at various scales have proved that however 'wise' a project appears to be it needs adoption in the longer term — we are back to political filters, this time at grass-roots level. The scientific inputs to such plans are notable in the field of water pollution. As Chapter 7 suggests, models of point-source pollutants are using the principle of pathways (including environmental capacities) and targets to set objectives for water quality management. In principle, therefore, additional point sources can be 'planned in' to the most advantageous location; in any other location the *strict* application of principles of environmental economics suggest that costs must rise, e.g. of discharge permits.

However, it is precisely the *limited* adoption of environmental economics in environmental management which curtails the ability rationally to 'plan in' projects spatially in this way. In the United Kingdom, a nation in which urban planning began with a concern for conditions for human beings *within the home* (see Cullingworth, 1985), there has been a long-lasting policy of leaving pollution control to the legislators rather than offering detailed opportunities to planners. Simplistically, therefore, planning deals with the general acceptability of development on the site proposed for it (within strategic zones set by occasional 'structure plans'). Because most communities have been generally inclined towards local economic development their politicians have generally approved. Central government agencies have then controlled the pollution performance of the development.

There are somewhat clearer, more direct roles for planners in allocating land for non-productive uses, including conservation, and in protecting wilderness reserves against development encroachment. It is of interest in this case to find public — or agency — pleading rather than rational application of biogeographical principles (see Chapter 3) to be the main factor guiding policy.

14.2.2 Planning and pollution: attitudes in the developed world

It is fortunate for English-speakers that a wealth of literature has emanated from the long history of town and country planning in the United Kingdom and that treatments of the relationship between planning and pollution have already emerged (e.g. Miller and

Wood, 1983). Some authors have compared attitudes to planning and pollution between a traditionally welfare-orientated society such as Britain and the 'free-market' conditions of the United States (Wood, 1989). Miller and Wood (1983) summarize four policy options for local planners under UK legislation; regulation of development with pollution implications can be achieved by:

(a) withholding planning permission;
(b) consenting with conditions;
(c) agreement of voluntary codes with the developer;
(d) strategic planning as a framework within which to control pollution from individual projects.

It is, however, rare for pollution to be cited as the sole or principal reason for the refusal of planning permission and it is difficult for local representatives to be unduly noble in denying growth to their community in respect of downstream or down-wind externalities. As a result dust and noise pollution are much more likely to affect planning decisions than are greenhouse gases. As a government circular quoted by Miller and Wood puts it: 'planning conditions should not be imposed in an attempt to deal with problems which are the subject of controls under separate environmental legislation.'

The 'mother of planning' as a nation is therefore left with local representatives able to decide a preferred balance of land use (amenity, agriculture, industry) and to curtail inappropriate juxtapositions of land uses but little more. There have been exceptions such as the establishment of 'smokeless zones' in Manchester ahead of national legislation, but in most cases planning professionals tend to see their role as collative rather than regulatory in pollution issues.

By contrast, the United States has a much more open and conflict-orientated system of environmental controls, and in obtaining land-use permits from a range of pollution control agents the American developer is likely to face higher costs and longer delays. He/she is dealing with agencies who wish to make a rational decision with a high public profile and in terms of their particular pollution medium; by contrast the land-use control issues are subordinate. As Wood (1989) concludes, the UK system of planning may appear *ad hoc* but it is integrative and local within a homogeneous national framework; it is also land-use-led. Wood compares the case of air pollution controls on both sides of the Atlantic, primarily because it was to air pollution that the British principle of 'best practicable means' was first applied. He uses the following points for comparison: the characteristics of the developer/project, the development agency, the objectors and finally characteristics of the processes of authorization/control and concludes, as others do, that neither the US nor the UK approach is inherently better; progress towards environmental improvement occurs at an identical pace. However, the American approach should clearly become more efficient and the British approach more transparent and accountable.

Clearly, land-use planning as a means of environmental management is bound to evolve (in that each of Wood's list of characteristics is dynamic) and there will clearly be a challenge in respect of the scale of and risk from major development projects such as those connected with nuclear power (Chapter 11). Industrial major accident hazards (Chapter 10) will also need to be given a more profound and co-ordinated treatment by planners than at present. Walker (1991) reveals a wide diversity of current approaches from the drawing of risk contours round sites in the Netherlands to the less formal advice provided by the UK Health and Safety Executive to planners (Table 14.2).

14.2.3 'NIMBY' — 'Not in My Back Yard'

Another feature of large developments in the developed world which has confused some politicians has been that a stirring of environmental feelings has spread far and

Table 14.2 Approximate consultation distances for major hazard sites as specified by the Health and Safety Executive

Substance	Largest tank size (%)	Consultation distance (m)
Liquefied petroleum gas	25–40	300
held at pressure >1.4 bar	41–80	400
absolute	81–120	500
	121–300	600
	>300	1000
in cylinders or small bulk tanks <5.1 capacity	>25	100
Liquefied petroleum gas held under refrigeration <1.4 bar absolute pressure	>50	1000
Phosgene	>2	1000
Chlorine	10–100	1000
	>100	1500
Hydrogen fluoride	>10	1000
Sulphur trioxide	>15	1000
Acryontrite	>20	250
Hydrogen cyanide	>20	1000
Carbon disulphide	>20	250
Liquid oxygen	>500	500
Sulphur dioxide	>20	1000
Bromine	>40	600
Ammonia (anhydrous or as solution containing >50 by weight)	>100	1000
Hydrogen	>2	500
Ethylene oxide	5–25	500
	>25	1000
Propylene oxide (atmospheric pressure)	>5	250
(under pressure)	5–25	500
	>25	1000
Methyl isocynanate (dependent on nature of mix and storage)	1	1000

Source: Smallwood (1990).

wide, largely from the front parlours of the middle classes. It has been particularly associated with the development of nuclear power and transport routes. It is predicated on owner-occupation and land values but it has been succoured by planning processes which present the need for development as irrefutable, allowing only debate on siting, which quickly becomes a debate about alternatives — hence 'NIMBY'.

'NIMBY' has a subculture of 'NODAM' ('No Development After Mine') but in its main variant attempts at 'protective altruism under uncertainty'; it is therefore very close to the nub of the current dilemma in applying environmental rationality. The lesson for planners is clearly that the best available information on options for locating new developments must be a core of professional progress in the coming decades. Geographic information systems are vociferous candidates for the technical facilitation of 'NIMBY-Aid' options (see Chapter 11, also Charlton and Openshaw, 1986; Rhind, 1986) but it is also clear that unless we have a coherent national philosophy of role of planning, one in which planners are not politically neutral and where information is spread equitably, no amount of GIS will yield success (Blowers, 1986).

14.2.4 Environmental assessment (EA)

As explained in Chapter 1, the principle of

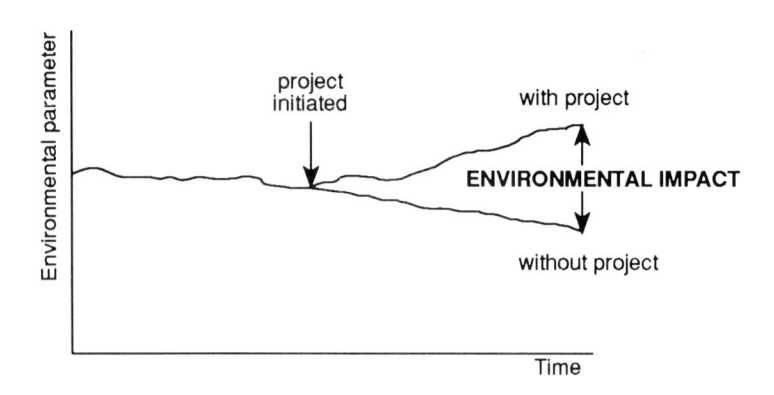

Figure 14.1 Environmental impact: a simple definition in relation to a development and a 'natural' trend. (Wathern, 1988)

EA was put into law first in 1967 in the United States (then known as Environmental Impact Assessment or EIA). The process of assessing the impacts of a proposed development — a close approach to rational planning — is in fact vested in two further processes, those of scoping and monitoring. Assessment is worthless unless all the potential impacts are considered — hence scoping. However, because environmental assessments also constitute 'decision-making under uncertainty', monitoring (or at least audits of the completed project) are also essential. In fact, the process of environmental assessment is becoming a test-bed for the reconciliation of political options, knowledge bases and development controls.

We may approach EA from the fundamental viewpoint of Figure 14.1, one which will pose familiar questions to environmental scientists: What will be the direction of future change? By how much? How certain are we? These need to be answered for the development against a dynamic, not static, national background. The immediate philosophical quandary is whether to register the marginal impact of the development and risk slightly enhanced environmental degradation or whether the mitigation techniques which can be forced on the developer by EA can and should be designed to cope with current levels of degradation — can the project lead to enhancement?

These questions are perhaps too 'heady' for a system which must be applied rigidly and bureaucratically if it is to be implemented at all. Whilst it is applied in many countries of the developed world (and increasingly by development agencies in the developing world), it is fair to say that EA takes on a different shape according to existing practices. In the United States its introduction was to a system of local, free-market development bargaining processes and in the view of many EA brought the system almost to a halt (Black, 1981). In Europe the European Communities Directive 337/85/EEC (OJ L175, 5.7.85) on EA has been less of a shock to systems with existing land-use planning structures (e.g. United Kingdom and the Netherlands) but has nevertheless spawned a new 'cottage industry' of carrying out assessments. Sapienza (1988) has summarized the process of EA in Europe as comprising:

(a) a description of the project and its main features;
(b) an outline of the main alternatives;
(c) a description of the aspects of the environment likely to be significantly affected by the project;
(d) a description of the main environmental effects;
(e) a description of measures to avoid, reduce and, if possible, compensate for

Planning, control or management?

Table 14.3 European Community guidance on the selection of developments for EA

Projects subject to mandatory environmental impactassessment (Annexe I)

Crude-oil refineries, installations for gasification and liquefaction of 500 tonnes or more of coal or bituminous shale per day
Thermal power stations and other combustion installations with a heat output of 300 megawatts or more, nuclear power stations, nuclear reactors (except research installations with less than 1 kilowatt continuous thermal load)
Permanent storage or final disposal installations for radioactive waste
Integrated works for the initial melting of cast-iron and steel
Certain installations for the extraction and processing of asbestos
Integrated chemical installations
Construction of motorways, express loads, long-distance railway lines, airports with runway length of 2100 metres or more
Ports and inland waterways for vessels of over 1350 tonnes
Waste disposal installations for the incineration, chemical treatment or landfill of toxic and dangerous wastes

Projects subject to voluntary environmental assessment (Annexe II)

Annexe II lists twelve categories of activity and identifies up to twelve subcategories of activity under each heading. They include:
Agriculture
Extractive industry
Energy industry
Processing of metals
Glass manufacture
Chemical industry
Food industry
Textile, leather, wood and paper industries
Rubber industry
Infrastructure projects
Miscellaneous projects
Modifications to development projects included in Annexe I and projects in Annexe I undertaken exclusively or mainly for the development and testing of new methods or products and not used for more than one year

the project's significant negative effects on the environment;
(f) a non-technical summary of the impact study.

Clearly, the penetration of EA depends largely upon the scale and type of development considered relevant for these processes. European legislation recognizes mandatory and discretionary types of development (see Table 14.3).

Lee and Wood (1988) welcomed the adoption of the Directive in the United Kingdom as improving

(a) integration across environmental media (air, land, water);
(b) anticipation of problems;
(c) information and consultation.

In (c) above there were clear signs of EA's American origins but Kennedy (1988) warned that Europe might fall into traps which still gape open in the United States — those of failing to consider EA as a *process* (not a document) and therefore of ignoring needs for early commencement and late monitoring of EA procedures. He also warned that 'EA will never be a purely technically predictive exercise'.

265

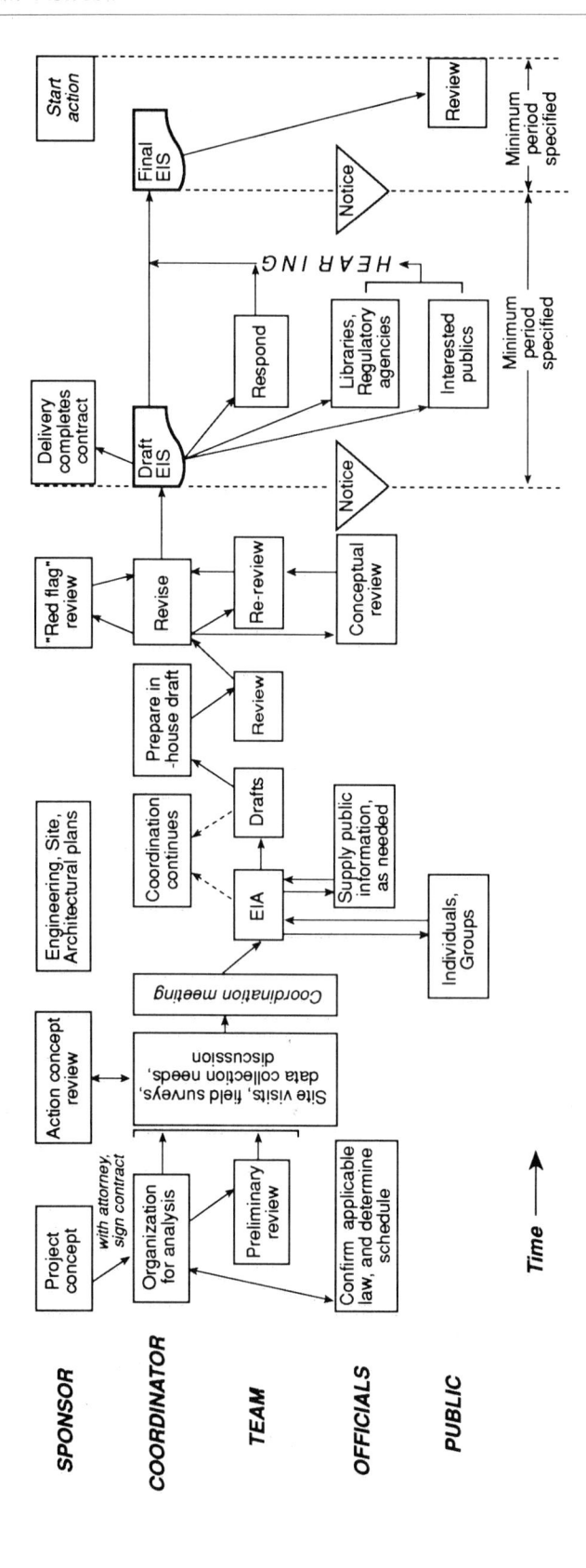

Figure 14.2 The progress of Environmental Impact Assessment and the preparation of the Statement (EIS) in the United States. (Black, P.E. 1981, *Environmental Impact Assessment*, New York, Praeger)

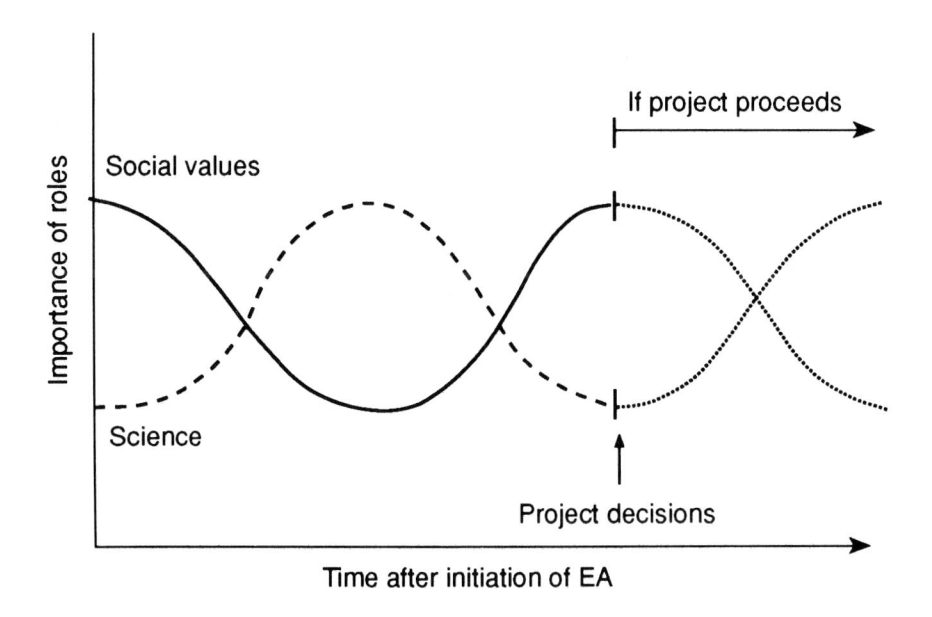

Figure 14.3 The changing dominance of social and scientific contributions through time in an ideal Environmental Assessment. (Beanlands, 1988)

Herrington (1988) is gloomy about the power of EA to protect the environment in the United Kingdom:

politicians and bureaucrats will continue to allow economic priorities to outweigh those of the environment in public decision-making ... EIA is likely to have nothing more than a peripheral or at most marginal impact on the quality of most planning decisions. (p. 159)

Of major concern in the first four years of EA in the United Kingdom have been issues of both policy and knowledge. The review of the first half of the period by Manchester University (Wood and Jones, 1991) suggests that many local authorities are allowing, often at junior level, relevant projects to escape the procedure; some feared that delays would put off developers whilst others were ignorant of the legislation. For those projects accorded EA, consultation with planners to ensure adequate scoping of the impacts was often inadequate and statutory consultees such as the conservation agencies were excluded.

As to the quality of knowledge incorporated in environmental statements, the Manchester study evaluated none as very good nor very poor but most as 'unsatisfactory'; a third of all statements lack the essential non-technical summary. Local authorities often need to take on the services of environmental consultants to audit the statements (compiled by other consultants); they cannot improve on the statements if by doing so they exceed the 16-week limit on the whole procedure of planning applications.

There is considerable professional anxiety in the United Kingdom about the quality of knowledge entering EA; environmental science has never been professionalized but the nation now has professional Institutes of Water and Environmental Management and of Ecology and Environmental Management. There is also an Insititute of Environmental Assessment for practitioners.

What established, professionally vindicated procedures can be applied to EA? The 'bare bones' of matrices, such as those devised by Leopold *et al.* (1971) and developed as dendritic networks by Canter (1985), seem

Table 14.4 A classification of assessment methods by task

Identification methods	— to assist in identifying the project alternatives, project characteristics and environmental parameters to be investigated in assessment
Data assembly methods	— to assist in describing the characteristics of the development and of the environment that may be affected
Predictive methods	— to predict the magnitude of the impacts which the development is likely to have on the environment
Evaluation methods	— to assess the significance of the impacts which the development will have on the environment
Communication methods	— to assist in consultation and public participation, and in expressing the findings of the study in a form suitable for decision-making purposes
Management methods	— to assist in managing the scoping of the study, the preparation of the impact study, the efficient conduct of the consultation process, etc.
Decision-making method	— to assist decision-makers in assessing and understanding the significance of environmental impacts relative to other factors relevant to a decision on the proposed development

Source: Newson (1992a) p. 149 and SI (1982) p. 210.

Table 14.5 Major activities and types of personnel involved in the EIA process

	Major activities	Types of personnel involved (examples only)
1.	Deciding if a study is necessary	— senior officers in competent authority, project leader in developer organization and certain of their support staff
2.	Scoping of study	— project leaders in competent authority and developer organization (with support staff), environmental control agency specialists, representatives of interest groups
3.	Managing the study	— developer/consultancy/competent authority project leader
4.	Preparing specialist consultancy/competent	— technical specialists employed by developer/consultancy/competent authority/other environmental control agencies (some of the work may be subcontracted to a number of different consulting groups
5.	Preparing the study	— developer/consultancy project leaders (with support assistance, possibly including professional writers)
6.	Reviewing the study	— competent authority project report leader (with technical support assistance), members of review panel (if in existence)
7.	Organizing the consultation process	— competent authority project leader (with support assistance), chairman of public inquiry
8.	Participating in the consultation process	— environmental control agency specialists, interest group representatives, general public
9.	Synthesising the findings of the consultation process	— competent authority project leader, chairman of public inquiry
10.	Using the findings of the study and consultation process to reach a decision on the proposed action	— competent authority senior officers and elected representatives
11.	Monitoring and post-auditing and environmental consequences arising from project implementation	— technical specialists employed by developer, competent authority and other environmental control agencies; competent authority senior officers.

Source: Newson (1992a) p. 149 and SI (1982) p. 210.

Table 14.6 Patterns in accuracy of impact predictions

Impact category	n*	(actual/predicted magnitude) log-transformed x̄**	s***	Mean accuracy as %
Emissions to air	11	0.941	1.187	39
Ambient air quality	17	0.881	1.294	41
Waste water discharge	7	0.663	0.266	51
Surface water quality	10	1.174	1.034	31
Groundwater	7	1.213	1.030	30
Cooling and marine discharge	7	0.37	30.282	69
Social	6	0.223	0.195	80
Atmospheric	28	0.803	1.167	45
Surface and groundwater	24	1.036	0.880	35
Primary	28	0.648	0.803	52
Secondary	40	0.939	1.086	39
Less severe	38	0.643	0.630	53
More severe	28	1.116	1.292	33
Mining	25	1.262	1.324	28
Mineral processing	16	0.694	0.700	50
Power generation	24	0.500	0.533	61
Overall	68	0.819	0.983	44

n* = number of predictions
x̄** = mean (actual/predicted)
s*** = standard deviation

Source: Buckley (1991).

to have progressed very little and only a third of the authors assembled by Wathern (1988) for the most concentrated statement on EA by practitioners pay detailed attention to methodology.

The most spectacular visual guidance to procedures is coming from those who wish to show how the complex interaction between environmental science and the political economy of development operates in practice (e.g. Figure 14.2 and its simplification in Figure 14.3). Inevitably, even within 16 weeks, the balance between social values and science shifts but perhaps not so harmonically (or harmoniously!) as Figure 14.3 pretends. Again in a social science vein, it is somewhat easier to tabulate the methodology of EA by general task and by personnel (Tables 14.4 and 14.5) than it is to bespeak the 'proper' conduct of EA under national legislation. Perhaps the best definition of 'proper' here is that given by Massam (1988) who suggests that Canada's EA procedures should move towards the 'Yes in my back yard' syndrome.

We end with a very optimistic development and a very pessimistic analysis of EA. The optimism is provided by the EC which is drafting a Directive to apply EA to 'policies, plans and programmes'. It is optimistic because land-use decisions mainly arise from agencies or free enterprise capitalizing on an opportunity provided by policy. Thus EA at local level will mesh in with higher-order EAs in the category of development involved, giving much greater thoroughness and consistency.

As a pessimistic conclusion, however, and in the light of environmental science as our salvation, Buckley (1991) has analysed up to a thousand environmental statements produced in Australia and, for those predictions which are numerical and which have been audited post-project, less than 50 per cent accuracy is revealed, with a discrepancy of over two orders of magnitude in some cases (Table 14.6).

14.3 Environmental control

We have used a long-established system of planning, that of the United Kingdom, to illustrate the restrictions inherent to, or forced upon planning; we therefore turn to less flexible controls on environmental behaviour. In this section we focus not on law as a means of environmental management (see Chapter 4) but on the technical aspects of law enforcement and, at a higher order of reasoning, on the principles which might be part of a co-ordinated legal programme for environmental protection.

14.3.1 Standards 'tight' and 'loose'

There are, in relation to pollution control, two general codes of practice in which elements of legislation can be wrapped; such 'wrapping' is essential since it represents what is attainable by monitoring for compliance: of the binary 'command and control' epithet this is the crucial second element.

By chance these two general codes are, respectively, those of the British tradition (environmental quality standards) and those of emerging European policies (uniform emission standards) — EQS and UES. The British tradition dates back to the Alkali Acts of 1863–1906 and encompasses four sub-principles:

(a) control is applied when scientific evidence justifies it
(b) pollution should be prevented at source;
(c) the best commercially viable technology should be used to abate pollution;
(d) the polluter should bear the costs of control.

These seem unquestionably wise but conceal elements liable to provoke extreme arguments with respect to public trust in expert judgement and to scientific judgement of long-term effects. A summary of opponents' views might be 'practical but weak and devious'. These opponents might point to the superiority of UES over EQS in respect of (a) its simplicity which allows it to be enforced rigidly; (b) its recognition that minimum pollution is the best aim for the planet.

The British traditionalist would then respond that the UES approach has three drawbacks:

(a) it fails to consider the impact of emissions on the receiving medium (deleterious in the case of high background contamination, harmless in the case of good dilution/dispersion capacities);
(b) it is not cost-effective because dischargers cannot obtain benefits from a favourable environmental location;
(c) it provides no framework for non-point sources; it cannot guarantee environmental quality.

Because of the long political battles between the two codes, the United Kingdom, which is forced to operate under EC rulings, has sought to develop hybrid legislation (Haigh, 1989) in which, for the case of water pollution control, certain substances are dangerous enough to warrant 'Red List' status and to be subject to UES within a general control framework of (statutory) EQS. In an attempt to 'go one better' than European legislators and to demonstrate that UES and EQS are best used together within an even more thorough framework, the UK government developed Integrated Pollution Control (IPC). Under this principle control is guided by selection of the Best Practicable Environmental Option (BPEO).

14.3.2 Integrated Pollution Control (IPC)

It could be agreed that IPC did not begin

Planning, control or management?

Table 14.7 Comparison of contamination potentials for three types of disposal

Contaminants	Effects of ocean disposal	Effects of land disposal	Effects of incineration
Pathogens (bacteria, viruses, and parasites)	Contamination of shellfish	No problem if sludge is stabilized; should not apply sludge on crops grown for raw consumption	Destroyed by incineration
Metals	Mostly non-toxic as a result of chemical changes after dumping	Absorption of cadmium zinc, nickel, and copper potentially toxic to plant growth; potential hazard to human health from eating crops contaminated with cadmium	Metals in air emissions and ash
Organic chemicals	Bioconcentration by fish, a potential health hazard	Potential absorption by cattle grazing on land where sludge has been spread; potential groundwater contamination by smaller, more soluble compounds	Largely degraded by incineration; possible formation of suspected carcinogens

Source: Newson (1992a) p. 149 and SI (1982) p. 210.

with a battle between UES and EQS but with another legislative development in the United States — the setting up of the unified Environmental Protection Agency (EPA) in 1970. Irwin (1990), however, traces the most recent upsurge in IPC policies to a statement by OECD in 1985:

There is an increasing awareness that the traditional approach of dealing with problems in only one medium at a time, such as air or water, may not be the most efficient way of dealing with all pollution problems: it may simply result in the transfer of the pollutant from one medium to another; it may fail to take account of the physical, chemical, biological, and commercial cycles that can affect the pollutant; or it may not adequately consider the multiple paths of exposure that can combine to create human health and environmental risks

Table 14.7 illustrates that the decision to dispose of potential pollutants raises very weighty issues both systematically/scientifically and regionally/geographically. Perhaps the United Kingdom's interest in IPC stems from its tradition of EQS since, for example, the choice between ocean, land or incineration disposal can be seen but in the capacity of the chosen environment to accept the waste and the desirable quality of those which 'escape' it.

The Royal Commission on Environmental Pollution (1988) advised the UK government that Integrated Pollution Control should operate under decisions favouring the Best Practicable Environmental Option (BPEO). The Commission defined these words as follows:

Best: doubtful if there is an absolute best for all time;
Practicable: refers to local conditions, technical knowledge and financial implications, including capital, revenue and public costs;

Table 14.8 Number of works and processes under IPC inspection (Alkali & Works Regulation Act, 1906, Health & Safety at Work etc. Act, 1974)

NUMBER OF WORKS UNDER INSPECTION

	England and Wales 1986*	Scotland 1986*	England 1987**	Wales 1987*
Alkali works	1	0	1	0
Scheduled works	1984	305	1762	181

NUMBER OF SEPARATE PROCESSES (PRESCRIBED PREMISES) UNDER INSPECTION

Process	England and Wales 1986	1987	Scotland 1986	Process	England and Wales 1986	1987	Scotland 1986
Acetylene	0	0	0	Hydrochloric acid	142	137	11
Acrylate	29	31	1	Hydrofluoric acid	14	14	2
Aldehyde	11	11	2	Hydrogen cyanide	18	17	3
Alkali	1	1	0	Iron and steel	66	64	8
Aluminium	40	41	11	Lead	99	97	5
Amines	86	86	11	Lime	31	31	0
Ammonia	54	52	6	Magnesium	5	5	0
Anhydride	18	17	1	Manganese	1	1	0
Arsenic	19	20	0	Metal recovery	27	25	2
Asbestos	82	65	3	Mineral	754	739	166
Benzene	29	25	4	Nitrate and chloride of iron	2	3	1
Beryllium	11	9	1	Nitric acid	142	138	9
Bisulphite	105	103	13	Paraffin oil	0	0	0
Bromine	38	36	4	Petrochemical	28	28	4
Cadmium	21	20	0	Petroleum	115	117	33
Carbon disulphide	18	18	2	Phosphorus	6	5	0
Carbonyl	1	1	0	Picric acid	1	1	0
Caustic soda	0	0	0	Producer gas	0	0	0
Cement	27	26	2	Pyridine	16	17	2
Ceramic	24	23	0	Selenium	14	13	1
Chemical fertilizer	19	18	4	Smelting	13	12	0
Chemical incineration	69	68	12	Sulphate of ammonia and chloride of ammonia	13	11	1
Chlorine	98	98	8	Sulphide	104	101	11
Chromium	5	5	0	Sulphuric acid (Class I)	0	0	0
Copper	39	36	1	Sulphuric acid (Class II)	24	23	3
Di-isocyanate	186	194	8	Tar and bitumen	73	70	75
Electricity	81	75	17	Uranium	3	3	0
Fluorine	18	18	0	Vinyl chloride	11	9	0
Gas and coke	42	37	6	Zinc	18	19	0
Gas liquor	6	5	1			2581	(England)
						258	(Wales)
Totals					2915	2839	455

* Figures prepared from data available on 31 December 1986.
** Figures prepared from data available on 31 December 1988.

Source: Newson (1992a) p. 149 and SI (1982) p. 210.

Table 14.9 Main UK pollution control arrangements

Type of Pollution	Enforcement Legislation	Agency
Air pollution (registered works) must apply bpm	Alkali Act, Health & Safety at Work etc Act	HMIP (IAPI) Department of the Environment
Air and noise (unscheduled works)	Control of Pollution Act, Part III, Clear Air Acts, Nuisance Provisions of Public Health Act	Environmental Health Dept of the local authority, i.e. the district council
Water	Control of Pollution Act, Part II	Regional Water Authority, [HMIP (Water Division) advises]
Marine	Food & Environment Protection Act 1985	Ministry of Agriculture, Fisheries and Food
Waste Disposal to land	Control of Pollution Act, Part I	Waste Disposal Authority: county councils in England [HMIP (HWI) advises]
Radioactive waste	Radioactive Substances Act 1960	HMIP (RCI) Department of the Environment
Control through land use planning	Town & Country Planning Acts 1971 etc.	Planning Department of local authority

Source: Owens (1991).

Environmental: essential to consider local/remote and short-/long-term issues afffecting all media. Malfunctioning, accidents and eventual decommissioning also to be included;

Option: to include not only waste disposal but minimization and recycling.

The 'nuts and bolts' of IPC concern agency structure and plant-by-plant procedures. From April 1991 the UK Environmental Protection Act brings around 100 categories of industrial process in around 5,000 plants under IPC. Her Majesty's Inspectorate of Pollution (HMIP) will receive applications from new and existing plants for an authorization to operate. Monitoring and publication are essential features of the new procedures (Table 14.8). Each process considered for authorization is considered not only within the BPEO principle but in terms of 'BATNEEC' (Best Available Technology Not Entailing Excessive Cost). This is a development of the long-standing 'Best practicable means' in UK policy on technical evaluation. Some would argue that this is an escape clause for parsimonious polluters and conservative scientists, but others stress both the stronger 'best available' of BATNEEC (implying dynamic quest for perfection) and the need to keep industry in business so that it can resource its own environmental protection under the 'polluter pays principle'.

Inevitably the operation of BATNEEC within IPC under BPEOs will depend strongly on the ability of HMIP to be proactive, i.e. to administer a consent-based system, to ensure compliance through monitoring and to manage efficiently the consultation processes. The burden of inspection is enormous as can be judged from Table 14.8 listing HMIP's current duties in respect of the Alkali and Health & Safety at Work Acts.

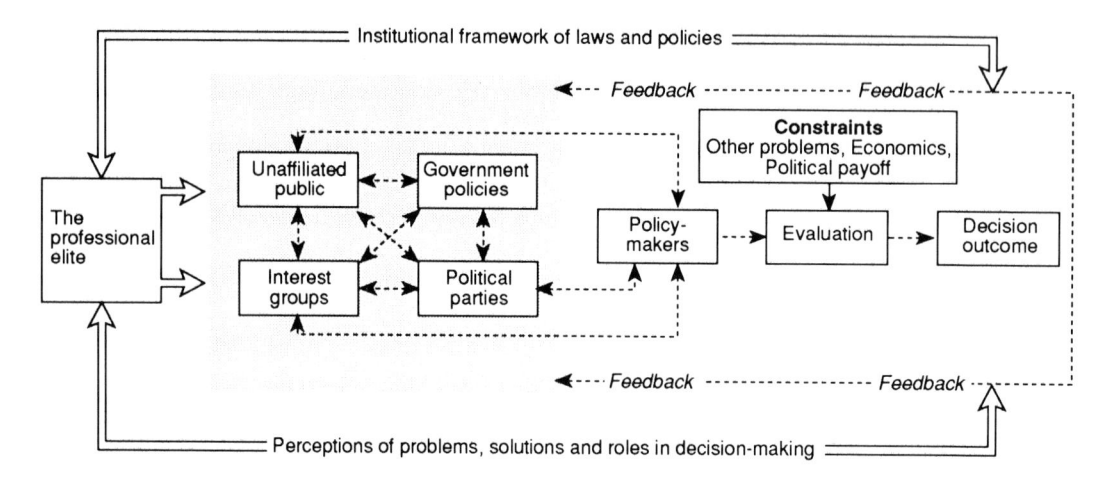

Figure 14.4 The players in environmental decision-making

Owens (1990) reveals a further problem in respect of a plethora of existing agencies involved with pollution control and operating at an appropriate geographical scale (see Table 14.9). Research at the University of Hull also reveals geographical scale problems of administration as compromising the best intentions of IPC to use geographical principles; in Humberside there are both overlaps and disjunctions in space as well as mismatches of technologies and control strategies. O'Riordan (1990) warns that without sufficient resources and power HMIP could operate IPC with inspectors who are 'box-tickers' rather than modellers and assessors of inter-pollutant trade-off. There seems little doubt that this warning is being heeded as politicians of all parties define their view of a powerful UK EPA.

14.4 Managing the environment

Figure 14.4 illustrates a simplistic pattern of decision-making in environmental policy, applied to the conditions prevalent in the United Kingdom. We have also discussed throughout the book the relative roles of moral persuasion, economics and the law in bringing about policy operations.

14.4.1 The need to consult

Management, therefore, has environmental options and it can raise or lower the status of the scale and locus of its operations, perhaps in relation to issue-attention cycles but hopefully in relation to longer-term objectives. Throughout, attention returns again and again to the role of 'the people' in bringing about sustainable patterns of development. We have learned this from many directions, such as:

(a) the apparent equilibrium of primitive peoples and their environment achieved via a simple but highly codified set of principles;
(b) the failure of development schemes, such as irrigated agriculture, if local people are not considered to be the agents of a new production pattern;
(c) the relevance of national tradition and the notion of 'the way we are' in relation to developed-world environmental policies;
(d) the increasing use of public consultation in policy instruments such as EA and IPC.

The decision contexts of two models for operating environmental management

A 'professionalized' decision-making process

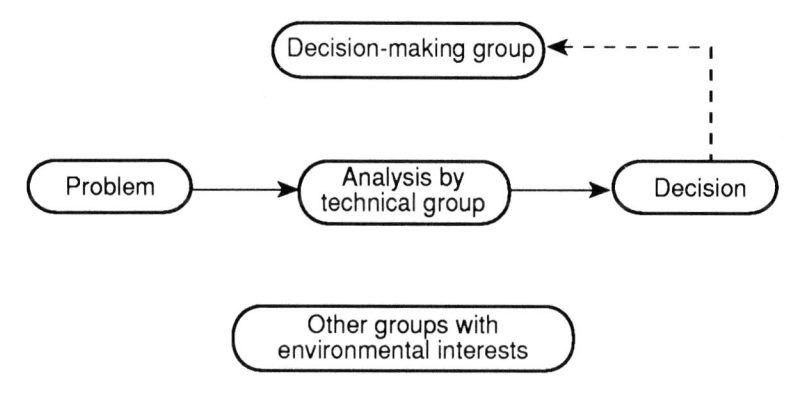

An interactive decision-making process

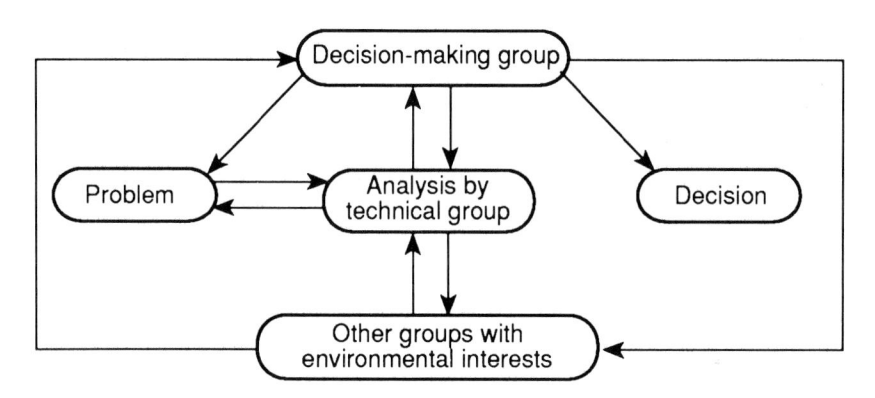

Figure 14.5 Routes to decision-making, technocratic (above), consultative (below). (Modified from Lee, 1989)

policies are shown in Figure 14.5; the paradox in the necessary growth of environmental information is that power is not conferred by possession of that information. Instead it becomes 'public knowledge' or is 'given away'. The cost of maintaining perfectly rational short-term policies through the agency of unwilling or uncommitted individuals, corporations or nations may lead to degradation in other sectors; consultation is cost-effective even when it produces semi-rational outcomes.

We are returned, once more, to the problems of decision-making under uncertainty. Table 14.10 presents a characterization of levels of knowledge on environmental issues and the equivalent role of that knowledge in decision-making. Compared with the characterization of policy processes in British administration (Table 14.11) we can observe that as we descend the former table we also descend the latter towards consultation and politico-rational categories.

Successive policy developments have sought to widen consultation on the basis of making environmental information widely available; this is costly, risks misunderstanding and can infringe industrial secrecy as

Table 14.10 A decision protocol to promote learning from experience

	Information gradients in impact analysis				Decision categories for project approval		
Confidence level	Data set ratings	Process knowledge	Approach permitted	Type of approval	Colour code	Terms and conditions of implementation	
1 High	Good	Proven cause-effect relationship	Statistical prediction	Unqualified	Green	Normal standards and regulations apply.	
2 Fairly high	Sufficient	Evidence for hypothesis	Quantitative simulation	Qualified	Yellow	Subject to special requirements, e.g. monitoring programme or post-project audit.	
3 Fairly low	Incomplete	Postulated linkages	Conceptual modelling	Conditional	Orange	Must conform to stringent environmental controls; additional information to be filed prior to construction, research management experiment imposed.	
4 Low	Poor	Speculation	Professional opinion	Deferral	Red	Abandon project, or moratorium on development until major redesign or special studies completed, e.g. contingency planning or pilot project for demonstration and evaluation sources.	

Source: Sadler (1988).

Table 14.11 Land-use planning: policy processes in contemporary British public administration

Bureaucratic-legal:	the determination of actions in terms of formal procedural and legal rules
Techno-rational:	the determination of actions in terms of the judgement of experts and scientific reasoning
Semi-judicial:	the determination of actions through formal hearings of the arguments of conflicting interests, with an assessor balancing the relative merits of the arguments
Consultative:	the determination of actions through negotiation and debate with, and among, concerned and affected groups; the forms of such consultation may include: corporatist: negotiation of a wide range of issues over a long period of time with specified representatives of specific groups; bargaining: negotiation with specific groups over a specific issue where mutual dependency between the group and the state is involved; pluralist politics: political debate among pressure groups, with politicians determining the balance of advantage in political terms; open democratic debate: 'rational' debate, where all affected parties discuss the advantages and disadvantages of particular courses of action and reach agreement without domination
Politico-rational:	the determination of action by the judgement of politicians in the formal arenas of representative democracy

Source: Healey *et al.* (1988).

has been proved by recent US policy. In the United Kingdom confidentiality has long been regarded as part and parcel of the technocratic framework for environmental management — 'leave it to the experts'. Despite the pleas of a Working Party on Public Access to Information held by Pollution Control Authorities (Department of the Environment, 1986) and the registers of information set up by the Control of Pollution Act (1974), it seems likely that an EC Directive on the Freedom of Environmental Information will eventually bring the Community (and with it the United Kingdom) into line with the United States and Japan, leaving those wishing for secrecy with the burden of proof for this exception (Clinton-Davis, 1990).

14.4.2 Environmental politics: not labour, not capital

If policy developments play into the hands of the individual in terms of environmental information, will it invariably be individuals who take action from it? Unlikely. The most likely outcome of this dissemination is a reorientation of politics towards representative groups (currently protest groups but in future political parties). The role of 'green' politics is currently marginal; even where green parties enjoy election success they are forced into largely conventional political axes. It seems inevitable that these axes will themselves bend in the next few decades. Paehlke (1989) examines the history of the relationship between environmentalism and politics and claims that 'In environmental politics, distributional issues take second place to choice of technology, design of technology and one of technology' (p. 189). He sees environmentalism and anti-environmentalism as axes normal to those of left and right in politics. It seems that, whatever the geometry, some adoption of the knowledge/participation issues raised in this chapter as central to environmental management will have a profound effect on politics, if not parties, in the next decades.

References

Black, P.E. (1981), *Environmental Impact Analysis*, Praeger, New York.
Blowers, A. (1986), 'Geographers, planners and politics: theory and practice', in P.T. Kevill and J.T. Coppock (eds), *Geography: Planning and Policy-making*, Geo Books, Norwich, pp. 25–44.
Buckley, R.C. (1991), 'How accurate are environmental impact predictions?' *Ambio*, 20(3–4), 161–2.
Canter, L. (1985), *Environmental Impact of Water Resources Projects*, Lewis, Chelsea, Michigan, 352 pp.
Charlton, M. and Openshaw, S. (1986), 'Planning and land use change in a UK metropolitan county: spatial data manipulation, policy analysis and computer modelling', in P.T. Kivell and J.T. Coppock (eds), *Geography: Planning and Policy-making*, Geo Books, Norwich, pp. 113–40.
Clinton-Davis, S. (1990), 'The freedom of Environmental Information Directive', *Environmental Law*, 14(1), 2–4.
Collingridge, D. and Reeve, C. (1986), *Science Speaks to Power: The Role of Experts in Policymaking*, Frances Pinter, London, 175 pp.
Cullingworth, J.B. (1985), *Town and Country Planning in Britain*, 9th edn, Allen and Unwin, London.
Darwin, C. (1859), *The Origin of Species*, John Murray, London.
Department of the Environment (1986), 'Public access to environmental information', Pollution Paper 23, HMSO, London, 48 pp.
Haigh, N. (1989), *EEC Environmental Policy and Britain*, 2nd edn, Longman, Harlow, 382 pp.
Healey, P., McNamara, P., Elson, M. and Doak, A. (1988), *Land Use Planning and the Mediation of Urban Change*, Cambridge University Press, 295 pp.
Henderson-Sellers, A. (1991), 'Policy advice on greenhouse-induced climatic change: the scientist's dilemma', *Progress in Physical Geography*, 15(1), 53–70.
Herrington, J. (1988), 'Environmental values in a changing planning system', in M. Clark and J. Herrington (eds), *The Role of Environmental Impact Assessment in the Planning Process*, Mansell, London, 140–62.
Hutton, J. (1795), *Theory of the Earth with Proofs and Illustrations*, William Creech, Edinburgh.
Irwin, F. (1990), 'Introduction to Integrated Pollution Control', in N. Haigh and F. Irwin (eds), *Integrated Pollution Control in Europe and North America*, Conservation Foundation, Washington, DC, 3–30.
Kennedy, W.V. (1988), 'What can Europe learn from 15 years of EIA in North America?' *EER*, 2(2), 17–19.
Lee, N. and Wood, C. (1988), 'Implementing the EC Directive on EIA', *EER*, 2(2), 12–17.
Leopold, L.B., Clark, F.E., Hanshaw, F.F. and Bolsley, J.R. (1971), *A Procedure for Evaluating Environmental Impact*, USGS Circular 645, Dept Interior, Washington, DC.
Lovelock, J.E. (1979), *Gaia: A New Look at Life on Earth*, Oxford University Press, Oxford, 157 pp.
Massam, B.A. (1988), 'Environmental assessment in Canada: theory and practice', Canada House, London, Lecture Series 39, 32 pp.
Miller, C. and Wood, C. (1983), *Planning and Pollution*, Clarendon Press, Oxford, 232 pp.
Newson, M.D. (1992a), 'Land and water: convergence, divergence and progress in UK policy', *Land Use Policy*, 9(2), 111–21.
O'Riordan, T. (1990), 'Integrated Pollution Control', in R. Macrory (ed.), *Environmental Challenges: The Institutional Dimension*, Imperial College Centre for Environmental Technology, London, pp. 19–27.
O'Riordan, T. and Rayner, S. (1991), 'Risk management for global environmental change', *Global Environmental Change*, 1(2), 91–108.
Owens, S. (1990), 'The unified pollution inspectorate and Best Practicable Environmental option in the United Kingdom', in N. Haigh and F. Irwin (eds), *Integrated Pollution Control in Europe and North America*, Conservation Foundation, Washington, DC, 169–208.
Paehlke, R. (1989), *Environmentalism and the Future of Progressive Politics*, Yale University Press, Newhaven and London, 325 pp.
Rhind, D.W. (1986), 'Remote sensing, digital mapping and geographical information systems: the formation of public policy', in P.T. Kivell and J.T. Coppock (eds), *Geography: Planning and Policy-making*, Geo Books, Norwich, pp. 277–90.
Ronan, C.A. (1983), *The Cambridge Illustrated History of the World's Science*, Cambridge University Press, Cambridge, 543 pp.
Royal Commission on Environmental Pollution (1988), *Twelfth Report: Best Practicable Environmental Option*, HMSO, London, 70 pp.

Sadler, B. (1988), 'The evaluation of assessment: post-EIS research and process development', in P. Wathern (ed.), *Environmental Impact Assessment, Theory and Practice*, Unwin Hyman, London, pp. 129-42.

Sapienza, G. (1988), 'Environmental Impact Assessment in the European Community', *EER*, 2(2), 8-12.

Smallwood, R.J. (1990), 'The UK experience in communicating risk advice to land-use planning authorities', in H. Gow and H. Otway (eds), *Communicating with the Public about Major Accident Hazards*, Elsevier Applied Science, London.

University of Manchester EIA Centre (1991), 'UK Environmental Statement 1988-90: an analysis', Occasional Paper 29.

Walker, G.P. (1991), 'Land use planning and industrial hazards: a role for the European Community', *Land Use Policy*, 9, 227-40.

Wathern, P. (ed.) (1988), *Environmental Impact Assessment, Theory and Practice*, Unwin Hyman, London, 332 pp.

Wood, C. (1989), *Planning Pollution Prevention: A Comparison of Siting Controls over Air Pollution Sources in Great Britain and the USA*, Heinemann Newnes, Oxford, 243 pp.

Wood, C. and Jones, C. (1991), *Monitoring Environmental Assessment and Planning*, HMSO, 51 pp.

Index

Note: Numbers in bold refer to figures

ALARA 220
ALARP 220
accidents 139
acidification 113–117, **115**
acid lakes **116**
acid rain **17**, 39–40, 114
ammonia 114
Arvill R. 5

BATNEEC 75, 272
BPEO 271, 272
Bhopal 197
biodiversity 38, 246
biogeography 49

CFC's 61–62, 233
CHEMET 204, 206–207
CIMAH 201–202, 206
Calder N. 10
Caldwell L.K. 3
capacities 251
carbon dioxide **239**
charges 92
chemicals 23, 151
chemical incidents 197
Chernobyl 20, 35, 60, 114, 200, **201**, 213, 215
Christianity 4, 5
Clean Air Act 109–110, 240
commons 81, 83–84, 251, 252–253
common law 63
conservation 38, 76, 89, 249, 250
consultation 263, 274–277
consumer surveys **99**
contingent valuation 89–90
Coto de Donana
contamination 21, 172
critical loads 120–126
culture 3

damage/costs 88, 91
'deep green' 3
dioxin 156
disasters 197

dust **112**

EQS 270
Earthrise 14
economics **86**
emergency planning 196–210
energy 232–241
energy reserves 234–237
energy supply 236
Engel J.R. 5
environmental assessment 88, 263–270, **266**, **267**
environmental audits 100
environmental capacities 141
environmental controls 270–274
environmental data 215
environmental decision-making 274
environmental impact 85, **264**
environmental indicators 9
environmental information 78
environmental politics 277–278
ethics 3
European Communities 60, 61–62, 69, 72, 265
eutrophication 40, **137**, 138
extinction 37
externalities 84

Flixborough 199, 200
fuels 235
fuel technology 238–240

GDP 98, 232
GIS 218, **223**, **227**, 226, **229**, 228, 248
GNP 95
Gaia 245, 250, 256
Garaet el Ichkeul 44, **46**, 47, 52
green economics 87
Green Parties 9, 10
'greenhouse' effect 41, 234, 246–250

heavy metals 40, 172–195
Herrington 267
holism 258–259

Index

Holocene 15, 245
household waste 152, **153**
hunter-gatherers 242–243

IPC 270–274
incineration 155–156, 161–163
indigenous peoples 253
integrated pollution control 170
island biogeography 49–50
issue-attention cycle 6, 100, 145

Japan 18
Johnston R.J. 8

LDC's 17
lake sediments 51
landfill 154–155
liability 71, 74
lifestyle effects 214
Love Canal 151, 164–167

marine pollution 19
market 81
Marx K. 4, 84
McLaren D.J. 14
metal accumulation **175**
methane **16**, 20, 161
monitoring 22, 30–35, 52, 139, 143, 144
municipal solid waste 152–153, 156

NIMBY 12, 153, 262–263
Naess A. 4
nature 253–254
Neolithic revolution 244
New Environmental Age 6, 11
Newcastle coal 233
Nitrate Sensitive Areas 145
nuclear power plants 222–225
nuclear regulation 219–221
nuclear waste 225–228
nuisance 63–66, 67

OECD 271
OPEC 236
ozone (ground-level) 117–120
   (stratospheric) 61, 233

PHAST 203
Paehlke R.C. 9, 17
palaeoecology 249
Pearse P.H.
pesticides **27**
planning law 64–65, 74
Pleistocene overkill **243**
pollution
   air 109–128
   control, UK 273
   fluoride 111–112
   groundwater 139, 158–160
   lead 180, 184
   nickel 180
   nitrate 138
   pathways 29, 30, 130, 178, 261
   pesticides 138
   river 70, 179–182
   silage 132, 134–137
   slurry 132, 134–137
   sources 28, 112, 130, 150, 174, 261
   targets 29, 130, 183, 261
   thresholds 185–186
precautionary principle 28
project cycle 97
property law 63, 67

radiation 213–217
   infrared 217
   ionizing 218–221
   microwaves 217
   non-ionizing 216–218
   optical 217
   radio frequency 217–218
   ultrasound 228–229
   units 219
   UV 216–217
radionuclides 221
Ramsar Convention 43–44
recycling 157
reductionism 10, 258–259
restoration (habitat) 254–256
risk 221–222, 224, 254
river quality classification 147–148
River Tyne 189–192
river water quality 32–34, 130–149
Roddick A. 101

Sandbach F. 9
Santos M. 28
satellite accounts 93
scale 8
scarcity 82
Sellafield 213
sewage 40, 130–132, **132**, 134, 142
sewerage **134**
Seveso 198–199
smog **110**
sulphur deposition **124–125**
'Silent Spring' 7, 9, 25, 28
site plan **205**
'Spaceship Earth' 80, 81
species-area 43
statute law 68

281

Index

surrogate values 88–89
surveillance 31
sustainability 83, 101–103
synergism 26

Tyneside 150, **163**, 173, 187–189
toxicity 22, 186–187
   testing 22–24
Toynbee A. 4

UES 270
uncertainty 260

Vorsorgeprinzip 28, 260

WAZAN 203
Warn P.R. 14
Waste Disposal Authorities 179
waste disposal costs 169
waste dumping 66, 154
water abstraction 57–58
wetlands 44–49, 50
wilderness 251–252
world trade 94, 236–237

YIMBY 168

zone of consequence **204**